DEVELOPMENTS IN SEDIMENTOLOGY 39

HYDRODYNAMICS AND SEDIMENTATION IN WAVE-DOMINATED COASTAL ENVIRONMENTS

FURTHER TITLES IN THIS SERIES
VOLUMES 1, 2, 3, 5, 8 and 9 are out of print

4   F.G. TICKELL
THE TECHNIQUES OF SEDIMENTARY MINERALOGY
6   L.VAN DER PLAS
THE IDENTIFICATION OF DETRITAL FELDSPARS
7   S. DZULYNSKI and E.K. WALTON
SEDIMENTARY FEATURES OF FLYSCH AND GREYWACKES
10 P.McL.D. DUFF, A. HALLAM and E.K. WALTON
CYCLIC SEDIMENTATION
11 C.C. REEVES Jr.
INTRODUCTION TO PALEOLIMNOLOGY
12 R.G.C. BATHURST
CARBONATE SEDIMENTS AND THEIR DIAGENESIS
13 A.A. MANTEN
SILURIAN REEFS OF GOTLAND
14 K.W. GLENNIE
DESERT SEDIMENTARY ENVIRONMENTS
15 C.E. WEAVER and L.D. POLLARD
THE CHEMISTRY OF CLAY MINERALS
16 H.H. RIEKE III and G.V. CHILINGARIAN
COMPACTION OF ARGILLACEOUS SEDIMENTS
17 M.D. PICARD and L.R. HIGH Jr.
SEDIMENTARY STRUCTURES OF EPHEMERAL STREAMS
18 G.V. CHILINGARIAN and K.H. WOLF, Editors
COMPACTION OF COARSE-GRAINED SEDIMENTS
19 W. SCHWARZACHER
SEDIMENTATION MODELS AND QUANTITATIVE STRATIGRAPHY
20 M.R. WALTER, Editor
STROMATOLITES
21 B. VELDE
CLAYS AND CLAY MINERALS IN NATURAL AND SYNTHETIC SYSTEMS
22 C.E. WEAVER and K.C. BECK
MIOCENE OF THE SOUTHEASTERN UNITED STATES
23 B.C. HEEZEN, Editor
INFLUENCE OF ABYSSAL CIRCULATION ON SEDIMENTARY
ACCUMULATIONS IN SPACE AND TIME
24 R.E. GRIM and GUVEN
BENTONITES
25A G. LARSEN and G.V. CHILINGAR, Editors
DIAGENESIS IN SEDIMENTS AND SEDIMENTARY ROCKS, I
26 T. SUDO and S. SHIMODA, Editors
CLAYS AND CLAY MINERALS OF JAPAN
27 M.M. MORTLAND and V.C. FARMER, Editors
INTERNATIONAL CLAY CONFERENCE 1978
28 A. NISSENBAUM, Editor
HYPERSALINE BRINES AND EVAPORITIC ENVIRONMENTS
29 P. TURNER
CONTINENTAL RED BEDS
30 J.R.L. ALLEN
SEDIMENTARY STRUCTURES
31 T. SUDO, S. SHIMODA, H. YOTSUMOTO and S. AITA
ELECTRON MICROGRAPHS OF CLAY MINERALS
32 C.A. NITTROUER, Editor
SEDIMENTARY DYNAMICS OF CONTINENTAL SHELVES
33 G.N. BATURIN
PHOSPHORITES ON THE SEA FLOOR
34 J.J. FRIPIAT, Editor
ADVANCED TECHNIQUES FOR CLAY MINERAL ANALYSIS
35 H. VAN OLPHEN and F. VENIALE, Editors
INTERNATIONAL CLAY CONFERENCE 1981
36 A. IIJIMA, J.R. HEIN and R. SIEVER, Editors
SILICEOUS DEPOSITS IN THE PACIFIC REGION
37 A. SINGER and E. GALAN, Editors
PALYGORSKITE-SEPIOLITE: OCCURRENCES, GENESIS AND USES
38 M.E. BROOKFIELD and T.S. AHLBRANDT, Editors
EOLIAN SEDIMENTS AND PROCESSES

DEVELOPMENTS IN SEDIMENTOLOGY 39

HYDRODYNAMICS AND SEDIMENTATION IN WAVE-DOMINATED COASTAL ENVIRONMENTS

Edited by

B. GREENWOOD and R.A. DAVIS, Jr.

Departments of Geography and Geology, Scarborough College, University of Toronto, Scarborough, Ont. M1C 1A4 (Canada)
Department of Geology, University of South Florida, Tampa, FL 33620 (U.S.A.)

Reprinted from Marine Geology, Vol. 60 (1—4)

ELSEVIER
Amsterdam — Oxford — New York — Tokyo   1984

ELSEVIER SCIENCE PUBLISHERS B.V.
1 Molenwerf
P.O. Box 211, 1000 AE Amsterdam, The Netherlands

Distributors for the United States and Canada:

ELSEVIER SCIENCE PUBLISHING COMPANY INC.
52, Vanderbilt Avenue
New York, N.Y. 10017

Library of Congress Cataloging in Publication Data
Main entry under title:

Hydrodynamics and sedimentation in wave-dominated
  coastal environments.

  (Developments in sedimentology ; 39)
  Papers from a symposium held at the Eleventh International Congress on Sedimentology which met at Hamilton, Ontario in Aug. 1982.
  "Reprinted from Marine geology, vol. 60 (1-4)."
  Bibliography: p.
  1. Coast changes--Congresses. 2. Sedimentation and deposition--Congresses. 3. Hydrodynamics--Congresses.
I. Greenwood, B. (Brian) II. Davis, Richard A. (Richard Albert), 1937-   . III. International Congress on Sedimentology. IV. Marine geology.
V. Series.
GB450.2.H96 1984   551.4'57   84-10321
ISBN 0-444-42400-8 (U.S.)

ISBN 0-444-42400-8 (Vol. 39)
ISBN 0-444-41238-7 (Series)

© Elsevier Science Publishers B.V., 1984
All rights reserved. No part of this publication may be reproduced, stored in a retrieval system or transmitted in any form or by any means, electronic, mechanical, photocopying, recording or otherwise without the prior written permission of the publisher, Elsevier Science Publishers B.V./Science & Technology Division, P.O. Box 330, 1000 AH Amsterdam, The Netherlands.

Special regulations for readers in the USA — This publication has been registered with the Copyright Clearance Center Inc. (CCC), Salem, Massachusetts. Information can be obtained from the CCC about conditions under which photocopies of parts of this publication may be made in the USA. All other copyright questions, including photocopying outside of the USA, should be referred to the publisher.

Printed in The Netherlands

CONTENTS

Preface VII

Waves, long waves and nearshore morphology
    A.J. Bowen and D.A. Huntley (Halifax, N.S., Canada) 1
Spatial and temporal variations in spectra of storm waves across a barred nearshore
    R.G.D. Davidson-Arnott and D.C. Randall (Guelph, Ont., Canada) 15
Waves, currents, sediment flux and morphological response in a barred nearshore system
    B. Greenwood and D.J. Sherman (Scarborough, Ont., Canada) 31
Landward migration of inner bars
    T. Sonamura and I. Takeda (Ibaraki, Japan) 63
Sediment flux and equilibrium slopes in a barred nearshore
    B. Greenwood (Scarborough, Ont., Canada) and P.R. Mittler (Vancouver, B.C., Canada) 79
Sediment transport and morphology at the surf zone of Presque Isle, Lake Erie, Pennsylvania
    D. Nummedal, D.L. Sonnenfeld (Baton Rouge, La., U.S.A.) and K. Taylor (Erie, Pa., U.S.A.) 99
Sedimentology and morphodynamics of a microtidal beach, Pendine Sands, SW Wales
    C.F. Jago (Gwynedd, U.K.) and J. Hardisty (Bristol, U.K.) 123
High-frequency sediment-level oscillations in the swash zone
    A.H. Sallenger, Jr. and B.M. Richmond (Menlo Park, Calif., U.S.A.) 155
Wave-formed structures and paleoenvironmental reconstruction
    H.E. Clifton and J.R. Dingler (Menlo Park, Calif., U.S.A.) 165
Boundary roughness and bedforms in the surf zone
    D.J. Sherman (Woods Hole, Mass., U.S.A.) and B. Greenwood (Scarborough, Ont., Canada) 199
Tidal-cycle changes in oscillation ripples on the inner part of an estuarine sand flat
    J.R. Dingler and H.E. Clifton (Menlo Park, Calif., U.S.A.) 219
Bedforms and depositional sedimentary structures of a barred nearshore system, eastern Long Island, New York
    R.C. Shipp (Orono, Me., U.S.A.) 235
Beach and nearshore facies: southeast Australia
    A.D. Short (Sydney, N.S.W., Australia) 261
Structures in deposits from beach recovery, after erosion by swell waves around the southwestern coast of Aruba (Netherlands Antilles)
    J.H.J. Terwindt (Utrecht, The Netherlands), C.H. Hulsbergen (Emmeloord, The Netherlands) and L.H.M. Kohsiek (The Hague, The Netherlands) 283
What is a wave-dominated coast?
    R.A. Davis, Jr. (Tampa, Fla., U.S.A.) and M.O. Hayes (Columbia S.C., U.S.A.) 313
Shoreface morphodynamics on wave-dominated coasts
    A.W. Niedoroda (Houston, Texas, U.S.A.), D.J.P. Swift (Dallas, Texas, U.S.A.), T.S. Hopkins (Upton, N.Y., U.S.A.) and Chen-Mean Ma (Houston, Texas, U.S.A.) 331
Control of barrier island shape by inlet sediment bypassing: East Frisian Islands, West Germany
    D.M. FitzGerald (Boston, Mass., U.S.A.), S. Penland and D. Nummedal (Baton Rouge, La., U.S.A.) 355

Coarse clastic barrier beaches: A discussion of the distinctive dynamic and morphosedimentary characteristics
R.W.G. Carter (Londonderry, Northern Ireland) and J.D. Orford (Belfast, Northern Ireland) 377

Shoreface translation and the Holocene stratigraphic record: Examples from Nova Scotia, the Mississippi Delta and eastern Australia
R. Boyd (Halifax, N.S., Canada) and S. Penland (Baton Rouge, La., U.S.A.) 391

Holocene sedimentation of a wave-dominated barrier-island shoreline: Cape Lookout, North Carolina
S.D. Heron, Jr. (Durham, N.C., U.S.A.), T.F. Moslow (Baton Rouge, La., U.S.A.), W.M. Berelson (Los Angeles, Calif., U.S.A.), J.R. Herbert (Houston, Texas, U.S.A.), F.A. Steele III (Casper, Wyo., U.S.A.) and K.R. Swanson (San Francisco, Calif., U.S.A.) 413

Reconstruction of paleo-wave conditions during the Late Pleistocene from marine terrace deposits, Monterey Bay, California
W.R. Dupré (Houston, Texas, U.S.A.) 435

Reconstruction of ancient sea conditions with an example from the Swiss Molasse
P.A. Allen (Cardiff, U.K.) 455

PREFACE

This volume evolved from a symposium held at the Eleventh International Congress on Sedimentology which met at Hamilton, Ontario in August 1982 under the auspices of the International Association of Sedimentologists. The symposium entitled "Coastal Environments Dominated by Waves" was convened jointly by Brian Greenwood (representing the primary sponsors, the Geological Association of Canada and Canadian Society of Petroleum Geologists) and Skip Davis (representing the Society of Economic Paleontologists and Mineralogists, the co-sponsor).

The objective of the symposium was to provide a forum for discussion of the links between hydrodynamics and sedimentation in both modern and ancient wave-dominated coastal environments. Of the 35 original papers 23 are presented in this volume: they range in content from theory and experiment on recent sedimentation under waves (papers 2—9), to the interpretation of bed forms and sedimentary structures under a range of environmental constraints (papers 10—15), to macro-scale sedimentation, morphodynamics and the stratigraphic record (papers 16—21) and finally to explicit reconstruction of ancient environmental constraints on sedimentation revealed in the rock record (papers 22 and 23). Each paper is an original research contribution and has been selected to indicate the current levels and directions of research on sedimentation in wave-dominated coastal environments.

We would like to thank the authors for their patience during the production of this volume and the many referees who gave their time freely to rigorously review the manuscripts.

BRIAN GREENWOOD and RICHARD A. DAVIS, Jr. (Editors)

WAVES, LONG WAVES AND NEARSHORE MORPHOLOGY

A.J. BOWEN and D.A. HUNTLEY

Department of Oceanography, Dalhousie University, Halifax, N.S. B3H 4J1 (Canada)

(Received August 4, 1983; revised and accepted November 1, 1983)

ABSTRACT

Bowen, A.J. and Huntley, D.A., 1984. Waves, long waves and nearshore morphology. In: B. Greenwood and R.A. Davis, Jr. (Editors), Hydrodynamics and Sedimentation in Wave-Dominated Coastal Environments. Mar. Geol., 60: 1—13.

Recent field measurements on beaches of different slopes have established that wave motion at periods substantially longer than the incident waves dominates the velocity field close to the shore. Analysis of a number of extensive data sets shows that much of this long wave motion is in the form of progressive edge waves, though forced wave motion, standing edge waves and free waves propagating away from the shore may also contribute to the energy.

Theoretically, the drift velocities in bottom boundary layers due to edge waves show spatial patterns of convergence and divergence which may move sediment to form either regular crescentic or cuspate features when only one edge wave mode dominates, or a bewildering array of bars, bumps and holes when several phase-locked modes exist together.

Convincing field demonstration of the link between nearshore topography and edge waves only exists for the special case of small-scale beach cusps on steep beaches, formed by edge waves at the subharmonic (twice the period) of the incident waves. At longer periods the link is proving more difficult to establish, due to the longer time-scales of topographic changes, the interaction between pre-existing topography and the water motion, and the observation of broad-banded edge wave motion which is not readily linked to topography with a well-defined scale.

These ideas are, however, central to the study of nearshore processes, as most of the plausible alternate hypotheses do not seem to lead to quantitative predictions. Clearly, further theoretical and observational work is essential.

INTRODUCTION

It is clear from the most superficial examination that the nearshore zone is a very active area of sediment transport. This activity is observed on both long time scales, seasonal changes in beach profiles being evident in many parts of the world, and on time scales of a few hours, during a major storm for example. It was natural that the early attempts to explain these processes focussed on the obvious driving mechanism, the waves moving in from deep water. The belief that these incident waves are the only important mechanism was not often explicitly stated, perhaps because it was already known that the strong, wave-driven currents occurred in and just outside the surf zone. Shepard et al. (1941) had pointed to the possible geological importance of rip currents. However, an expectation that the waves themselves, not the nearshore currents, are the dominant process was implicit in the large number of laboratory experiments in long, narrow wave flumes that sought to explain the shape of the beach (Rector, 1954) or the existence of longshore bars (Johnson, 1949; Iwagaki and Noda, 1963).

An interesting consequence of the assumption that the incoming waves are the dominant process was the very limited set of parameters that seemed to be involved in determining characteristics of the beach such as the beach profile. In the idealized laboratory experiment, a single wave frequency σ (or wavelength L) and height H characterized the waves, while the sand was apparently adequately described by a density and a typical grain size d. Any further property of the system was then only dependent on a very few non-dimensional numbers, H/L and H/d particularly. The criteria for the formation of a longshore bar given by Iwagaki and Noda (1963) are typical of this approach.

An interesting question is why these very sensible ideas did not lead to a satisfactory model for on-offshore processes on a beach. One way of focussing on this question is to look at conditions well inside the surf zone. To a very good approximation, breakers inside the surf zone are "saturated", that is, the wave height at any point is limited by the local depth (Fig.1). This is what might be expected on dimensional grounds (Longuet-Higgins, 1972) and what is found both in laboratory experiments using monochromatic waves and in detailed field measurements in conditions of both broad and narrow spectral distributions of wave energy (Thornton and Guza, 1982).

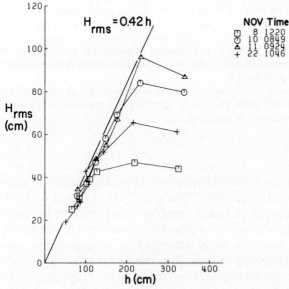

Fig.1. Saturation: the limitation of wave height H in the surf zone by the local depth, h (after Thornton and Guza, 1982).

If breakers are saturated, the local wave height is proportional to the depth, regardless of conditions offshore. How then does the beach face "know" that there are large waves offshore? To address this question it is useful to look more closely at the nature of the wave motion very close to the shore, in the run-up for example.

WAVES AND LONG WAVES

There has recently been considerable interest in the direct measurement of the swash movement, the run-up, at the shoreline using both run-up meters, essentially wave staffs laid parallel to the beach, and time-lapse photography (Holman and Guza, 1984). The observations show that in the region very close to the shore the dominant wave motion is not normally at frequencies typical of the incident waves. On steep beaches the run-up shows significant energy at the subharmonic frequency of the incident waves (Huntley and Bowen, 1973). On gently sloping beaches the frequencies are generally much lower and not obviously related to those of the incoming waves. However, the amplitude of the low-frequency motion is very clearly related to size of the incident waves. Figure 2 shows that the amplitude of the movement at the shoreline at low frequencies increases linearly with

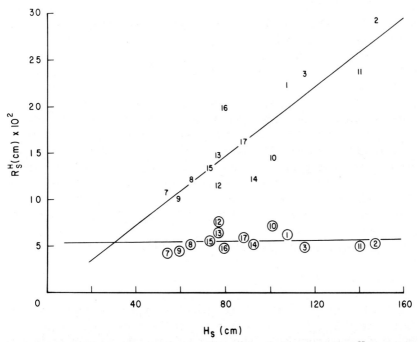

Fig. 2. The horizontal excursion in the swash at the shoreline R_s^H as a function of the significant waveheight offshore H_s. The open numbers show the magnitude of the significant excursion at periods longer than 25 s, those circled at periods less than 25 s (after Guza and Thornton, 1982).

increasing incident wave height while the amplitude at the incident frequency remains constant, as would be expected from the idea of saturation. The amplitude of the incident waves at the shoreline increases with breaker height for very small incident waves but reaches a constant value at a small value of ϵ (Guza and Bowen, 1976), where

$$\epsilon = \frac{H\sigma^2}{2g\beta^2} \tag{1}$$

and β is the beach slope. Thus we would expect shoreline amplitudes to be independent of offshore wave conditions when ϵ is large, which will generally be the case at incident wave frequencies where the waves are normally large enough to break. ϵ is, in fact, an important dimensionless parameter, a surf similarity measure, characterizing many properties of the system including, for example, the type of breaking.

The occurrence of low-frequency waves is not unexpected. Munk (1949) observed similar waves outside the surf zone and suggested that they were associated with the wave groups which naturally occur with the beating of several incident wave frequencies. This led to the adoption of the generic term "surf beat" for all motions in a frequency range from 30 s to several minutes. Despite the early evidence for the existence of such waves, they did not attract much interest. In retrospect this is surprising because it should have been evident that the wavelengths associated with these motions are of the same order as those of major multiple bar structures. There seems to have been a general impression that these waves were not only long, but also of low amplitude.

However, the results of Guza and Thornton (1982) shown in Fig.2 suggest that such low-frequency motions are dominant over a significant region in the surf zone and that this dominance increases in very severe conditions, precisely the conditions in which the most active sediment transport is expected to occur. It may be useful to emphasize that this dominance is not of one small quantity relative to another. The orbital velocities associated with these low frequency motions are of the same order as the currents associated with the incident waves and the wave-induced nearshore circulation, with typical values of the order of 1 m s^{-1}. This is clearly an important contribution to the total velocity field. To understand the way in which this motion influences the nearshore morphology, we need to know more about the onshore and longshore structure of this low-frequency activity.

SURF BEAT AND EDGE WAVES

Figure 3 schematically illustrates the principle types of waves that may contribute to the low-frequency motion in terms of the wave frequency σ and the longshore wavenumber λ. The solid lines denote a set of edge-wave modes whose amplitude is largest at the shoreline and dies away seaward so that the wave is trapped to the shoreline. Mathematically, these solutions require the dispersion relation:

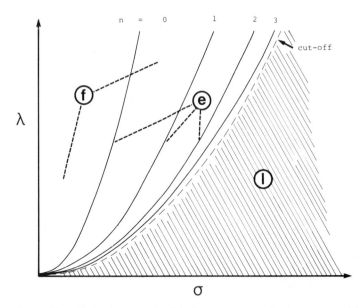

Fig.3. The dispersion relationship between wave frequency σ and longshore wave number, λ, for a beach slope $\beta = 0.12$. n is the modal number of the edge waves (e). l indicates the region of leaky modes and f illustrates areas of forced modes.

$$\sigma^2 = g\lambda \sin(2n + 1)\beta, \qquad n = 0, 1, 2 \ldots \qquad (2)$$

where n is the mode number shown in Fig.3. If $(2n + 1)\beta > \pi/2$ there are no trapped solutions, and waves having these (σ, λ) values may propagate to, or from, deep water and are consequently known as "leaky" modes. They include the normal incident waves.

The spaces between the edge wave modes are combinations of (σ, λ) values which do not satisfy the dispersion relation, eq.2. Any motion here is a forced wave and the response will theoretically be weaker than at the free modes for which the response is "resonant". There is an additional set of forced modes which do not satisfy the necessary dispersion relationship for free modes in the on-offshore direction. These waves have a frequency and wavenumber that do not satisfy the normal conditions for surface gravity waves. Examples are the harmonics of the incident waves and the set-down under groups of incident waves (Longuet-Higgins and Stewart, 1962).

There have been a number of observations of wave conditions near the shore that suggest that both leaky waves and edge waves are important components of this low-frequency motion (Suhayda, 1974; Huntley, 1976; Wright et al., 1979; Holman, 1981). However, in order to separate modes when a number of different modes, trapped and leaky, occur at the same frequency a large longshore array is necessary. Using data from the 520 m long longshore array at the first Nearshore Sediment Transport Study site, Torrey Pines Beach, California, Huntley et al. (1981) were able to determine the observed low-frequency energy spectra as functions of both frequency and longshore wavenumber, and hence plot the observed energy on a diagram like Fig.3. Figure 4 shows an example for longshore currents. Clearly most of the energy here occurs along the expected edge-wave dispersion curves. Their analysis also suggests that other forms of motion, possibly forced waves driven by incoming wave groups, also contribute to the energy. However, results like those shown in Fig.4 provide unambiguous confirmation that progressive edge waves are indeed present at surf-beat frequencies.

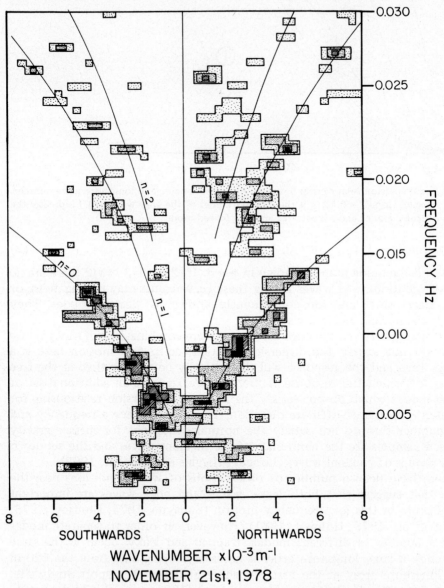

Fig.4. Contours of energy in the longshore component of velocity plotted on a (σ, λ) diagram, with negative and positive wavenumbers representing waves moving southwards or northwards, respectively, along the beach (after Huntley et al., 1979).

LONG WAVES AND NEARSHORE MORPHOLOGY

As previously mentioned, surf beat, whether a leaky mode or an edge wave, provides length scales which are very much of the same order as many of the morphological features actually observed (Bowen and Inman, 1971; Short, 1975; Bowen, 1980). Leaky modes are primarily seen in the on-off shore direction and are therefore most likely to play a role in the formation of shore parallel features such as longshore bars. Edge waves, on the other hand, have a well-defined longshore length scale and are therefore a possible cause of rhythmic longshore features. However, a single progressive edge wave provides no longshore variability on the average. It is seen, at different locations in the longshore direction, as exactly the same wave, simply arriving at slightly different times. To produce longshore variation at least two modes of the same frequency are necessary. Bowen and Inman (1971) showed that a standing edge wave, which can be regarded as two edge waves of the same frequency, mode and amplitude propagating in opposite directions, might generate a nearshore morphology of cusps and crescentic bars whose wavelength would be half that of the edge waves (Fig.5). The drift velocities, calculated from Hunt and Johns (1963), seemed to provide a pattern which would form the observed structure.

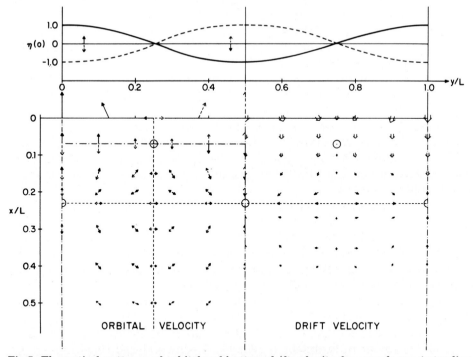

Fig.5. Theoretical patterns of orbital and bottom drift velocity for a mode $n = 1$ standing edge wave whose shoreline elevation $\eta(0)$ is shown above. L is the edge wave wavelength. The solid arrows indicate the orbital velocity at the time when $\eta = 0$ everywhere, the dashed arrows the velocity half a wave period later (after Bowen and Inman, 1971).

Holman and Bowen (1982) have recently extended these ideas to look at any combination of edge-wave modes of the same frequency. The waves can be of different modes, different amplitudes and may propagate in the same direction or in opposite directions. In every case, provided there is some coherence between the modes, a longshore pattern will result. For example, the combined elevation of two modes m and r is given by

$$\eta = a_m X_m(x)\sin(\lambda_m y - \sigma t) + a_r X_r(x)\sin(\lambda_r y - \sigma t + \alpha) \tag{3}$$

where a_m, a_r are the shoreline amplitudes of the two modes, α is the phase angle between the waves and X_m and X_r give the dependence of the wave amplitudes on the offshore co-ordinate x. If we assume that α is constant, i.e. that the waves are "phase-locked", and look at a region in which $X_m(x) \sim X_r(x)$, then eq.3 can be rewritten as

$$\eta = (a_m - a_r)X(x)\sin(\lambda_m y - \sigma t) + 2a_r X(x)\sin[0.5(\lambda_m + \lambda_r)y - \sigma t + \alpha/2]$$
$$\cos[0.5(\lambda_m - \lambda_r)y - \alpha/2] \tag{4}$$

If $a_m = a_r$ and $\lambda_m = -\lambda_r$, this result reduces to the standing edge wave studied by Bowen and Inman (1971), the negative wavenumber showing a wave propagating in the negative y direction. However, for any two modes the cosine term in eq.4 produces a longshore modulation which is fixed in space, provided α is constant. This modulation has wavelength $\pi/(\lambda_m - \lambda_r)$ so the wavelength tends to be longer if the waves move in the same direction than if they move in opposite directions. Again, using expressions derived by Hunt and Johns (1963), the drift velocity pattern can be computed for any combination of modes. Holman and Bowen (1982) introduced a simple sediment transport model to compute the equilibrium slope for any pattern of drift velocities. Figure 6 shows both the drift velocities and the computed topography generated by the interaction of edge waves of modes 1 and 2 moving in the same direction. The waves were assumed to be of equal amplitude at the shoreline for the purposes of the illustration. In this case the drift velocity pattern generates both topographic features and a residual transport of sediment up the coast.

There is an important difference between the pattern of transport and the shape of the morphology. The morphology shown is assumed to be in equilibrium with the wave and drift velocity fields. The condition which must then be satisfied is that the divergence of the sediment transport vector is zero. Sediment is moving through the system, but everywhere the transport into any small area is balanced by the transport out. Even, for the case of crescentic bars where there is no net longshore drift velocity, there remains a local circulation of the sediment over the bar system (Fig.7), a circulation very similar to that deduced from studies of sedimentary structure and the movement of sediment tracers by Greenwood and Davidson-Arnott (1979) and Greenwood and Hale (1982).

One interesting result found by Holman and Bowen (1982) was that the theoretical topography computed in their model did not look familiar. It

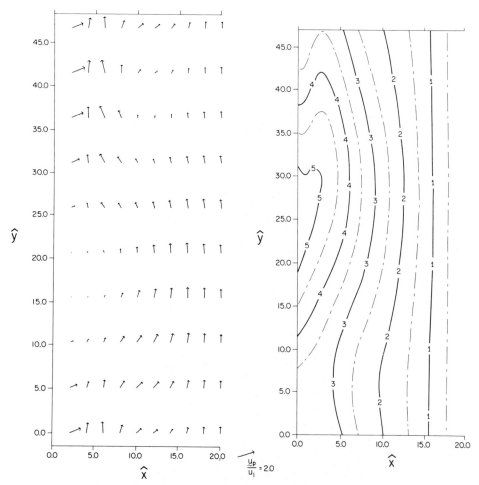

Fig.6. The dimensionless drift velocity, u_p/u_l, and derived equilibrium bottom contours for the interaction of two edge waves, modes 1 and 2, of the same frequency, both propagating in the direction of $\hat{y}$ positive. One complete wavelength of the pattern is shown. The coast is to the left, and $\hat{x}$ is the offshore coordinate (after Holman and Bowen, 1982).

only took on a familiar form, a crescentic bar for example, if it was added on to an overall beach slope. This has important implications when making measurements of sand bars in the field. It is natural to think of a bar crest as being at the point where the slope of the bottom is zero, i.e. where the local depth is a minimum. However the position of this point is dependent on the general slope of the beach in relation to the size of the perturbation creating the bar.

Furthermore, in the theoretical model there is no difference conceptually between a perturbation which gives a clear bar form and a smaller perturbation which produces a low tide terrace, a flattening of the profile rather than a bar per se. In fact it is clear that if the slope of the perturbation is

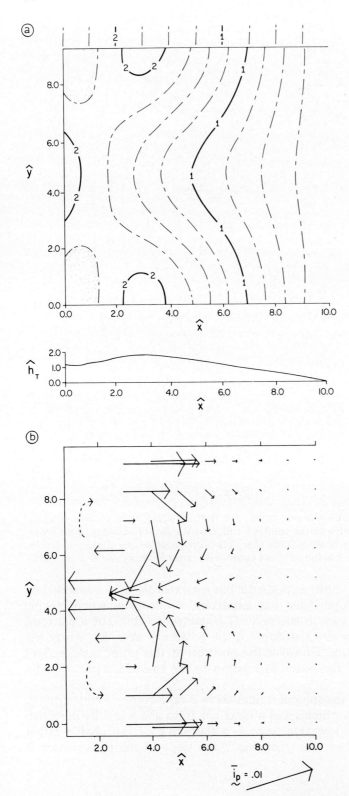

nowhere as large as the beach slope, no "bar" will be formed, while on a gentler beach the same perturbation will produce a bar.

These points may be readily illustrated by a simple model of a bar on a linearly sloping beach. Consider, for example, a Gaussian-shaped bar, centred at x_0, on a beach of slope β. The depth h can be written

$$h = -\beta x + a \exp -(x - x_0)^2 \tag{5}$$

The crest of the bar, defined to be where $\partial h/\partial x = 0$, will be where

$$\beta/2a = (x - x_0) \exp -(x - x_0)^2 \tag{6}$$

The minimum value of bar amplitude, a, for which a crest will occur will be where

$$\partial(\beta/2a)/\partial(x - x_0) = 0 \quad \text{i.e. } (x - x_0) = -0.71 \tag{7}$$

and this occurs for $\beta/2a = 0.43$. Larger values of this ratio will not cause a bar crest to be formed, merely a change in beach slope. Figure 8 shows some examples and illustrates how the bar "crest" can be significantly displaced from the location of maximum perturbation. This kind of displacement can make theoretical interpretation of measured bar locations difficult.

This clearly illustrates how low-frequency motion may influence the

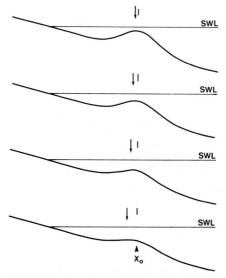

Fig.8. The beach profile described by eq.5, showing the effect of reducing the amplitude of the bar a. The position of the point of minimum depth, $dh/dx = 0$, is shown by the arrow.

Fig.7. a. Nearshore topography in the form of a crescentic bar generated by the interaction of two mode 1 edge waves illustrated in Fig.5. b. The residual sediment flux, $\bar{i}_p$, over this bar system. Transport is in at the horns of the bar, out over the cusp (after Holman and Bowen, 1981).

beach profile in a subtle way without necessarily providing obvious length scales. Nevertheless, there is a conceptual link between increasing waveheight, leading to increasing low frequency activity, larger perturbations in the topography and the formation of bars. This is entirely consistent with the conclusions of Iwagaki and Noda (1963) that large waves make bars, but the physical model is very different.

CONCLUSIONS

It is now clear that the velocities dominating the motion in the inner surf zone and at the shoreline are often those at surf-beat frequencies rather than incident wave frequencies. Recent measurements also show that the amplitude of this surf beat increases essentially linearly with the amplitude of the incident waves in deep water. The magnitude of these surf-beat velocities is on the order of 1 m s^{-1}.

Complete identification of this surf-beat energy has not yet been achieved, but it is clear that edge waves, trapped to the shoreline and propagating parallel to the shore, form a significant proportion of the total energy.

Numerical models of drift velocities and sediment transport show that regular topographic features are expected to be formed under a single long-period edge-wave mode, while complex patterns of bars, bumps and holes can occur when several modes, phase-locked together, occur simultaneously. Thus many of the observed topographic features on natural shorelines could be the result of edge-wave activity.

There remain, of course, a number of questions which are the subject of continuing research. Firstly, how do growing bars influence the edge waves which might produce them? Holman and Bowen (1982) ignore any feedback between the topography and the edgewaves but recent work by Kirby, Dalrymple and Liu (1981) suggests one approach to this problem. Secondly, do edge waves occur in nature over a narrow enough frequency range to explain the specific topographic length scales observed? Equation 2 shows that, on plane beach slopes, the wavelength depends strongly on the frequency. However, there are several possible mechanisms which might lead to a narrowing of the wavelength range, amongst them the feedback mechanism just mentioned. Thirdly, are different edge-wave modes sufficiently phase locked in nature to allow the range of features predicted by Holman and Bowen? Analysis is continuing in an attempt to answer this.

Despite these remaining uncertainties, the observations and conceptual ideas presented here lead to quantitative predictions of many features of bar shapes and scales and these must be tested against accurate measurements in the field. The simple model of a bar on a sloping beach (Fig.8) shows that this will require more than simply measuring the location of the bar "crest".

REFERENCES

Bagnold, R.A., 1963. Mechanics of marine sedimentation. In: M.N. Hill (Editor), The Sea, Vol. 3. Wiley-Interscience, New York, N.Y., pp.507—528.
Bowen, A.J., 1980. Simple models of nearshore sedimentation: beach profiles and longshore bars. In: S.B. McCann (Editor), The Coastline of Canada. Geol. Surv. Can., Ottawa, Ont., Pap. 80-10: 1—11.
Bowen, A.J. and Inman, D.L., 1971. Edge waves and crescentic bars. J. Geophys. Res., 74: 8662—8671.
Greenwood, B. and Davidson-Arnott, R.G.D., 1979. Sedimentation and equilibrium in wave formed bars: a review and case study. Can. J. Earth Sci., 16: 312—332.
Greenwood, B. and Hale, P.B., 1982. Lagrangian sediment motion in a crescentic nearshore bar under storm-induced waves and currents. Can. J. Earth Sci., 19: 424—433.
Guza, R.T. and Bowen, A.J., 1976. Resonant interactions for waves breaking on a beach. In: Proc. 15th Conf. on Coastal Engineering. Am. Soc. Civ. Eng., New York, N.Y., pp.560—579.
Guza, R.T. and Thornton, E.B., 1982. Swash oscillations on a natural beach. J. Geophys. Res., 87: 483—491.
Holman, R.A., 1981. Infragravity energy in the surf zone. J. Geophys. Res., 86: 6442—6450.
Holman, R.A. and Bowen, A.J., 1982. Bars, bumps and holes: models for the generation of complex beach topography. J. Geophys. Res., 87: 457—468.
Holman, R.A. and Guza, R.T., 1984. Measuring run-up on a natural beach. Coastal Eng., 8: 129—140.
Hunt, J.N. and Johns, B., 1963. Currents produced by tides and gravity waves. Tellus, 15: 343—351.
Huntley, D.A., 1976. Long period wave motion on a natural beach. J. Geophys. Res., 81: 6441—6449.
Huntley, D.A. and Bowen, A.J., 1973. Field observations of edge waves. Nature, 243: 160—162.
Huntley, D.A., Guza, R.T. and Thornton, E.B., 1981. Field observations of surf beat, 1, progressive edge waves. J. Geophys. Res., 83: 1913—1920.
Iwagaki, Y. and Noda, H., 1963. Laboratory study of scale effects in two dimensional beach processes. In: Proc. 8th Conf. on Coastal Engineering. Am. Soc. Civ. Eng., New York, N.Y., pp.194—210.
Johnson, J.W., 1949. Scale effects in hydraulic models involving wave motion. Trans. Am. Geophys. Union, 30: 517—525.
Kirby Jr., J.T., Dalrymple, R.A. and Liu, P.L.-F., 1981. Modification of edge waves by barred beach topography. Coastal Eng., 5: 35—49.
Longuet-Higgins, M.S., 1972. Recent progress in the study of longshore currents. In: R.E. Meyer (Editor), Waves on Beaches and Resulting Sediment Transport. Academic Press, New York, N.Y., pp.203—248.
Longuet-Higgins, M.S. and Stewart, R.W., 1962. Radiation stress and mass transport in gravity waves, with application to surf beats. J. Fluid Mech., 13: 481—504.
Munk, W.H., 1949. Surf beats. EOS, Trans. Am. Geophys. Union, 30: 849—854.
Rector, R.L., 1954. Laboratory study of equilibrium profiles of beaches. U.S. Army Corps Eng., Beach Erosion Board, Tech. Memo, 41, 38 pp.
Shepard, F.P., Emery, K.O. and Lafond, E.C., 1941. Rip currents: a process of geological importance. J. Geol. 49: 337—369.
Short, A.D., 1975. Multiple offshore bars and standing waves. J. Geophys. Res., 80: 3838—3840.
Suhayda, J.N., 1974. Standing waves on beaches. J. Geophys. Res., 79: 3065—3071.
Thornton, E.B. and Guza, R.T., 1982. Energy saturation and phase speeds measured on a natural beach. J. Geophys. Res., 87: 9499—9508.
Wright, L.D., Chappell, J., Thom, B.G., Bradshaw, M.P. and Cowell, P., 1979. Morphodynamics of reflective and dissipative beach and inshore systems: Southeastern Australia. Mar. Geol., 32: 105—140.

SPATIAL AND TEMPORAL VARIATIONS IN SPECTRA OF STORM WAVES ACROSS A BARRED NEARSHORE

ROBIN G.D. DAVIDSON-ARNOTT and DAVID C. RANDALL

Department of Geography, University of Guelph, Guelph, Ont. N1G 2W1 (Canada)

(Received April 16, 1983; revised and accepted December 15, 1983)

ABSTRACT

Davidson-Arnott, R.G.D. and Randall, D.C., 1984. Spatial and temporal variations in spectra of storm waves across a barred nearshore. In: B. Greenwood and R.A. Davis, Jr. (Editors), Hydrodynamics and Sedimentation in Wave-Dominated Coastal Environments. Mar. Geol., 60: 15—30.

Wave staffs and electromagnetic current meters were deployed on a profile across a two-bar system at Wendake Beach, southern Georgian Bay. This paper examines spatial and temporal changes in the characteristics of wave form, and the spectra of surface elevation and on-offshore current motion, during one storm. Non-linear effects of wave shoaling and breaking across the bars result in the appearance of secondary waves and both the wave and on-offshore current spectra have significant harmonic peaks during most of the storm. Significant low-frequency energy occurs only during the peak of the storm. While the peak frequency remains constant across the bar system, the proportion of energy in the primary peak is greatest in the troughs and lowest over the bar crests and there are similar changes in the proportion of energy in the first harmonic. However, in both surface elevation spectra and on-offshore current spectra, the greatest proportion of energy is found in frequencies related to the incident wind waves.

INTRODUCTION

On many gently sloping sandy coastlines the nearshore area is characterized by the presence of sand bars which assume a variety of forms, with most being aligned parallel to the shoreline (Greenwood and Davidson-Arnott, 1979). In some areas these are essentially permanent features of the nearshore profile (e.g. Greenwood and Davidson-Arnott, 1975; Davidson-Arnott and Pember, 1980; Hale and McCann, 1982) while in others they appear to be related to the occurrence of steep storm waves and may be destroyed during other times (e.g. Wright et al., 1979). It is evident that the existence of a barred, in contrast to a planar, profile reflects the establishment of a condition of dynamic equilibrium, resulting from a particular combination of processes controlling the rate and pattern of sediment transport in the breaker and surf zones. A number of controlling mechanisms have been suggested to account for the occurrence of the bars and their overall morphology and location, including standing waves (e.g. Short, 1975; Bowen, 1980), edge waves (Bowen and Inman, 1971; Holman and Bowen, 1982),

0025-3227/84/$03.00 © 1984 Elsevier Science Publishers B.V.

interaction between breaking waves and rip cell circulation (Greenwood and Davidson-Arnott, 1979) and harmonics generated during wave shoaling (Boczar-Karakiewicz et al., 1981). Ultimately, the occurrence of a non-barred or a barred profile is probably controlled by the characteristics of the incident waves and factors determining the type of shoaling deformation, breaking, and the amount of wave reflection.

Numerous studies have reported on the characteristics of wave transformation during shoaling, breaking and in the surf zone on planar slopes (e.g. Galvin, 1968; Divoky et al., 1970; Longuett-Higgins, 1976; Suhayda and Pettigrew, 1977; Shemdin et al., 1980) and on water motion associated with breaking waves (e.g. Thornton et al., 1976; Thornton, 1979; Battjes and Van Heteren, 1980). There have been fewer studies of wave transformation over barred profiles (e.g. Byrne, 1969; Wood, 1970; McNair and Sorensen, 1970; Busching, 1976; Wright et al., 1979; Mizuguchi, 1980) where the processes are complicated by multiple breaking and an irregular surf zone profile, with deeper water in the troughs leading to zones of wave reformation.

This paper reports on the form and spectral characteristics of waves over a barred system. In particular, attention is focussed on changes in the number and form of waves in the zones of wave shoaling, breaking and reformation, on spatial and temporal changes in the energy distribution within the wave spectra, and on the relationship between spectra from the water surface profile and the horizontal component of water motion below the surface. While consideration is given primarily to the effects of the bar topography on the incident waves, it is evident that some of the phenomena described here must in turn be considered in modelling overall nearshore sediment transport and, thus, ultimately in the formation of the bars themselves.

STUDY AREA

The study was carried out at Wendake Beach at the southeast end of Georgian Bay, Lake Ontario (Fig.1). The study site is located near the middle of a 5 km stretch of sandy beaches which form part of the head of Nottawasaga Bay. The beach and nearshore zone consist primarily of well-sorted fine to medium sand with occasional outcrops of till or bedrock, and is characterized by the presence of two or three bars. A nearly continuous, straight to sinuous outer bar extends parallel to the shore for most of the stretch of beach and there are one or two inner bars which are generally crescentic in form and broken by rather poorly defined rip channels. Three bars were present along the profile instrumented for the study (Fig.2). The outer bar was located about 100 m offshore with water depth over the crest of 1.6 m and in the trough 2.0 m. The second bar was located about 50 m offshore with water depth 1 m over the crest and there was a small, poorly developed bar just lakeward of the step. The outer bar was nearly symmetric in shape while the lakeward slope of the second bar was much steeper than the shoreward slope. Over the study period the outer bar

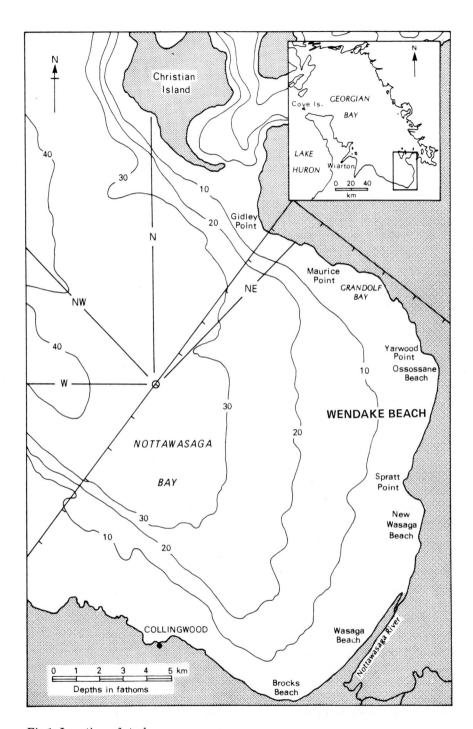

Fig.1. Location of study area.

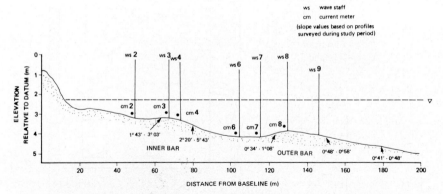

Fig.2. Profile across nearshore bars showing location of wave staffs and current meters used in study. Because of a failure in the electronics only frequency data is available for wave staff 8.

remained relatively stable, while the second bar grew in height and migrated about 15 m landward.

Prevailing winds are offshore from the southwest, with waves affecting the bars being generated by winds from the W, NW, and N, blowing over fetches of 40, 150 and 5 km, respectively, during the passage of depressions. During these storms, waves reaching Nottawasaga Bay have significant wave heights of 1—2 m and periods of 4—7 s (Davidson-Arnott and Pollard, 1980). The presence of shorefast ice and ice in Georgian Bay restricts wave processes for a 3—4 month period from mid-December. More details of the nearshore morphology and wave climate are given in Randall (1982).

METHODOLOGY

Water surface elevations and water motion were measured using continuous resistance wave staffs and biaxial electromagnetic flow meters. For the measurements reported here, nine wave staffs and six flow meters were placed along a shore normal profile in the configuration shown in Fig.2. Due to instrumentation problems data were not obtained from wave staffs 5 and 8. Pre-storm calibration indicated that wave staff linearity was good except for the lower 25 cm. Wind speed and direction were recorded using a D.C. wind generator and vane mounted on a 4 m high mast jetted into the sand on the beach. Analogue signals from each sensor were hard-wired to a Hewlett-Packard Data Aquisition System with analogue to digital scanner converter, and micro computer. Data were collected over 9-min spans with a sampling interval of 0.5 s. Spectral analysis was carried out for the first 1024 data points using a lag of 50. More details on instrumentation can be found in Greenwood and Sherman (1982).

STORM CHARACTERISTICS

A meteorological depression moving northeastward above Georgian Bay during May 31 and June 1 1980 generated storm waves affecting the study area over a period of about 19 h. During this time a total of ten sets of records were taken. Wind speed and direction recorded at the study site are presented in Fig.3 along with the significant wave height and peak period recorded at the outer wave staff (staff 9). Wave build-up occurred primarily in response to increasing wind speed and a clockwise shift in wind direction from the SW at about 18.00 h on May 31 to the NW at 22.00 h. At the peak of the storm average wind speeds recorded at the beach exceeded 9 m s^{-1} with corresponding wave height greater than 1.0 m and wave period greater than 5.0 s. After 1.00 h on 1 June, wind speed began to decrease gradually while wind direction shifted slowly towards the north, and correspondingly shorter fetch lengths, resulting in a slow decrease in both wave height and period.

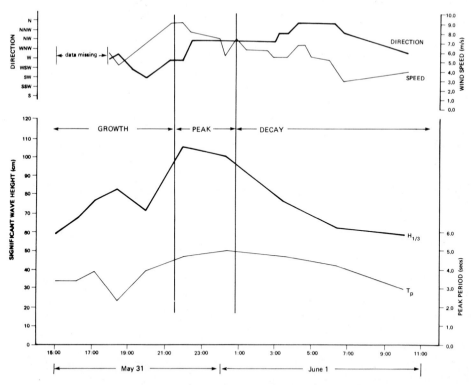

Fig.3. Trends for wind speed, wind direction, significant wave height ($4V$, where V^2 is the variance of the wave record) and peak period during the storm monitored here. Wave data are derived from spectra for wave staff 9.

SPATIAL VARIATION IN WAVE CHARACTERISTICS

Wave height

Changes in significant wave height across the bars are shown in Fig.4 for two runs (15.00 and 22.00 h on May 31) which reflect conditions during the growth and peak of the storm, respectively. The dashed lines indicate changes in wave height from one staff to the next over the bar and trough systems while the vertical lines show the changes in height at each staff in the interval between the two runs. Also shown in Fig.4 are values for the ratio of wave height to water depth (γ) changes in which reflect primarily the increase in wave height but also include the effects of increased water depth due to set-up.

The data for both runs show a similar spatial pattern with greatest heights occurring over the bar crest and a considerable reduction occurring in the trough landward of the bar, reflecting height loss through breaking on the bar, and the effect of deeper water in the trough. Even during the growth stage there is considerable breaking over the inner bar because of its shallow depth and steep lakeward slope, though during this time breaker type was

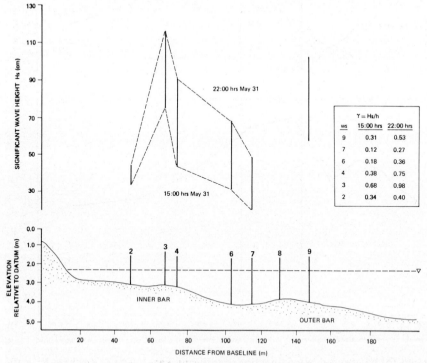

Fig.4. Spatial variations in significant wave height at the beginning and peak of the storm. The solid lines indicate the increase in wave height through time and the dashed lines indicate changes in height across the bar trough system.

observed to be primarily in the spilling range. At the peak of the storm there is a large increase in wave height on the inner bar, breaking is intense and occurs as a mixture of spilling and plunging types. The inshore breaker type parameters of Galvin (1968) and Battjes (1974) were calculated using significant wave height measured at staffs 9 and 3, and average slope values for the lakeward slope of the outer and inner bars, respectively. Values for the outer bar were well within the spilling range, but those for the inner system at the peak of the storm are close to the transition from spilling to plunging [0.79—0.85 compared to the transition value of 0.68 using Galvin's (1968) parameter]. If H 1/10 is used rather than H 1/3 then the values at the peak of the storm fall into the plunging category. Values of γ at staff 3 range from 0.68 to 1.06 over the storm and are thus in the range conventionally assumed for a highly dissipative surf zone. The intensity of breaking over the inner bar is reflected in the large reduction in significant wave height between staffs 3 and 2. The fact that there is little change in height and the value of γ at staff 2 over most of the storm is indicative of a saturated inner surf zone; and observations show the formation of surf bores rather than wave reformation in many instances.

Over the outer bar, while there is considerable peaking due to shoaling, breaking is confined to the largest waves during the growth phase and is of the spilling variety. During the peak of the storm breaking was more intense but still primarily within the spilling range with breaking ceasing as the waves moved into the trough. Thus values of γ recorded at staff 7 range from 0.12 to 0.27 and are much less than recorded for the saturated conditions landward to the inner bar. A puzzling feature is the magnitude of the height decrease in the trough (staffs 7 and 6) landward of the outer bar, particularly during the growth phase. It is difficult to account for this decrease on the basis of the observed breaker intensity and comparatively deep water over the bar crest, particularly as staffs 9 and 7 are located in approximately the same water depth.

Wave form and number

The change in wave characteristics across the nearshore profile can be seen visually by comparison of wave records at the peak of the storm (Fig.5). The water surface profiles from each staff have been time shifted so that the transformation of individual waves can be traced vertically. The effects of wave shoaling and breaking on the outer (staff 9) and inner (staffs 4 and 3) bars is seen in the increased wave height and steepness, and the peaked form of both the trough and the crest. A number of small crests are present, particularly on the crests and backs of larger waves. Some of these are small wind waves, reflecting continuous wave generation up to the beach during the storm, while others are probably secondary waves (solitons) resulting from breakdown of the primary waves during and just before breaking (e.g. Galvin, 1968; Byrne, 1969; Thornton et al., 1976). In the outer trough (staffs 7 and 6) the considerable reduction in wave height is readily apparent,

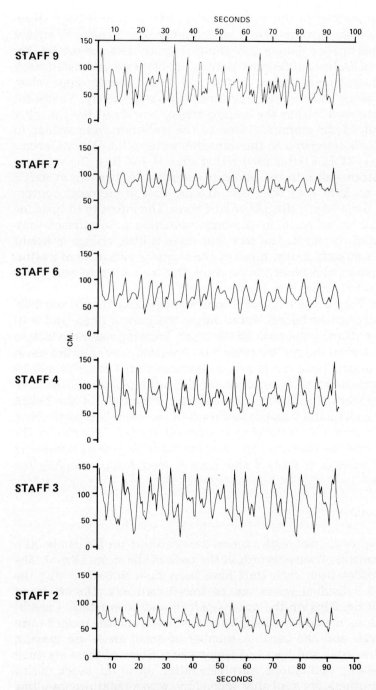

Fig. 5. Portion of the analogue record from wave staffs 9, 7, 6, 4, 3 and 2 taken at 00.30 h on June 1. The records have been time shifted so that individual waves can be traced vertically.

waves are less steep and troughs are more rounded. There is also, particularly at staff 7, a noticeable reduction in the number of smaller peaks and these re-appear as the waves shoal and break across the inner bar. However, landward of the inner bar at staff 2, while there is an obvious height reduction due to breaking, there does not appear to be the same reduction in the number of secondary crests. Actual counts of both visible peaks and zero crossings for two sets of records at the peak of the storm confirm this general pattern and show approximately a 30% reduction in wave crests between staffs 9 and 7. There is, however, little change in the number of wave crests between staffs 3 and 2.

Wave and current spectra

Spatial and temporal changes in the magnitude and distribution of energy associated with both water surface elevations and with the onshore-offshore component of water motion were examined by comparison of their spectra. Examples of the spectra recorded during the peak of the storm are shown in Fig.6.

The spectra for water surface elevation are typical of those associated with storm waves in fetch limited conditions, with most of the energy in a relatively narrow primary peak, a rapid decay towards lower frequencies and a more gentle slope towards the higher frequencies reflecting saturation in the equilibrium range (Thornton, 1979). There is considerable temporal and spatial variation in the shape of individual spectra both in terms of the size and number of peaks. Most of the spectra contain a single significant primary peak and many exhibit one or more secondary peaks at frequencies both higher (harmonic) and lower (sub-harmonic) than the primary peak frequency. With very low waves at the beginning and end of the storm a prominent primary peak is absent. Otherwise it ranged from 0.16 to 0.49 Hz (2.0—6.0 s) with changes reflecting the changes in wind speed and fetch described earlier. As has been found elsewhere (e.g. Thornton et al., 1976) there was no detectable change in the peak frequency across the breaker zones.

Secondary peaks at frequencies approximately twice that of the primary peak, the first harmonic, are most pronounced at the height of the storm. This peak is generally associated with non-linearities generated during shoaling and breaking and with the appearance of solitons. As would be expected, therefore, these peaks are best developed on the lakeward slopes and bar crests (staffs 9, 4 and 3) and are generally less apparent in the troughs (staffs 7, 6, 2; see Fig.6). They also occur for a greater proportion of the storm in the inner system. No significant peaks are present at higher order harmonics, but this is not surprising in view of the relatively high frequency of the primary peak.

In general there appears to be only limited energy at frequencies lower than the peak frequency in the wave record. However, during the height of the storm when wave breaking is most intense, a significant peak does develop at a frequency of about 1/4 of the primary peak. This is most

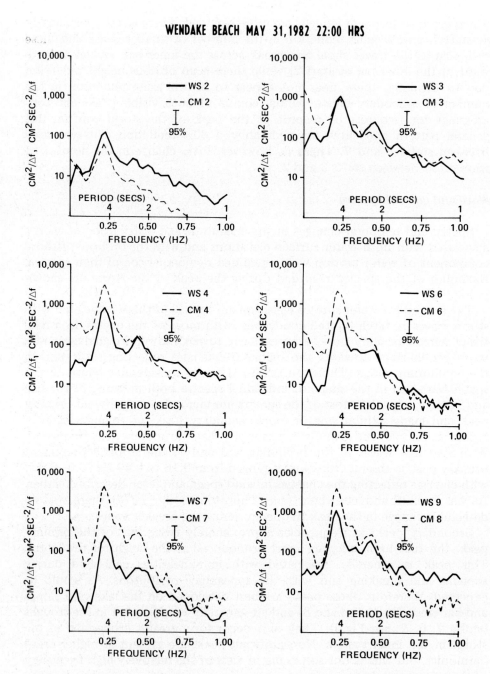

Fig.6. Spectra from wave staffs and X (onshore-offshore) component of electro magnetic current-meter records.

evident in the outer trough (staffs 7 and 6) and in the inner system (staffs 3 and 2) and is not significant in the breaker zones on the lakeward side of the bar crests (staffs 9 and 4). The low-frequency peak is discussed in more detail by Greenwood and Sherman (1984, this volume).

The spectra of onshore/offshore water motion measured at mid-depth are similar in general shape to that of surface water elevations, with most of the energy centred around the primary peak (see Fig.6). The secondary peak at the first harmonic is also well-developed. Thus, as has been noted in other areas, (e.g. Thornton et al., 1976) with spilling breakers water motion below the wave trough appears to be dominated by the incident wave field rather than by turbulence due to breaking. However, there is generally more energy at low frequencies in the current meter record and instead of a pronounced peak at 1/4 of the primary peak, the energy density increases linearly.

Temporal and spatial changes in energy distribution

Temporal changes in peak period are shown in Fig.2 and, as noted earlier, the peak period remained constant across the bar-trough system at all times. However, there were noticeable changes in the distribution of energy within the spectra both spatially and temporally, and an attempt was made to examine changes in the proportion of energy within the peak frequency (Vf) and in the first harmonic (V_2f). Several methods of defining the limits of the peak (e.g. Huntley, 1980) were tested but did not prove consistent, and instead a fixed bandwidth of 0.067 Hz centred on the primary peak was used. The upper and lower limits of the first harmonic were then determined at twice the upper and lower cut-off frequencies of the primary peak. Except at the height of the storm, where a small portion extended beyond the fixed bandwidth, the technique appeared to fit the primary peak well, and where a significant harmonic peak was present in the record, this also appeared to coincide well with the defined limits. These results are summarised in Table I.

The proportion of energy in the primary peak ranges from under 30% at the beginning and end of the storm to about 45% at the height of the storm when the primary peak is particularly well defined. Temporal variations in the proportion of energy in the first harmonic are much smaller but there is a similar, if not as well defined, pattern.

There appears to be a distinct pattern to spatial variations in the energy in the primary peak. Except at the height of the storm the proportion of energy is much greater in the trough (staffs 7 and 6) than on the outer bar (staff 9), and it again falls over the inner bar. Landward of the inner bar, the proportion is lower during most of the storm but higher than on the preceding crest at the beginning and end when waves are lowest. The spatial variations in V_2f are again much smaller and the pattern is less distinct. Over much of the storm the proportions of energy in the first harmonic are greatest at staff 2, the inner trough, and they are frequently higher in the outer trough as well although this is not consistent and the differences are small.

TABLE I

Variance in primary peak and first harmonic as a percentage of the total variance within each spectrum through the storm

Wave staff	Time									
	15:00	16:15	17:15	18:30	20:00	22:00	00:25	02:20	06:20	10:15
Primary peak										
9	*	26.2	30.5	22.7	24.4	45.0	35.0	26.8	16.0	*
7	*	42.0	40.1	28.9	34.1	43.1	37.3	32.9	30.2	*
6	*	43.5	36.0	30.1	36.0	44.3	36.2	34.2	35.5	*
4	31.8	38.1	31.9	26.2	31.2	42.8	39.2	28.9	29.1	24.5
3	32.0	34.9	25.4	22.3	30.1	38.6	34.7	27.8	26.0	27.2
2	36.3	40.8	30.6	19.3	25.9	33.8	29.8	30.5	34.0	28.0
First harmonic										
9	*	7.2	11.8	8.6	14.8	14.7	14.5	13.7	14.6	*
7	*	8.5	10.4	9.8	12.7	14.8	16.3	13.5	16.8	*
6	*	5.4	10.0	7.7	16.0	15.0	18.3	11.1	12.9	*
4	10.2	8.5	12.8	9.3	12.7	13.5	12.6	11.0	14.9	6.1
3	10.3	12.7	14.3	11.0	15.3	15.5	14.7	11.7	18.9	5.4
2	7.4	9.0	16.6	14.3	19.2	12.7	20.3	15.8	11.5	7.4

*No significant primary peak.

DISCUSSION

It is evident that, under storm conditions, the portion of the nearshore zone studied is typically dissipative rather than reflective (e.g. Guza and Bowen, 1976; Chappell and Wright, 1978; Wright et al., 1979). Waves break some distance offshore as spilling and plunging breakers and there is considerable energy transfer to higher harmonics (Thornton, 1979). The presence of the bars results in a segmentation of the surf zone into areas of high turbulence (the bar crests) and areas of lower turbulent dissipation (the troughs) with the intensity varying with incident wave height and between the outer and inner bar system.

The lakeward slope and crest of both bars is an area of wave shoaling and breaking. The shallower water over the inner bar crest and landward trough, however, results in greater intensity of wave breaking over the inner bar and some differences between conditions in the outer and inner troughs. At staff 2, landward of the inner bar, conditions appear similar to that of a saturated inner surf zone with frequent generation of surf bores and wave height controlled by turbulent dissipation (e.g. Horikawa and Kuo, 1966; Suhayda and Pettigrew, 1977). The inner bar acts as an effective filter for wave height over much of the storm and this is reflected in the nearly constant value for γ. The value of 0.4 recorded for γ is similar to that reported by Wright et al. (1982).

Over the outer trough the greater water depth and reduced breaker intensity leads to much lower values for γ. Wave breaking generally decreases as the wave moves into the trough and surf bores are not formed. This zone is frequently referred to as one of wave reformation though, under these conditions it is perhaps better described in terms of cessation of breaking since the primary wave form is never actually destroyed. The large decrease in wave height landward of the inner bar crest can be accounted for by wave breaking. However, the decrease over the outer bar between wave staffs 9 and 7, particularly during the early and late stage of the storm when there was little breaking there, is more difficult to explain. Since staffs 9 and 7 are in approximately the same water depth the changes cannot be explained purely on the basis of shoaling and changes in celerity. Some energy loss may be due to bed friction over the shallow bar crest. Another possible explanation may be related to spatial changes in the relative position of primary waves and the secondary waves generated during breaking and intense shoaling. In the trough these would be out of phase leading to an overall decrease in the amplitude of water surface fluctuations. As incident waves catch up with secondary waves generated by the preceding wave the overall amplitude of water motion would increase. The general increase in wave height at staff 6, although there is a slight increase in water depth suggests that this may in fact be occurring.

The transfer of energy to higher harmonics, which can be seen in the well developed peaks in the spectra of both water surface elevations and the on-offshore water motion, has been reported from a number of field and laboratory studies. Examination of actual wave profiles such as those shown in Fig. 5, suggests that the secondary waves generated are not phase locked, and thus travel more slowly through the trough than the primary waves from which they were derived. Interaction of these secondary waves with succeeding waves can produce spatial variations which are dependent on the recurrence length and which can therefore influence the location of breaker zones and the spacing of bars (e.g. Hulsbergen, 1974; Bijker et al., 1976; Boczar-Karakiewicz et al., 1981). Because of the presence of considerable incident wind wave energy at the first harmonic frequency and the generation of a second set of secondary waves as waves shoal over the inner bar, the behaviour of the secondary waves seen here is not easily traced and further work is necessary to determine their importance in this bar system.

Although the spectra of water surface elevations and on-offshore water motion show close agreement at the peak frequency and at higher harmonics, they differ somewhat at the low-frequency end of the spectrum. The surface elevation spectra show a fairly pronounced peak at about 1/4 the incident frequency while the current meter spectra show an essentially continuous increase in energy at frequencies below 0.125 Hz. The low-frequency peak is best developed in the trough and landward edge of the bar (staffs 8, 7, 6 and 3, 2), and occurs only at the peak and beginning of the decay phase of the storm when there is fairly intense breaking on the outer bar. There is a number of possible mechanisms for generating this sub-harmonic peak. Numerous investigations have reported on the generation of standing edge waves (e.g. Guza and Davis, 1974; Bowen and Guza, 1978; Huntley and

Bowen, 1978; Huntley et al., 1981). Chappell and Wright (1978) found standing edge waves at about 1/4 of the incident frequency were dominant in a bar trough system at the highest energy levels but were subordinate to edge waves at 1/2 of the incident frequencies with lower incident waves. However, the peak here is not well-developed in the current meter spectra and does not decrease offshore very rapidly as would be expected for a zero mode standing edge wave [see Greenwood and Sherman, (1984, this volume) for further discussion]. A second possibility is a resonant response of water levels in the trough to intense wave breaking over the bar crest, possibly related to groupiness in the incident wave record (e.g. Symonds and Huntley, 1980; Symonds et al., 1982).

Finally, a number of mechanisms has been proposed to explain bar formation, location and form which are related to phenomena associated with both higher harmonics (e.g. Boczar-Karakiewicz et al., 1981) and sub-harmonics (e.g. Holman and Bowen, 1982). From the analysis to date it is not possible to comment on the validity of these mechanisms in relation to the bar-trough systems studied. However, three points are worthy of note: (1) in both surface elevation spectra and on-offshore spectra the greatest proportion of energy is found in frequencies related to incident wind waves; (2) there are considerable temporal variations in the proportion of harmonic and sub-harmonic energy present and changes in the significance of these is much greater than the change in bar morphology during the storm; and (3) it seems likely that the bar morphology is the primary control on the spacing of the breaker zones and on the nature of the resonant interaction generating sub-harmonic energy (e.g. Huntley, 1980). The bars studied here are therefore macroscale features reflecting the cumulative effects of a number of storm events of different magnitudes rather than a bedform responding instantaneously to changing wave conditions. It also seems likely that no single mechanism, such as edge waves of a particular frequency, or recurrence length, can be used to explain the final morphological form. Rather, it may reflect an equilibrium resulting from the interaction of several mechanisms controlling the spatial and temporal patterns of sediment transport with the primary processes being related to the incident wind waves.

ACKNOWLEDGEMENTS

This paper results from a study carried out jointly with Brian Greenwood and Doug Sherman of Scarborough College, University of Toronto. Paul Cristilaw, Leslie Joynt and Ross Sutherland provided valuable field assistance. We are grateful to the support staff in the Geography Department, University of Guelph, for their technical, secretarial, cartographic and computing help. The paper has benefited from comments by David Huntley, Ed Thornton and Brian Greenwood. The work was supported by a N.S.E.R.C. operating grant (R.D.A.) and a University of Guelph Graduate Scholarship (D.R.).

REFERENCES

Battjes, J.A., 1974. Surf similarity. Proc. 14th Coastal Engineering Conference, Copenhagen, pp.466—480.
Battjes, J.A. and Van Heteren, J., 1980. Field measurements of wind wave kinematics. Proc. 17th Coastal Engineering Conference, Sydney, N.S.W., pp.347—362.
Bijker, E.W., Van Hijum, E. and Vellinga, P., 1976. Sand transport by waves. Proc. 15th Coastal Engineering Conference, Honolulu, Hawaii, pp.1149—1167.
Boczar-Karakiewicz, B., Paplinska, B. and Winiecki, J., 1981. Formation of sand bars by surface waves in shallow water. Laboratory experiments. Rozpr. Hydrotech., 41: 111—125.
Bowen, A.J., 1980. Simple models of nearshore sedimentation: beach profiles and longshore bars. In: S.B. McCann (Editor), The Coastline of Canada. Geol. Surv. Can., Pap. 80-10: 1—11.
Bowen, A.J. and Guza, R.T., 1978. Edge waves and surf beat. J. Geophys. Res., 83(4): 1913—1920.
Bowen, A.J. and Inman, D.L., 1971. Edge waves and crescentic bars. J. Geophys. Res., 76(36): 8662—8671.
Busching, F., 1976. Energy spectra of irregular surf waves. Proc. 15th Coastal Engineering Conference, Honolulu, Hawaii, pp.539—559.
Byrne, R.J., 1969. Field occurrences of induced multiple gravity waves. J. Geophys. Res., 74(10): 2590—2596.
Chappell, J. and Wright, L.D., 1978. Surf zone resonance and coupled morphology. Proc. 16th Coastal Engineering Conference, Hamburg, pp.1359—1377.
Davidson-Arnott, R.G.D. and Pember, G.F., 1980. Morphology and sedimentology of multiple bar systems, Southern Georgian Bay, Ontario. In: S.B. McCann (Editor), The Coastline of Canada. Geol. Surv. Can., Pap. 80-10: 417—428.
Davidson-Arnott, R.G.D. and Pollard, W.H., 1980. Wave climate and potential longshore sediment transport patterns, Nottawasaga Bay, Ontario. J. Great Lakes Res., 6(1): 54—67.
Divoky, D., Le Mehaute, B. and Lin, A., 1970. Breaking waves on gentle slopes. J. Geophys. Res., 75(9): 1681—1692.
Galvin, C.J., 1968. Breaker type classification on three laboratory beaches. J. Geophys. Res., 73(12): 3615—3659.
Greenwood, B. and Davidson-Arnott, R.G.D., 1975. Marine bars and nearshore sedimentary processes, Kouchibouguac Bay, New Brunswick. In: J. Hails and A. Carr (Editors), Nearshore Sediment Dynamics and Sedimentation. Wiley, New York, N.Y., pp.123—150.
Greenwood, B. and Davidson-Arnott, R.G.D., 1979. Sedimentation and equilibrium in wave-formed bars: a review and case study. Can. J. Earth Sci., 16(2): 312—332.
Greenwood, B. and Sherman, D.J., 1983. Shore-parallel flows in a barred nearshore. Proc. 18th Coastal Engineering Conference, Cape Town, pp.1677—1696.
Greenwood, B. and Sherman, D.J., 1984. Waves, currents, sediment flux and morphological response in a barred nearshore. In: B. Greenwood and R.A. Davis, Jr. (Editors), Hydrodynamics and Sedimentation in Wave-Dominated Coastal Environments. Mar. Geol., 60: 31—61 (this volume).
Guza, R.T. and Bowen, A.J., 1976. Resonant interactions from waves breaking on a beach. Proc. 15th Coastal Engineering Conference, Honolulu, Hawaii, pp. 560—579.
Guza, R.T. and Davis, R.E., 1974. Excitation of edge waves by waves incident on a beach. J. Geophys. Res., 79: 1285—1291.
Hale, P.B. and McCann, S.B., 1982. Rhythmic topography in a meso-tidal, low-wave-energy environment. J. Sediment. Petrol., 52: 415—429.
Holman, R.A. and Bowen, A.J., 1982. Bars, Bumps and Holes: Models for the generation of complex beach topography. J. Geophys. Res., 87: 457—468.

Horikawa, K. and Kuo, C., 1966. A study on wave transformation inside the surf zone. Proc. 10th Coastal Engineering Conference, Tokyo, pp.217—233.
Hulsbergen, C.H., 1974. Origin, effect and suppression of secondary waves. Proc. 14th Coastal Engineering Conference, Copenhagen, 392—411.
Huntley, D.A., 1980. Edge waves in a crescentic bar system. In: S.B. McCann (Editor), The Coastline of Canada. Geol. Surv. Can., Pap. 80-10: 111—121.
Huntley, D.A. and Bowen, A.J., 1978. Beach cusps and edge waves. Proc. 16th Coastal Engineering Conference, Hamburg, pp.1378—1393.
Huntley, D.A., Guza, R.T. and Thornton, E.B., 1981. Field observations of surf beat 1: Progressive edge waves. J. Geophys. Res., 85: 6451—6466.
Longuett-Higgins, M.S., 1976. Recent developments in the study of breaking waves. Proc. 15th Coastal Engineering Conference, Honolulu, Hawaii, pp.441—460.
McNair, E.D. and Sorensen, R.M., 1970. Characteristics of waves broken by a longshore bar. Proc. 13th Coastal Engineering Conference, Washington D.C., pp.415—434.
Mizuguchi, M., 1980. An heuristic model of wave height distribution in surfzone. Proc. 17th Coastal Engineering Conference, Sydney, N.S.W., pp. 278—289.
Randall, D.C., 1982. Changes in the form and energy spectra of storm waves across a nearshore bar system, Wendake Beach, Nottawasaga Bay, Ontario. Unpubl. M.Sc. thesis, Univ. of Guelph, Guelph, Ont., 133 pp.
Shemdin, O.H., Hsiao, S.V., Carlson, H.E., Hasselmann, K. and Schulze, K., 1980. Mechanisms of wave transformation in finite-depth water. J. Geophys. Res., 85(9): 5012—5018.
Short, A.D., 1975. Multiple offshore bars and standing waves. J. Geophys. Res., 80(27): 3838—3840.
Suhayda, J.N. and Pettigrew, N.R., 1977. Observations of wave height and wave celerity in the surf zone. J. Geophys. Res., 82(9): 1419—1424.
Symonds, G. and Huntley, D.A., 1980. Waves and currents over a nearshore bar system. Proc. of the Canadian Coastal Conference. Natl. Res. Council Can., Burlington, Ont., pp. 64—78.
Symonds, G., Huntley, D.A. and Bowen, A.J., 1982. Two-dimensional surf beat: Long wave generation by a time-varying break point. J. Geophys. Res., 87: 492—498.
Thornton, E.B., 1979. Energetics of breaking waves within the surf zone. J. Geophys. Res., 84: 4931—4938.
Thornton, E.B., Galvin, J.J., Bub, F.L. and Richardson, D.P., 1976. Kinematics of breaking waves. Proc. 15th Coastal Engineering Conference, Honolulu, Hawaii, pp.461—476.
Wood, W.L., 1970. Transformation of breaking wave characteristics over a submarine bar. Dept. of Natural Science, Michigan State University, Tech. Rep. 4.
Wright, L.D., Chappell, J., Thom, B.G., Bradshaw, M.P. and Cowell, P., 1979. Morphodynamics of reflective and dissipative beach and inshore systems: Southeastern Australia. Mar. Geol., 32: 105—140.
Wright, L.D., Guza, R.T. and Short, A.D., 1982. Dynamics of a high-energy dissipative surf zone. Mar. Geol., 45: 41—62.

WAVES, CURRENTS, SEDIMENT FLUX AND MORPHOLOGICAL RESPONSE IN A BARRED NEARSHORE SYSTEM

BRIAN GREENWOOD and DOUGLAS J. SHERMAN*

Departments of Geography and Geology, Scarborough Campus, University of Toronto, Scarborough, Ont. M1C 1A4 (Canada)
Department of Geography, Scarborough Campus, University of Toronto, Scarborough, Ont. M1C 1A4 (Canada)

(Received April 16, 1983; revised and accepted August 31, 1983)

ABSTRACT

Greenwood, B. and Sherman, D.J., 1984. Waves, currents, sediment flux and morphological response in a barred nearshore system. In: B. Greenwood and R.A. Davis, Jr. (Editors), Hydrodynamics and Sedimentation in Wave-Dominated Coastal Environments. Mar. Geol., 60: 31—61.

A shore-normal array of seven, bi-directional electromagnetic flowmeters and nine surface piercing, continuous resistance wave staffs were deployed across a multiple barred nearshore at Wendake Beach, Georgian Bay, Canada, and monitored for a complete storm cycle. Time-integrated estimates of total (ITVF) and net (INVF) sediment volume flux together with bed elevation changes were determined using *depth-of-activity rods*.

The three bars, ranging in height from 0.10 to 0.40 m accreted during the storm (0.03 m), and the troughs were scoured (0.05 m). Sediment reactivation depths reached 0.14 m and 12% of the nearshore control volume was mobilized. However, the INVF value for the storm was less than 1% of the control volume revealing a near balance in sediment volume in the bar system. Landward migration of the inner, crescentic and second, sinuous bars occurred in association with an alongshore migration of the bar form itself; the outermost, straight, shore-parallel bar remained fixed in location.

The surf zone was highly dissipative throughout the storm ($\epsilon = 3.8 \times 10^2$—$192 \times 10^2$) and the wave spectrum was dominated by energy at the incident frequency. Spectral peaks at frequencies of the first harmonic and at one quarter that of the incident wave were associated with secondary wave generation just prior to breaking and a standing edge wave, respectively. The former spectral peak was within the 95% confidence band for the spectrum while the latter contributed not more than 10% to the total energy in the surface elevation spectrum even near the shoreline.

During the storm wave height exceeded 2 m (H_s) and periods reached 5 s (T_{pk}): orbital velocities exceeded 0.5 m s^{-1} (u_{rms}) and were above the threshold of motion for the medium-to-fine sands throughout the storm. Shore-parallel flows in excess of 0.4 m s^{-1} were recorded with maxima in the troughs and minima just landward of the bar crest.

The rate and direction of sediment flux is best explained by the interaction of antecedent bed slopes with spatial gradients in the mean and asymmetry of the shore-normal velocity field. These hydrodynamic parameters represent "steady" flows superimposed on the dominantly oscillatory motion and assumed a characteristic spatial pattern from the

---

*Present address: Department of Geography, University of Southern California, Los Angeles, CA 90007, U.S.A.

storm peak through the decay period. Increases spatially in the magnitudes of both the mean flows and flow asymmetries cause an increasing net transport potential (erosion); decreases in these values spatially cause a decreasing net transport potential and thus deposition. These transport potentials are increased or decreased through the gravity potential induced by the local bed slope. Shore-parallel flow was important in explaining sediment flux and morphological change where orbital velocities, mean flows and flow asymmetries were at a minimum.

INTRODUCTION

Since the first scientific description of wave-formed bars by Elie de Beaumont in 1845, considerable effort has been devoted to understanding their origins and dynamics. A large literature exists on the morphology, and more recently the sedimentology, of these ubiquitous geomorphological features. Many questions remain, however, concerning the complex interactions between the nearshore fluid and sediment motions, and the topography itself (for a review see Greenwood and Davidson-Arnott, 1979; and Greenwood, 1982). Knowledge of these interactions is important, since bars frequently form stable, equilibrium, bathymetric configurations in a wide range of coastal environments, often under conditions of high alongshore sediment transport (Greenwood and Davidson-Arnott, 1979; Greenwood and Mittler, 1984, this volume). In other places, in contrast, bars are formed and destroyed as conditions change (Goldsmith et al., 1982; Bowman and Goldsmith, 1983).

Most studies of the hydrodynamics and sediment dynamics of barred nearshores have been restricted to theoretical and laboratory research with the associated problems of scale in the latter. Only recently has instrumentation been available allowing comprehensive experiments in the prototype, capable of providing data for the testing of theoretical models. Since sediment flux and bar morphodynamics are a direct response to wave-induced currents and since the latter (at least at present) are the more easily measured, then examination of such fluid motions should provide insights into sediment and bar dynamics. This paper documents an experiment designed to monitor the incident and secondary wave characteristics, wave-induced orbital and shore-parallel flows, and the local sedimentary response during a single storm event in a barred nearshore in the Canadian Great Lakes.

LOCATION OF STUDY

The study site is at Wendake Beach in Southern Georgian Bay, Ontario (Fig.1). It is tideless, and waves are generated over discrete time intervals during the passage of meteorological depressions, causing rapid changes in the magnitude and direction of incident wave energy. Fetch lengths vary, with maxima to the WNW (84 km) and W (51 km). With limited fetch and local storm winds, wind forcing of the waves is continuous to the shoreline under most wave states.

In the Wendake Beach system 3 bars are present on a mean slope of 0.015 in medium to fine sands (0.43—0.13 mm mean diameter). The outer bar is

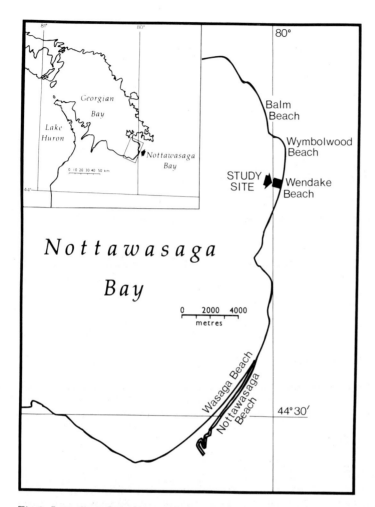

Fig.1. Location of study area.

110 m from the shoreline, approximately 0.50 m high, relatively straight, near-symmetrical and shore-parallel (Fig.2). The inner bars, at approximately 54 and 10 m from the shoreline, respectively, range in height from 0.10 to 0.35 m and are asymmetrical in section and sinuous to crescentic in planform. There is a reduction in spacing of the bars landward and, although some alongshore periodicity in form is evident in the inner bars, it is difficult to isolate: the dominant periodicity in form is in the shore-normal direction and thus the process(es) controlling sedimentation are most probably periodic in the same direction. While the bars are distinct bathymetric configurations, the nearshore slope deviates little from a planar configuration: a least squares linear fit explains more than 90% of the observed variance in relief.

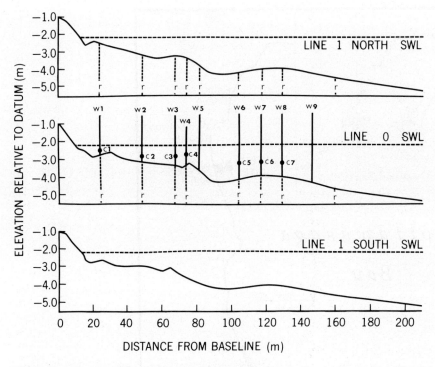

Fig.2. Nearshore profiles and instrument deployment prior to the storm, Wendake Beach, 1980:05:31. Line zero is the instrument array transect; lines 1 North and 1 South are spaced 30 m either side of the zero line. W = wave wire; C = current meter; r = depth of activity rod. Note that along the instrument transect the depth of activity rods were offset 1 m to avoid interference by the wave wire supports.

EXPERIMENTAL DESIGN

The field experiment was designed specifically to examine the two-dimensional nature of the nearshore hydrodynamics and sedimentation in a plane normal to the shoreline. It involved the deployment of a shore-normal array of wave and current sensors cable-linked to a high-speed data acquisition and mass storage system. Morphological changes were monitored using conventional profiling with level and staff in shallow water and echo sounding offshore; more detailed measurements of sediment flux and bed elevation change were made using *depth-of-activity* rods. More detail on the fluid monitoring system can be found in Greenwood (1982) and Greenwood and Sherman (1983); the depth-of-activity rods are described fully in Greenwood and Hale (1980), Greenwood et al. (1980) and Greenwood and Mittler (1984, this volume). Only a synopsis of methods will be presented here.

Wave and current sensors

Two types of instruments were used to monitor the fluid motions. Water surface elevation changes associated with waves were measured by surface piercing, continuous resistance wave staffs following the basic design of Truxillo (1970). Helically wound, steel wires set in 2 or 3 m long, 18.8 mm diameter PVC pipes were mounted by insulated brackets on 37.5 mm galvanized steel pipes. The latter were mounted on two metre long bases jetted into the bed. Staffs were field calibrated individually to specific oscillator-detector circuits and were linear except for the lowermost 0.25 m.

Shore-normal and shore-parallel horizontal flows were monitored using biaxial electromagnetic flow meters designed by Marsh-McBirney. These instruments have a time constant of 0.2 s and measure flows up to 3 m s^{-1}. Considerable work on the accuracy of these meters has been undertaken (see review of Huntley, 1980) and, while some problems still exist, the level of accuracy is thought to be significantly better than 10%. The meters were mounted in a specially designed bracket that allowed rotation in both the vertical and horizontal planes for accurate orientation of the probes. This assembly was mounted on a stainless steel support. A galvanized steel pipe, with fin to prevent rotation, was jetted into the bed to provide the base support.

Figure 2 documents the instrument locations during the experiment. Of particular importance were the locations of the current meters, since near-bed flows were to be related to sediment transport. Unfortunately our knowledge of the structure of turbulent oscillatory boundary layers over rough beds is still rudimentary, especially with respect to combined waves and currents (see Grant and Madsen, 1979). In any case the thickness of the bottom boundary layer will be highly variable depending upon local flow and bed roughness conditions. With the necessity of a fixed position in the shallow water depths of the Wendake Beach surf zone, and the likelihood of a logarithmic law being important even in the oscillatory boundary layer (Jonsson, 1980) it was decided to place the sensors at the same relative water depth ($h/2$) in all cases. In this way at least the same segment of the water column was monitored.

Data acquisition and analysis

All sensors were hardwired to the shore-based data acquisition system, which consisted of a mini-computer controlled, high-speed multiplexer and voltmeter giving potential analogue-to-digital conversion rates up to 1000 channel readings per second. Typically the instrument array was scanned every 0.5 s over a period of 9 min and the data stored on magnetic tapes. Record lengths were, of necessity, short to maintain time series stationarity in such a volatile environment. This did, however, restrict the scale of any low-frequency effects which could be detected.

The 1100 data points per sensor per sampling were truncated to 1024 points for spectral analysis and the computation of descriptive statistics, using Biomedical Computer Packages (BMDO3T and BMDP2D), after Dixon (1971) and Dixon and Brown (1979). Wave and current spectra were computed with a unit bandwidth of 0.016 Hz (64 lags), a detrended but unfiltered time series, and a fast fourier transformation to determine the spectral densities. Phase and coherence of cross spectra were also calculated. Simple descriptive statistics of the velocity field were computed using moments; for on-offshore, for example:

$$\text{Mean} = \bar{u} = \frac{1}{N} \sum_{i=1}^{N} u_i \qquad (1)$$

$$\text{Standard deviation} = u_{\text{rms}} = \left[\frac{1}{N} \sum_{i=1}^{N} (u_i - \bar{u})^2 \right]^{1/2} \qquad (2)$$

$$\text{Skewness} = u_{\text{sk}} = \frac{1}{N} \sum_{i=1}^{N} (u_i - \bar{u})^3 / (u_{\text{rms}})^{3/2} \qquad (3)$$

where u is the on-offshore velocity and N the sample size.

Monitoring sediment flux and local morphological response

Although sediment monitoring equipment has advanced rapidly recently, no single instrument can accurately measure both suspended load and bedload and associated bed elevation changes on a continuous basis. A simple, well-tested technique is used in this study to estimate sediment flux and morphological response. This involves *depth-of-activity rods*: these are 1 or 2 m long steel rods (0.5 cm diameter) with a free sliding washer that migrates vertically down the rod during sediment erosion and is buried at the reactivation limit by any subsequent deposition. Measurements by Scuba divers are taken prior to and at the end of the transport event to record the thickness of the active layer and the bed elevation changes. This allows estimates to be made of the time-integrated total and net volume flux through a storm as well as the net bed elevation changes for a series of discrete points. At Wendake Beach, *depth-of-activity rods* were deployed across the nearshore zone at the locations shown in Fig.2. Owing to incomplete installation prior to the storm, only bed elevation changes were recorded along line 0 (the instrument transect), but both depth-of-activity and bed elevation changes were recorded along line IN.

WAVE-GENERATED CURRENTS AND SEDIMENT TRANSPORT POTENTIAL

Sediment motion in the nearshore bars must be related to both shore-normal currents induced directly by primary wave oscillations, and to shore-parallel currents (longshore currents) resulting from the radiation stress associated with wave breaking and additional stresses induced by wind and alongshore pressure gradients. Secondary waves such as leaky mode or edge waves may be superimposed on these primary components. Since orbital velocities under waves are generally larger than any unidirectional current within the surf zone, the component of boundary shear stress due to waves will be significantly greater (Grant and Madsen, 1979) and therefore most important in the initiation of sediment movement. However, waves are inefficient transporters of sediment unless the oscillation is asymmetrical and/or combined with a superimposed current. Since the primary periodicity in sediment accumulation is normal to shore at Wendake Beach, then transport by shore-normal currents induced by primary and secondary waves is likely of greatest importance in both bar generation and maintenance in this system.

Waves within the nearshore are reasonably approximated using linear wave theory, where sinusoidal water motion is implicit (e.g. Gaughan and Komar, 1975). Waves in the surf zone, however, are markedly non-linear and water motions no longer symmetric. Sediment transport in this zone will be a function not only of the absolute velocity near the bed (a function of wave height), but also the frequency distribution of the velocity vectors. Of paramount importance in the oscillatory flow field, where the range of magnitudes of current vectors is large, is the asymmetry of the distribution of these current vectors, since sediment transport is not related linearly to the velocity, but more probably to the cube or fourth power of the absolute velocity. Quite small asymmetries can therefore be more important than any net (Lagrangian type) steady drift, even where the latter may be quite high. The importance of this was clearly recognized by Inman and Bagnold (1963), Inman and Frautschy (1966) and Wells (1967) and most recently in the transport models proposed by Bowen (1980), Bailard (1981) and Bailard and Inman (1981).

STORM EVENT 1980: 05:31—1980:06:01

Figure 3 illustrates the temporal pattern of wind speed and direction and the angle of wave approach through the storm. Storm waves were initiated by 1500 h (EDST) on May 31 with winds of 3 m s^{-1} from the WSW, which increased to a maximum of 8 m s^{-1} from the W at 2200 h. With the passage of the frontal system, winds veered to the NW at 2230 h reaching speeds of 7 m s^{-1}; a gradual reduction in speed occurred as a further switch to the N took place at 0320 h. The storm winds dropped below 3 m s^{-1} at 0650 h as the winds backed towards the westerly direction again and the storm ended by 1015 h on June 1. This dramatic change in wind and wave direction is typical of the passage of meteorological depressions in this region.

Incident and secondary waves

The general pattern of growth of the incident wave spectrum is best illustrated by characteristics measured at wave wire nine at the lakeward margin of the three-bar-system (see Fig.2). It should be stressed, however, that during the storm peak, the surf zone width reached some distance to lakeward; theory indicated that depth controlled breaking occurred up to 229 m from the shoreline at this time.

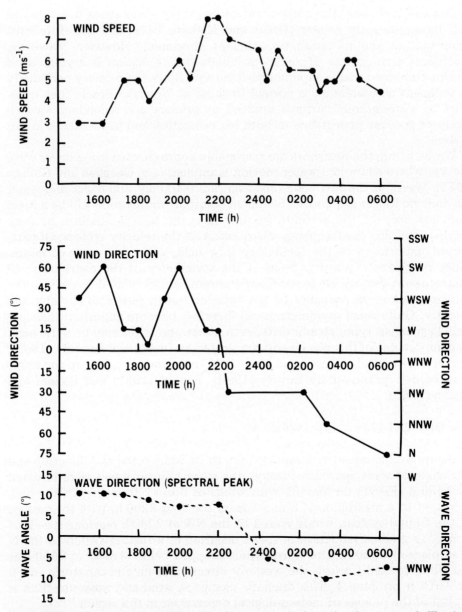

Fig.3. Wind speed, direction and angle of wave approach to the shore normal, Wendake Beach, 1980:05:31:15:00—1980:06:01:10:15.

Wave heights, periods and directions of approach

Table I summarizes the important incident wave parameters. Significant wave heights in excess of 2 m and with periods of 5 s occurred at the storm peak. Spilling breakers were the dominant wave form, with most intense breaking on the bar crests. Although wave reformation did occur in the

TABLE I

Incident wave parameters, Wendake Beach, 1980:05:31—06:01

Time (h)	Height[1] H_{rms} (m)	H_s (m)	Period T_{pk} (s)	Angle[2] θ (°)	Breaker[3] index B_b	Scaling[4] parameter ϵ ($\times 10^2$)	Surf zone[5] width X_b (m)
1500	0.59	0.82	3.4	10	0.40—0.56	8	74
1615	0.68	0.95	3.4	10	0.46—0.64	6.5	124
1715	0.76	1.06	4.0	10	0.37—0.52	6.4	143
1830	0.85	1.19	2.3	9	1.26—1.76	192	133
2000	0.78	1.09	4.3	7	0.33—0.46	5.9	127
2200	1.52	2.12	5.0	8	0.48—0.67	5.9	202
0025	0.98	1.37	5.0	5	0.31—0.43	4.8	190
0320	0.76	1.06	5.0	10	0.24—0.33	3.8	142
0620	0.62	0.87	4.0	6	0.30—0.43	4.8	121

[1]Wave heights determined from the total variance of the record; [2]angle based upon measured orbital vector (after Sherman, 1983); [3]Breaker Index, $B_b = H_b/gmT^2$ (after Galvin, 1972) where beach slope $(m) = 0.013$ and values for both H_{rms} and H_s are shown; [4]Scaling Parameter $(\epsilon) = a_b\omega^2/g\tan^2 m$ (after Guza and Inman, 1975) where a_b is taken as the root-mean-square amplitude at breaking following Wright et al. (1979); [5]estimated based upon solitary wave breaking criterion ($\alpha_b = 0.78$).

troughs observations indicated that during the storm peak and well into the decay period the whole of the surf zone was saturated with spilling breakers propagating across the troughs. Plunging breakers were occasionally observed on the second bar during the later stages of the storm and only rarely were true bores developed, occurring within a few metres of the beach face. Waves dissipated finally by collapsing at the foot of the beach.

The surf zone was in a highly dissipative state at all times and the scaling parameter (Table I) varied over two orders of magnitude. Large angles of wave approach (5°—10°) throughout the storm (Fig.3) gave rise to significant shore-parallel currents.

WAVE AND CURRENT SPECTRA

Figure 4 indicates the form and temporal changes evident in the incident spectra. In an environment such as this, where wind forcing is continuous into the surf zone, the spectra exhibit considerable energy at a wide range of frequencies above the peak. In all cases the latter is marked by a very sharp truncation at the lower frequency end. As expected, spectral growth was accompanied by a consistent shift in the spectral peak to lower frequencies (Fig.4a and b), which remained right through the decay phase.

During the most intense part of the storm a markedly bimodal spectrum appeared (Fig.4b), with a second peak at twice the frequency of the incident peak. Energy at this frequency (the first harmonic) has long been recognized in active surf zones. It can appear as an artifact of harmonic analysis of the strongly non-linear wave form, but also through the generation of secondary waves as very rapid shoaling and energy conservation takes place just prior to

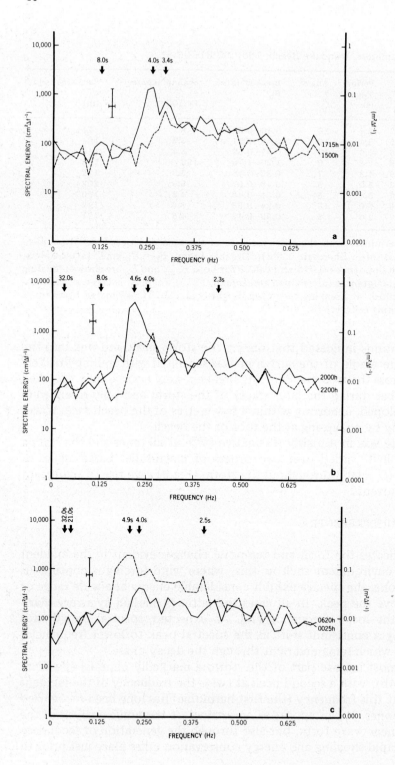

breaking (Thornton et al., 1974). The presence of secondary waves at this frequency was confirmed in the surface elevation analogue record (see also Davidson-Arnott and Randall, 1984, this volume), but overall the energy at this frequency was low (Table II).

There is very little evidence at the outer limits of the surf zone for structure in the spectra at frequencies lower than that of the incident waves (Fig.4), although under highly dissipative conditions many workers have noted a dominance of low-frequency energy (e.g. Holman et al., 1978; Holman, 1981; Wright et al., 1982). At Wendake this may reflect the distance offshore of wave wire 9 (136 m). Examination of spectra closer inshore (wave wire 4, 65 m offshore; see Fig.2) reveals an energy peak at a frequency one fourth that of the incident wave (Fig.5). It was, however, statistically significant only at the peak of the storm but remained significant until the very end. It was associated with a highly dissipative state (Table I) and the largest incident wave heights and periods. It does not, however, contribute more

TABLE II

Contributions of differing frequencies of oscillation to shore-normal and shore-parallel flows in the surf zone, Wendake Beach, 1980:01:00:25

Location offshore (m)	Frequency (Hz)	Current speeds*	
		Shore-normal flows u_{rms} (cm s^{-1})	Shore-parallel flows v_{rms} (cm s^{-1})
14	0.188	4.2	4.4
	0.047	5.0	5.4
38	0.188	8.0	5.6
	0.047	4.5	3.2
57	0.188	8.7	4.4
	0.047	5.5	2.6
64	0.188	12.0	3.6
	0.047	4.5	3.4
94	0.188	8.3	2.7
	0.047	2.6	1.5
106	0.188	8.7	3.1
	0.047	2.8	1.3
118	0.188	10.2	2.6
	0.047	3.6	1.3

Simple calculations based upon variances over one unit bandwidth at the respective spectral peaks. The peaks were identified using surface elevation spectra.

Fig.4. Wave spectra at outer margin of the surf zone through the early part of the storm (a), during the storm peak (b) and during the period of storm decay from the peak (c). These and all following spectra are based on 1024 data points. Unit bandwidth is 0.016 Hz and is given by the horizontal bar: the 95% confidence band around the spectral estimates is given by the vertical bar. All spectra have 32° of freedom.

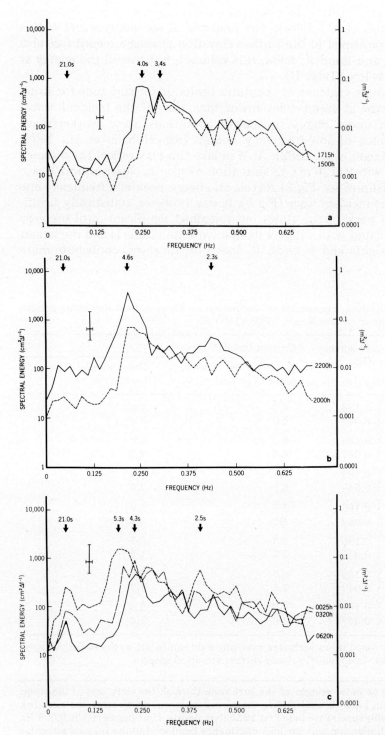

Fig.5. Wave spectra from the inner part of the surf zone through the early part of the storm (a), the storm peak (b), and the period from storm peak to decay (c).

than 10% to the total variance of the surface elevation even at its maximum (Table III). It appears of more importance in the velocity field: the maximum contribution of this low frequency band to the total energies of the on-offshore and alongshore motions was found to be 22 and 25%, respectively. Although energy at the incident frequency dominated the on-shore motions, the low frequency component was of equal importance in alongshore oscillations (Table III). Conclusive identification of the mode of this oscillation is not possible with the present dataset. Nevertheless, there does appear to be a spatial pattern to the energy at this frequency. Figure 6 documents the shore-normal variability in the spectra from a number of wave staffs and Fig.7 illustrates the spatial distribution of peak energies at both incident period and the longer period component (0.047 Hz) for on-offshore (u) and alongshore (v) velocity spectra. While there is a general tendency for energy at 0.047 Hz to decrease with distance from the shoreline (Fig.7 and Table II) it is important to note that a strong peak in u occurs over the crest of bar 2 in coincidence with a trough in the values of v. This suggests the trapping of energy near the shore as might be expected if the 21 s oscillation were an edge wave, and also the presence of a nodal point at the bar 2 location if the edge wave was in a standing mode. Evaluation of phase relationships between surface elevation (n) and the two components of the horizontal velocity field (u and v), for the incident and low-frequency spectral peaks at positions close to the shoreline, support the presence of edge wave motions at 0.047 Hz (Table IV). In no case however, did this low-frequency component dominate the spectrum. In all cases incident wave energy was more important and only in two cases were the low-frequency peaks in the velocity spectra statistically significant (Fig.8).

TABLE III

Surf zone spectral energies, Wendake Beach, 1980:06:00:25

Location offshore (m)	Total energy ($\times 10^3$)			Percentage of total*					
	η (cm^2)	u (cm^2·s^{-2})	v (cm^2 s^{-2})	η		u		v	
				21s	5s	21s	5s	21s	5s
14 (T1)**	—	15.0	31.2	—	—				
38 (T2)	4.9	41.1	13.6	8	41	15	48	24	10
57 (B2)	26.9	44.0	23.1	10	47	22	43	23	24
64 (B2)	15.0	61.4	24.8	6	46	14	57	25	18
94 (T3)	8.0	33.8	6.5	4	43	9	53	23	27
106 (B3)	4.3	3.8	10.1	3	43	7	54	22	36
118 (B3)	—	59.1	7.8	—	—	3	53	20	11
136 (B3)	20.6	—	—	9	45	—	—	—	—

*Values determined using the variance in unit bandwidth at these peak frequencies.
**T1, T2, T3 refer to the trough locations, and B2, B3 to the bar locations.

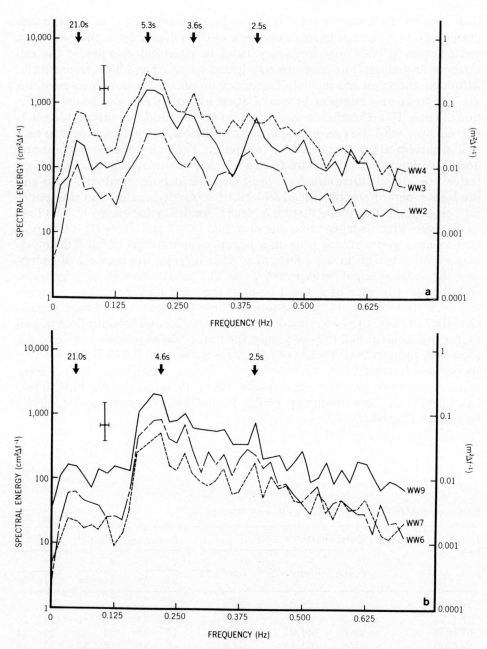

Fig.6. Wave spectra across the surf zone at the storm peak (0025 h). Note the strength of the low frequency oscillation at 21s in the inner surf zone (a) compared to further offshore (b). Wave wire numbers refer to the locations in Fig.2.

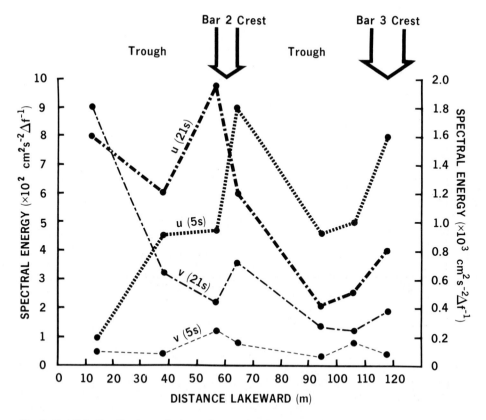

Fig.7. Spatial distribution of spectral energies for the incident peak, 5 s (0.188 Hz) and the low-frequency peak, 21 s (0.047 Hz) in the shore-normal (u) and shore-parallel (v) velocity spectra. For simplicity the value of the spectral density over one unit bandwidth at the peak is shown. Right-hand scale refers to the low-frequency peak.

TABLE IV

Phase relationships between surface elevation (η), onshore velocity (u) and longshore velocity (v), Wendake Beach, 1980:06:00:25 and 06:25

Time (h)	Location offshore (m)	Frequency (Hz)	Phase (coherence) in degrees	
			η vs. u	u vs. v
0025	14	0.047	+80 (0.50)	+14 (0.47)
		0.188	−140 (0.60)	−15 (0.75)
	38	0.047	+120 (0.78)	157 (0.30)
		0.188	−178 (0.98)	+94 (0.58)
0620	14	0.047	+88 (0.36)	+3 (0.32)
		0.250	−130 (0.75)	0 (0.18)
	38	0.047	+60 (0.30)	−180 (0.12)
		0.250	−165 (0.98)	−20 (0.08)

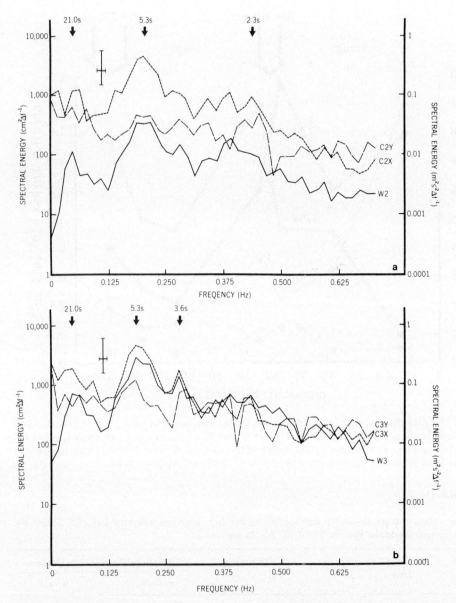

Fig. 8. Wave and current spectra at the storm peak (0025 h): (a) 38 m offshore; (b) 57 m offshore. C2 and C3 are current meter designations and X refers to the shore-normal and Y to the shore-parallel directions. W2 and W3 are wave wire designations. Locations are shown in Fig. 2.

Spatial and temporal variability of wave-orbital motions

Early in the storm (1500 and 1715 h), wave orbital velocities were already sufficient to initiate sediment motion (Fig.9). Taking a conservative velocity of 0.2 m s^{-1} for the threshold of entrainment, it is evident that the Wendake Beach sands would be in motion across the whole of the three-bar system. Maximum instantaneous currents at these times were in excess of 0.7 and 1.3 m s^{-1}, respectively. The mean flows (Fig.10) were offshore, but did not exceed 0.05 m s^{-1} and were therefore close to the instrument accuracy. At

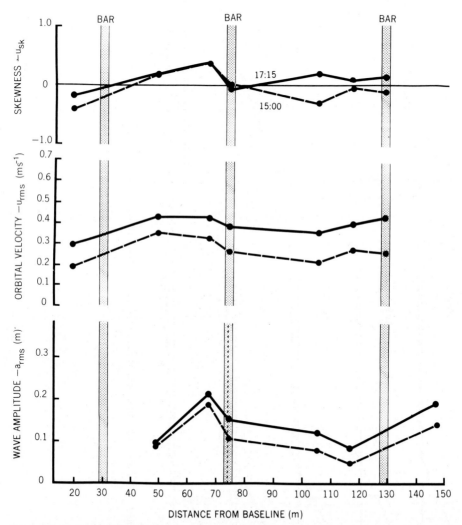

Fig.9. Spatial variability of wave amplitude, orbital velocity and orbital asymmetry across the surf zone, Wendake Beach, 1980:05:31:15:00 and 1980:05:31:17:15. Negative skewness is offshore. Dots mark bar crest locations and the vertical dashed line indicates the first breaker line based upon the depth-controlled breaking criterion.

48

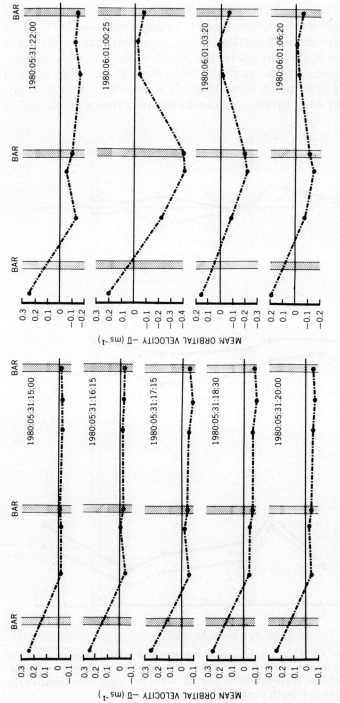

Fig.10. Spatial variability of mean orbital velocities across the surf zone, Wendake Beach, 1980:05:31:15:00—1980:06:01:06:20.

1500 h wave breaking was initiated on the second bar and by 1715 h the surf zone width had expanded lakeward of the third bar (Fig.9). Relatively small variation in orbital velocity was measured across the surf zone, but strong orbital asymmetries were present in the troughs. Near symmetrical motion occurred on the bar crests. The basic pattern was one of lakeward asymmetry (negative skewness) in the innermost trough and landward asymmetry (positive skewness) in the second trough; an initial lakeward asymmetry changing to a landward asymmetry was evident in the outer trough (Fig.9).

As wave height increased during the storm so did the orbital velocities: at the storm peak (2200—0025 h) u_{rms} reached maxima in excess of 0.60 m s^{-1} (Fig.11), with peak orbital flows of 1.7 m s^{-1}. As expected, maximum currents were found in the shallow water over the bar crests. However, since height loss through breaking was now occurring well-lakeward of the outer bar and across the full surf zone width, the largest velocities were on the outer bar with decreasing values landward (Fig.11).

After 2200 h a strong, steady lakeward drift was superimposed on the orbital velocities across virtually the whole of the surf zone. This drift reached values of 0.4 m s^{-1} in the second trough and across the crest of the second bar at 0025 h (Fig.10). Only in the very shallow zone landward of the first bar were mean flows landward (Fig.10), as an apparent response to bore development near the beach face. At 2200 h asymmetry in the oscillatory motion was landward across the whole of the surf zone (Fig.12), and could be capable of producing net transport opposed to that of the mean drift. Thus sediment transport differentials could be controlled by this asymmetry superimposed on the steady drift.

During the peak of the storm (0025 h), a pattern of mean flow and flow asymmetry developed which remained for the duration of wave activity (Figs.11 and 13). It was still present at 0620 h when wave heights and absolute orbital velocities had dropped to values almost identical to those at the beginning of the storm (cf. Figs.9 and 11).

On-offshore sediment transport, net sediment flux and morphological change will depend upon the interactions of mean flows with the flow asymmetries, provided flows are great enough to initiate transport. If, therefore, one can assume that most sediment transport and morphological change will be associated with the highest orbital velocities when bedload transport would be the highest (the storm peak), and with the decaying limb of the storm as any suspension settling accelerates, then the period 0025—0620 h becomes critical in any explanation of sediment flux.

The primary flow characteristics at this time were (Figs.10 and 13):

(a) Mean flows were offshore everywhere except landward of the first bar.

(b) Mean flows reached maxima on the landward slope and across the bar crest of the second bar (0.4 m s^{-1}), decreasing both landward into the trough and lakeward across the outer trough and bar crest. In the outer part of the surf zone, flows were less than 0.1 m s^{-1}.

(c) Flow asymmetries were onshore (positive u_{sk}) landward of the central part of the second trough and lakeward of the centre of the outer trough.

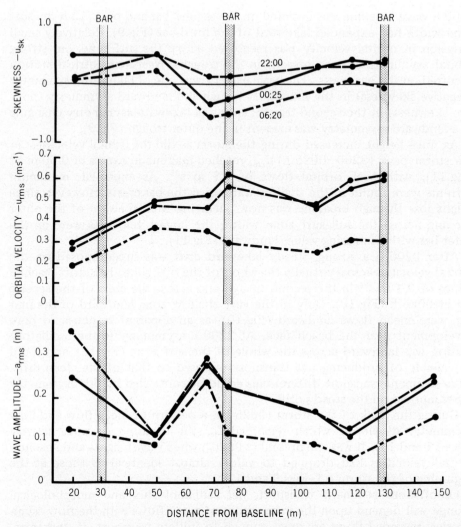

Fig.11. Spatial variability of wave amplitude, orbital velocity and orbital asymmetry across the surf zone, Wendake Beach, 1980:05:31:22:00, 1980:06:01:00:25 and 1980:06:01:06:20.

Across the landward slope and crest of bar two, however, flow asymmetries were offshore with largest negative skewness values occurring in the part of the trough immediately landward of the bar slope and crest.

(d) Symmetrical flow fields must therefore have occurred in the second and outer troughs.

With respect to the contribution of secondary oscillations to the fluid motion described above it has already been noted that a very large difference exists between the spectral densities of the water surface elevation at the incident frequency and the low-frequency peak at 21 s; this difference is not as great with respect to the two components of the velocity field especially

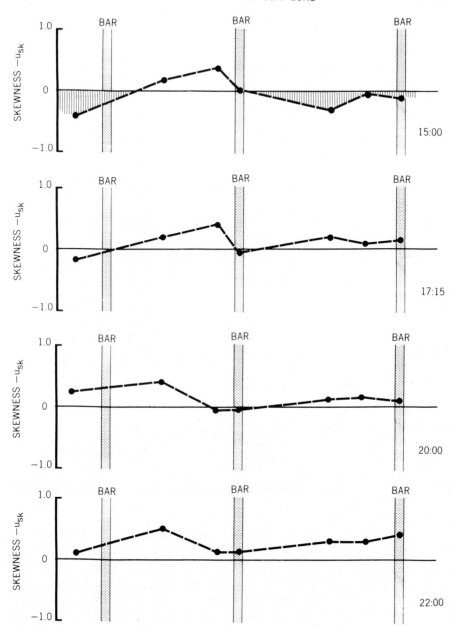

Fig.12. Spatial variability of orbital asymmetry across the surf zone, Wendake Beach, 1980:05:31:15:00—22:00.

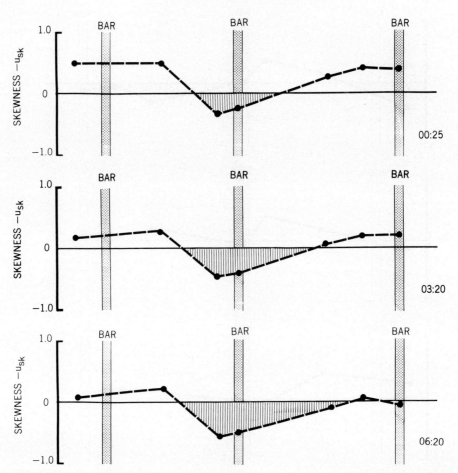

Fig.13. Spatial variability of orbital asymmetry across the surf zone, Wendake Beach, 1980:06:01:00:25—06:20.

close to the shoreline (Fig.8). However, the low-frequency peaks in the velocity structure are poorly defined and overall the surf zone is dominated by oscillatory currents at the incident wave frequency. Table II shows the relative contribution to nearshore currents of energy at the peak incident frequency and at 0.047 Hz during the time of maximum total energy and most significant low frequency spectral peak. Except very close to the shoreline, velocities associated with the incident waves are higher by a factor of 1.5 to 3. If the total energies in the two peaks were considered this difference would be greater still. Contrary to the results of similar studies, the currents at Wendake Beach are dominated by oscillations at the incident frequency.

Spatial and temporal variability of shore-parallel currents

During the early part of the storm (Fig.14) the time-averaged longshore component of motion in the surf zone was somewhat non-coherent, generally less than 0.20 m s^{-1} and thus close to or below the threshold of sediment motion. However, these flows were superimposed on the shore-normal oscillations, which would have set sediment in motion at this time and thus longshore advection of sediment could occur.

By 1830 h a strongly coherent flow pattern was established and maintained until the peak of the storm. Maximum velocities occurred on the lower lakeward slopes of the bars and reached a value in excess of 0.40 m s^{-1} in the

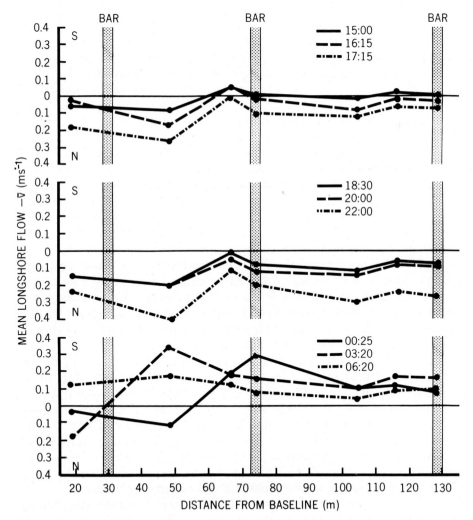

Fig.14. Spatial variability of shore-parallel currents across the surf zone, Wendake Beach, 1980:05:31:15:00—06:20.

second trough at 2200 h. Minima occurred just lakeward of the bar crests in both second and third trough-bar systems. Between 2200 and 0025 h a change in the direction of wind and wave approach caused a bi-directional, disequilibrium response in the longshore current (Fig.14). It will be recalled, however, that a strongly coherent pattern of shore-normal flows existed at this time. Upon reversal of the longshore current, maximum flows occurred in the second trough and the lowest velocities were now occurring in the outer trough. As waves decayed so did the shore-parallel flows until velocities across the surf zone were again less than 0.20 m s^{-1} by 0620 h, although the orbital velocities (u_{rms}) were still well above this value.

Sediment flux and morphological response

From the preceding discussion it is apparent that the potential for significant sediment motion was present over most of the nearshore zone throughout the storm and that net transport in the on-offshore direction could have been forced by both the mean flow and asymmetries associated with the orbital motions. Furthermore, spatial gradients in these flows (giving rise to net water and sediment flux) and local slopes (controlling the gravity potential on transported sediment) would interact to determine whether erosion (accelerating transport rates), deposition (decelerating transport rates) or no change (constant transport rates) would occur. Shore-parallel flows would become increasingly important where orbital velocities were low, near-symmetrical, or both. Any shore-normal gradient in these longshore currents could then produce transport differentials leading to shore-normal patterns in sediment erosion and accretion.

Examination of pre- and post-storm profiles along the instrument transect and at 30 m either side revealed bars present at both times, and with relatively small differences in morphology. Thus the bars are not totally destroyed by a single episode of storm-waves, but rather are in some form of equilibrium, either with a single storm or a series of storm events. Recent analysis of depth of activity and bed elevation data using a grid of 51 rods for a series of six storms at this location suggest strongly the existence of a steady state, at least in the overall sediment budget (Greenwood, in prep.).

In this storm, morphological changes were evident both from profile data and the depth-of-activity rods. In general there was a landward shift of the first and second bar crests (5 and 10 m, respectively), while the outer bar crest remained stable. Accretion of both second and third bars occurred, with erosion of the adjacent troughs. Landward movement of bars 1 and 2 was associated with a southerly alongshore shift in the sinuous form of the two bars. The latter is important, since southward flowing currents only occurred at the storm peak and during the decay period, after the major switch in the direction of wave approach at approximately 0000 h.

Figure 15 illustrates the local depths of activity and bed elevation change across the bar system together with the spatial distribution of the primary hydrodynamic parameters at the peak of the storm (0025 h). This particular data set was used because: (1) at this time both orbital velocities and asymmetries were greatest and thus likely to reflect the most important sediment transport phase of the storm; and (2) this general pattern was consistently present for the longest part of the storm (at least six hours).

Sediment reactivation was, as expected, controlled strongly by the antecedent morphology: in the shallow water over the bar crests, where orbital velocities were greatest, depth of activity reached a maximum. In contrast, the outer trough and lakeward slopes of the third bar with deeper water, lower wave heights and lower orbital velocities were marked by minima. In the trough landward of bar 2 a relatively high value for sediment reactivation was obtained because of the shallowness of the water.

It is rather surprising, however, given the measured currents, that the depth of activity relative to the pre-storm surface was not larger. Maxima were only 14 cm on both second and third bars. This indicates that with high, relatively uniform, bed stress, sediment entrainment reaches a maximum, which cannot be increased regardless of the duration of applied stress unless considerable net transport of sediment occurs. The latter was clearly not the case: although considerable volumes of sediment were set in motion during the storm (the average depth of activity was 9 cm), the net flux was less than two percent of the nearshore control volume (Greenwood, in prep.).

Bed elevation change indicated accretion of the bars and scouring of the troughs, and thus bar growth during storms: however, these changes were small, reaching maxima of only a few centimeters (Fig.15). This would suggest that conditions of near-equilibrium transport existed over the locally varying nearshore slopes for much of the storm. This could have been achieved through a continuous, but uniform, sediment transport rate over the bar system or, more probably, by some form of oscillating equilibrium, where a balance existed between the net oscillatory flows (mean and asymmetry), the local slope and sediment inertia.

Accretion on the landward slope and crest of bar 2 combined with erosion of the adjacent trough would account for the observed landward migration of the bar crest (Fig.15). This is best explained by increasing landward flows from the centre of this trough landward, and a similar increase in lakeward flows from the centre of the trough lakeward (Fig.15). Such a pattern was evident in both the mean flow ($\bar{u}$) and flow asymmetry (u_{sk}). The trough thus represents an erosional node. Decreases in both the mean flows lakeward and the lakeward asymmetries toward and across the landward slope and crest of bar 2 would explain the deposition here; the relative steep landward slope ($3-7°$) would provide further restraint on the lakeward transport. Enhancement of this pattern would be unlikely as a result of shore-parallel flows since the shore-parallel currents were lower in this second trough than on the landward slope and crest of the adjacent bar (see Fig.15).

Decreasing lakeward mean flows and flow asymmetries lakeward of the crest of the middle bar, would suggest continuing accretion into the centre of the outer trough (albeit somewhat less than on the landward side and crest of the bar). Also orbital asymmetries either side of the outer trough

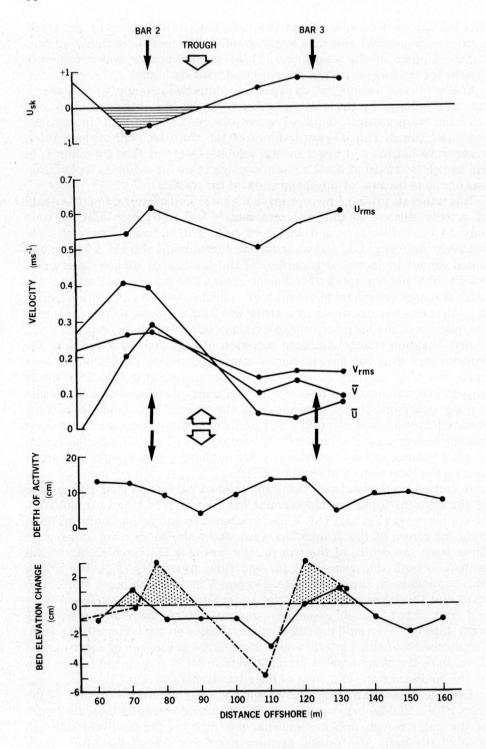

suggest convergence of sediment transport paths towards a depositional node in the centre of this outer trough. In fact erosion (3—5 cm) characterizes this location. It would appear, therefore, that with a reduction in u_{rms}, $\bar{u}$ and u_{sk}, and the consequent reduction in total load (corroborated in part by the depth of activity data), the longshore currents become the primary advecting agents (Fig.15) and caused the observed erosion.

Across the crest of the outer bar, bed elevation changes were positive on the landward side indicating accretion and negative on the lakeward side indicating erosion (Fig.15). At this location mean orbital flows were low, as indeed were mean longshore flows; it would seem that the best correlation is between the strong landward asymmetries in the orbital velocity field and a landward movement of sediment across the bar crest (Fig.15).

DISCUSSION AND CONCLUSIONS

Sediment transport and morphological change in the multiple bar system at Wendake Beach are most satisfactorily explained by the interrelationship of spatial gradients in the means and asymmetries of shore-normal oscillatory currents with the local slopes. Orbital asymmetries integrated over time represent a *steady* sediment transport potential which, together with the mean flows, is superimposed on the initial entrainment induced by the orbital motions themselves.

Figure 16 summarizes the patterns of fluid flow, erosion, deposition and morphological response that can be inferred from the field measurements. Sediments accumulate on the bars and are scoured from adjacent troughs during storms to cause bar growth: the alignment of scour and accretion relative to the pre-storm topography may cause a shift in bar location lakeward or landward. Sediment deposition occurs, paradoxically, in areas with the highest absolute orbital speeds: it results from spatial decreases in the orbital asymmetry and thus in the *steady* sediment drift. The gravitational resistance to sediment in transport is enhanced wherever local bed slopes oppose the *steady* drift and in some cases may aid accretion, as for example on the landward slope of bar 2 (Fig.16). Erosion results also form gradients in the orbital asymmetries, but in this case the asymmetries increase spatially and thus the sediment transport potential increases. Reversals (in the directional sense) of the orbital asymmetries can create either erosional nodes, as in the case of the middle trough, or depositional nodes, as in the case of outer trough (Fig.16).

Shore-parallel *steady* currents contribute to the net alongshore advection of sediment. They are of prime importance to the shore-normal patterns of

---

Fig.15. Spatial distribution of bed elevation change, depth of activity, orbital velocity (u_{rms} and $\bar{u}$), orbital asymmetry (u_{sk}) and longshore current velocity (v_{rms} and $\bar{v}$) across the outer 2 bar-trough systems, at Wendake Beach. Hydrodynamic data are from the storm peak, 1980:06:01:00:25. $\bar{u}$ is offshore at all points and $\bar{v}$ is to the south at all points. With respect to bed elevation change and depth of activity, the solid line represents data from line IN and the dotted line, data from line 0.

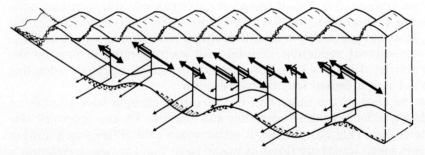

Fig.16. Generalized fluid flows, sediment flux and morphological response over the near-shore bars at Wendake Beach inferred from measurements 1980:05:31—1980:06:01. The relative magnitude of near bed, shore-normal oscillatory flow is shown by the solid double arrow and the direction and magnitude of the net flows (resulting from mean flows and flow asymmetries) by the broad open arrow. The relative magnitude of the shore-parallel flow is shown by the pairs of thin arrows. Cross-hatching represents areas of erosion and dots areas of accretion. A solid line indicates the post-storm profile.

sedimentation where absolute orbital velocities are small and/or symmetrical. The relatively strong longshore current in the outer trough is the most likely cause of erosion here since the orbital asymmetries indicate a depositional node.

The energy driving both shore-normal and shore-parallel currents at Wendake Beach was primarily from incident, wind-generated waves. Secondary waves were present during the peak of the storm, but contributed relatively little energy to the total spectrum. Waves at the frequency of the first harmonic of the incident wave were associated with breaking, and a low-frequency oscillation at a period four times that of the incident period was significant very close to the shoreline. The latter would appear to be a standing edge wave formed during the storm peak and remaining for the duration of the storm. The strong mean on-offshore flows and orbital asymmetries were coincident in time with the appearance of the edge wave and it is tempting to conclude a causal relationship as suggested by previous workers (e.g. Bowen and Inman, 1971; Holman and Bowen, 1982). If indeed the latter is correct then the edge wave node located over bar 2 would be consistent with Symonds and Huntley's (1980) observation of a resonant edge wave controlled by bar spacing. Furthermore, it would suggest that bars form under nodes of long waves (Carter et al., 1973) and thus result primarily from bed-load transport rather than suspension load (Bowen, 1980). Evidence from primary sedimentary structures (Greenwood, in prep.) indicates a predominance of bedding produced by oscillation ripples, megaripples and sheet flow. No good example of massive bedding as might be produced by suspension fallout was ever observed. Secondary waves at a frequency of the first harmonic also coincided with the storm peak, in association with extensive wave breaking. If, as seems probable, such waves are dispersive (Thornton, 1979) then interactions with the primary wave may be capable of producing the orbital patterns observed. Certainly such wave—wave interactions have been proposed for bar genesis (e.g. Hulsbergen, 1974; Bijker et al., 1976; Van de Graaf and Tilmans, 1980; Boczar-Karakiewicz and Bona, 1982).

The present paper provides only a qualitative interpretation of the relationships between fluid and sediment motions, and their integration into a quantitative model remains for future work. At this time a plethora of theories exist for predicting shore-normal transport (some purely analytical others based on results from model experiments) and only detailed comparisons along the lines of those of Bowen (1980), Seymour and King (1982) and Bailard (1983) will determine the most appropriate predictor for sediment transport in a barred system such as Wendake Beach.

ACKNOWLEDGEMENTS

This study forms part of a continuing research programme in Nearshore Hydrodynamics and Sedimentation supported by equipment (E39218, E6614) and operating (A7956) grants awarded to B.G. from the Natural Sciences and Engineering Research Council of Canada. A University of Toronto Open Fellowship, an Ontario Graduate Scholarship and a Postdoctoral Scholarship at Woods Hole Oceanographic Institution assisted D.J.S. Computer costs were defrayed by both Scarborough College and the Department of Geography, University of Toronto. The Academic Workshops at Scarborough College provided instrument design and construction and the Graphics and Photography Department assisted with the illustrations. Invaluable field and laboratory assistance was provided by M. Rollingson, P. Christilaw and R. Sutherland (Scarborough College) and R.G.D. Davidson-Arnott and D.C. Randall (University of Guelph). The writing of this paper was greatly assisted by the facilities and hospitality received by the senior author during research leave visits to the Universities of Uppsala, Sweden and Sydney, Australia. In particular Professor J.O. Norrman, Goran and Bodil Albjar (Uppsala), and Professor M. Daly and Dr. Andy Short (Sydney) are to be thanked. Drs. J. Bailard and A.H. Sallenger reviewed the manuscript and are thanked for their comments. The results and interpretations presented in this paper are, however, solely the responsibility of the authors.

REFERENCES

Bailard, J.A., 1981. An energetics total load sediment transport model for a plane sloping beach. J. Geophys. Res., 86: 10,938—10,954.
Bailard, J.A., 1983. Modeling offshore transport in the surf zone. Proc. 18th Coastal Engineering Conf., Cape Town, pp.1419—1438.
Bailard, J.A. and Inman, D.L., 1981. An energetics bedload model for a plane sloping beach: local transport. J. Geophys. Res., 86: 2035—2043.
Bijker, E.W., Van Hijum, E. and Vellinga, P., 1976. Sand transport by waves. Proc. 15th Coastal Engineering Conf., Honolulu, Hawaii, pp.1149—1167.
Boczar-Karakiewicz, B. and Bona, J.L., 1982. Shallow water waves over mildly varying topography. Abstracts, 18th International Conf. on Coastal Engineering, Cape Town, pp.275—276.

Bowen, A.J., 1980. Simple models of nearshore sedimentation; beach profiles and longshore bars. In: S.B. McCann (Editor), The Coastline of Canada. Geol. Surv. Can., Pap. 80-10, pp.1—11.
Bowen, A.J. and Inman, D.L., 1971. Edge waves and crescentic bars. J. Geophys. Res., 76; 8862—8871.
Bowman, D. and Goldsmith, V., 1983. Bar morphology of dissipative beaches: an empirical model. Mar. Geol., 51: 15—33.
Carter, T.G., Liu, P.L.-F. and Mei, C.C., 1973. Mass transport by waves and offshore sand bedforms. J. Waterways, Harbors Coasts, Am. Soc. Civ. Eng., WW2: 165—184.
Davidson-Arnott, R.G.D. and Randall, D.C., 1984. Spatial and temporal variations in spectra of storm waves across a barred nearshore. In: B. Greenwood and R.A. Davis, Jr. (Editors), Hydrodynamics and Sedimentation in Wave-Dominated Coastal Environments. Mar. Geol., 60: 15—30 (this volume).
De Beaumont, E., 1845. Lecons de Géologique Practique. Paris.
Dixon, W.J., 1971. BMD: Biomedical Computer Programs. University of California Press, Berkeley, Calif., 600 pp.
Dixon, W.J. and Brown, M.B., 1979. BMD P-79: Biomedical Computer Programs, P-Series, University of California Press, Berkeley, Calif.
Galvin, C.J., 1972. Wave breaking in shallow water. In: R.F. Meyer (Editor), Waves on Beaches and Resulting Sediment Transport. Academic Press, New York, N.Y., pp.413—455.
Goldsmith, V., Bowman, D. and Kiley, K., 1982. Sequential stage development of crescentic bars: HaHoterim beach, southeastern Mediterranean. J. Sediment. Petrol., 52: 233—249.
Grant, W.P. and Madsen, O.S., 1979. Combined wave and current interaction with a rough bottom. J. Geophys. Res., 84: 1797—1808.
Greenwood, B., 1982. Bars. In: M.L. Schwartz (Editor), Encyclopaedia of Beaches and Coastal Environments. Dowden, Hutchinson and Ross, Stroudsberg, Pa., pp.135—139.
Greenwood, B., 1983. Hydrodynamics and sedimentation in barred nearshores: a review of work by the Scarborough Campus Coastal Research Group. In: N.A. Rukavina (Editor), Proc. Third Workshop on Great Lakes Coastal Erosion and Sedimentation. Natl. Water Research Institute, Burlington, Ont., pp.45—48.
Greenwood, B. and Davidson-Arnott, R.G.D., 1979. Sedimentation and equilibrium in wave-formed bars: a review and case study. Can. J. Earth Sci., 16: 312—332.
Greenwood, B. and Hale, P.B., 1980. Depth of activity, sediment flux and morphological change in a barred nearshore environment. In: S.B. McCann (Editor), The Coastline of Canada: Littoral Processes and Shore Morphology. Geol. Surv. Can., Pap. 80-10, pp.89—109.
Greenwood, B. and Mittler, P.R., 1984. Sediment flux and equilibrium slopes in a barred nearshore. In: B. Greenwood and R.A. Davis, Jr. (Editors), Hydrodynamics and Sedimentation in Wave-Dominated Coastal Environments. Mar. Geol., 60: 79—98 (this volume).
Greenwood, B. and Sherman, D.J., 1983. Shore-parallel flows in a barred nearshore. Proc. 18th Coastal Engineering Conf., Cape Town, pp.1677—1696.
Greenwood, B., Hale, P.B. and Mittler, P.R., 1980. Sediment flux determination in the nearshore zone: prototype measurements. In: Workshop on Instrumentation for Currents and Sediments in the Nearshore Zone, National Research Council of Canada, Associate Committee for Coastal Erosion and Sedimentation, Ottawa, Ont., pp.99—120.
Guza, R.T. and Inman, D.L., 1975. Edge waves and beach cusps. J. Geophys. Res., 80: 2997—3012.
Holman, R.A., 1981. Infragravity energy in the surf zone. J. Geophys. Res., 86: 6442—6450.
Holman, R.A. and Bowen, A.J., 1982. Bars, bumps and holes: models for the generation of complex topography. J. Geophys. Res., 87: 457—468.
Holman, R.A., Huntley, D.A. and Bowen, A.J., 1978. Infragravity waves in storm conditions. Proc. 16th Coastal Engineering Conf., Hamburg, pp.268—284.

Hulsbergen, C.H., 1974. Origin, effect and suppression of secondary waves. Proc. 14th Coastal Engineering Conf., Copenhagen, pp.392—411.
Huntley, D.A., 1980. Electromagnetic flow meters in nearshore field studies. In: Workshop on Instrumentation for Currents and Sediments in the Nearshore Zone, National Research Council of Canada, Associate Committee for Coastal Erosion and Sedimentation, Ottawa, Ont., pp.47—60.
Inman, D.L. and Bagnold, R.A., 1963. Littoral processes. In: M.N. Hill (Editor), The Sea, Vol. 3. Wiley-Interscience, New York, N.Y., pp.529—533.
Inman, D.L. and Frautschy, J.D., 1966. Littoral processes and the development of shorelines. Proc. Coastal Engineering Speciality Conference, Santa Barbara, Calif., pp.511—536.
Jonsson, I.G., 1980. A new approach to oscillatory rough turbulent boundary layers. Ocean Eng., 7: 109—152.
Seymour, R.J. and King Jr., D.B., 1982. Field comparisons of cross-shore transport models. J. Waterway, Port, Coastal and Ocean Div., 108: 163—179.
Symonds, G. and Huntley, D.A., 1980. Waves and currents over a nearshore bar system. Proc. Canadian Coastal Conf., Natl. Res. Council Can., pp. 64—78.
Thornton, E.B., 1979. Energetics of breaking waves within the surf zone. J. Geophys. Res., 84: 4931—4938.
Thornton, E.B., Galvin, J.J., Bub, F.L. and Richardson, D.P., 1974. Kinematics of breaking waves. Proc. 15th Coastal Engineering Conf., Honolulu, Hawaii, pp.461—476.
Truxillo, S.G., 1970. Development of a resistance-wire wave gauge for shallow-water wave and water level investigations. Unpublished manuscript, Coastal Studies Institute, Louisiana State University, Baton Rouge, La.
Van de Graaf, J. and Tilmans, W.M.K., 1980. Sand transport by waves. Proc. 17th Coastal Engineering Conf., Sydney, N.S.W.
Wells, D.R., 1967. Beach equilibrium and second-order wave theory. J. Geophys. Res., 77: 497—504.
Wright, L.D., Chappell, J., Thom, B.G., Bradshaw, M.P. and Cowell, P., 1979. Morphodynamics of reflective and dissipative beach and inshore systems: Southeastern Australia. Mar. Geol., 32: 105—140.
Wright, L.D., Guza, R.T. and Short, A.D., 1982. Dynamics of a high-energy dissipative surf zone. Mar. Geol., 45: 41—62.

LANDWARD MIGRATION OF INNER BARS

TSUGUO SUNAMURA and ICHIROU TAKEDA

Institute of Geoscience, University of Tsukuba, Ibaraki (Japan)

(Received December 30, 1982; revised and accepted April 12, 1983)

ABSTRACT

Sunamura, T. and Takeda, I., 1984. Landward migration of inner bars. In: B. Greenwood and R.A. Davis, Jr. (Editors), Hydrodynamics and Sedimentation in Wave-Dominated Coastal Environments. Mar. Geol., 60: 63—78.

A field investigation was carried out to collect data of inner bar migration. Profiles were measured once or twice a week for a two-year period at Naka Beach, Ibaraki Prefecture, Japan. It was found that the onshore migration of inner bars could be described by two dimensionless quantities as: $5D/(H_b)_{max} < (H_b)_{max}/gT_{max}^2 < 20D/(H_b)_{max}$ where $(H_b)_{max}$ is the maximum value of daily average breaker height during one interval between surveys, T_{max} is the average wave period of the day giving $(H_b)_{max}$, D is the mean size of the beach sediment, and g is the acceleration due to gravity. Analyses based on surfzone sediment dynamics yields $\bar{v}/(wD/b) = 2 \times 10^{-11} (\bar{H}_b/D)^3$, where $\bar{v}$ is the average speed of onshore bar-migration, b is the bar height, $\bar{H}_b$ is the average breaker height, and w is the fall velocity of the beach sediment. Nomographs for the speed of landward migrating bars are also presented.

INTRODUCTION

Storm waves transport beach material offshore causing beach erosion and form a sand bar in the surfzone as a temporal sediment reservoir. Post-storm waves gradually move the sand bar onshore. The bar eventually emerges from the water level and welds onto the beach face. Such migrating bars have been called "ridge and runnel" topography by North American sedimentologists (e.g., Davis et al., 1972; Hayes, 1972; Owens and Frobel, 1977), but, because the application of this terminology has been questioned by Orford and Wright (1978), the present paper uses the term "inner bar".

Many researchers have considered the landward migration of inner bars (e.g., Evans, 1939; King and Williams, 1949; Sonu, 1969, 1973; Davis and Fox, 1972a, b, 1975; Davis et al., 1972; Hayes, 1972; Greenwood and Davidson-Arnott, 1975; Owens and Frobel, 1977; Fox and Davis, 1978; Short, 1978, 1979; Hine, 1979; Sasaki, 1982). However, few quantitative studies have been performed on this problem in connection with surfzone sediment dynamics. The purpose of the present study is to relate the landward migration of the inner bar to nearshore wave parameters and sediment properties.

STUDY AREA AND DATA ACQUISITION

Naka Beach, Ibaraki Prefecture, Japan, facing the Pacific Ocean was selected as the study area (Fig.1). The beach, located between Tokaimura and Ajigaura, is an approximately straight, north—south oriented, open coast with a 5-km shoreline length. The shoreline is stable on a long-term basis (Tanaka et al., 1973). The beach sediment is composed of coarse sand in the northern part of the beach and nearly fine sand in the southern part.

An outer bar is located 200—300 m offshore, nearly parallel to the shoreline; the water depth at the outer bar crest is 2—3 m below MSL. Nearshore bottom contours are approximately parallel to the shoreline (Fig.1). The average bottom gradient is 0.011 to a water depth of 20 m, and is almost constant along the beach. No significant alongshore difference in incident wave characteristics has been observed. The study area experiences a maximum tidal range of 1.4 m and a mean of 1 m.

Three monitoring sites were established along this beach (Fig.1). They are North, Central, and South sites, the first having an alongshore length of 300 m and the other two a length of 500 m each. Beach profile surveys were conducted at each site at an interval of once or twice per week for a period

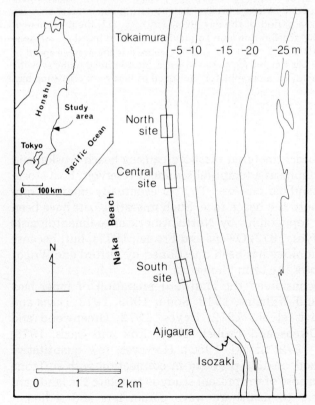

Fig.1. Study area and three monitoring sites.

of two years beginning August 28, 1980. Survey lines, drawn perpendicular to the general shoreline trend, were established at an interval of 20 m at the North site and 50 m at the Central and the South sites. The survey, conducted using a telescopic level, a surveyor's rod, and tape, was extended to the limit of wading, approximately 1.5 m deep.

Sediment samples were collected from the mid-foreshore at each site on October 23, 1980, April 5, 1981, and September 19, 1981. The time-averaged, mean grain size of the beach sediment was found to be 0.76 mm (0.39 ϕ) at the North site, 0.66 mm (0.60 ϕ) at the Central site, and 0.26 mm (1.9 ϕ) at the South site (Takeda, 1983).

Twenty-minute wave measurements were made every two hours by an ultrasonic-type wave gage installed 10 km south of the study area (at a water depth of 21 m). Daily averages of deep-water significant wave height and period are obtained.

RESULTS AND DISCUSSION

Condition for landward bar-migration

Figure 2 shows some representative examples of onshore migration sequences of inner bars. The bar migrates onshore with a marked slip-face, as reported widely from the coasts of other parts of the world (e.g., Hoyt, 1962; Davis et al., 1972; Hayes, 1972; Greenwood and Davidson-Arnott, 1975; Owens and Frobel, 1977; Hine, 1979; Hunter et al., 1979; Dabrio and Polo, 1981). The onshore migration is brought about by onshore sediment transport (e.g., Komar, 1976, p.298). Such distinct beach accretion as shown in Fig.2b, d, and e usually took place during shoreward bar-migration, although minor beach erosion was sometimes observed.

Figure 3 shows a few examples of offshore migration of inner bars. Large waves often hindered the nearshore profile survey. On such occasions, the position of the bar crest was roughly determined by wading when it was possible. The dashed line in Fig.3 shows the approximate bar profile. Davis and Fox (1975) have also observed offshore migration under stormy conditions. Offshore sediment transport causes the seaward bar-migration, giving rise to beach erosion (Fig.3).

Because the direction (onshore or offshore) of bar migration seems to be closely related to the shoreline change (accretion or erosion), the relationship which allows a demarcation of beach accretion and erosion could be applied to the delimitation of bar migrating directions. Such an accretion-erosion relationship (Sunamura, 1980), originally derived from wave-tank experiments (Sunamura and Horikawa, 1974), is described by:

$$\frac{H_0}{L_0} = C(\tan \bar{\beta})^{-0.27} \left(\frac{D}{L_0}\right)^{0.67} \tag{1}$$

where H_0 and L_0 are the deep-water wave height and wavelength, respectively, tan $\bar{\beta}$ is the average nearshore bottom slope to a water depth of 20 m, D is

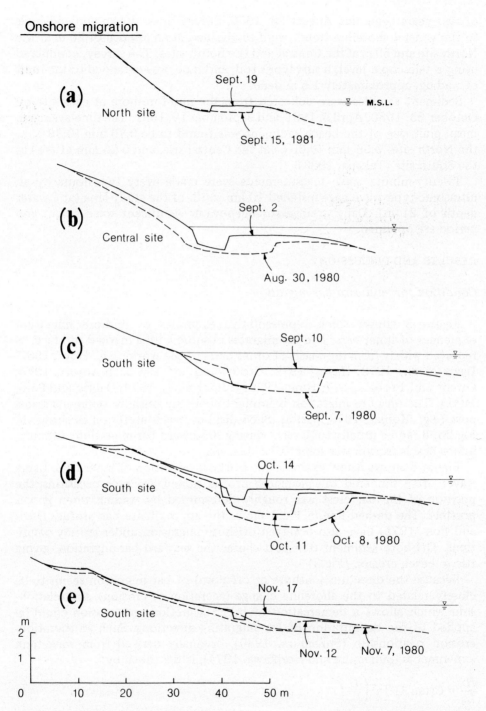

Fig.2. Onshore movement of inner bars.

Offshore migration

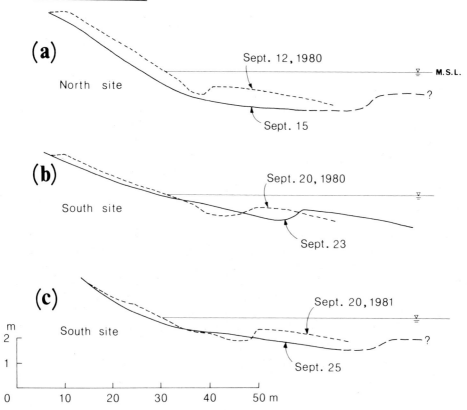

Fig.3. Offshore movement of inner bars.

the grain size of beach sediment, and C is an empirically determined constant ($C = 18$). Accretion (or erosion) takes place when the right-hand side of eq. 1 is greater (or less) than the left-hand side.

A relationship among wave breaker height, deep-water wave parameters, and nearshore bottom slope is approximated explicitly by (Sunamura and Horikawa, 1974; Sunamura, 1982):

$$\frac{H_b}{H_0} = (\tan \bar{\beta})^{0.2} \left(\frac{H_0}{L_0}\right)^{-0.25} \qquad (2)$$

where H_b is the breaker height. The deep-water wavelength, L_0, is related to the wave period, T, by linear theory:

$$L_0 = \frac{gT^2}{2\pi} \qquad (3)$$

where g is the acceleration due to gravity.

Using these three equations, the following relationship can be derived:

$$\frac{H_b}{gT^2} = \kappa \frac{D}{H_b} \qquad (4)$$

where κ is a dimensionless coefficient. In this derivation, such approximations as $1.35 \simeq 4/3$, $0.34 \simeq 1/3$, and $0.67 \simeq 2/3$ were used. Equation 4 is basically the same as eq. 1, but eq. 4 is convenient to use in the field, because H_b and T can be easily measured using a surveyor's rod and stopwatch (e.g., Bascom, 1964, p.173; Hoyt, 1971, p.21), although these measurements give only approximate values. In this study, H_b was estimated by substituting the data of the daily average wave parameters and tan $\bar{\beta} = 0.011$ into eq. 2.

Because temporal changes in wave climate exist even in a day, and because the topographic change is most sensitive to larger waves during one interval between surveys, the demarcation of bar migrating directions is better indexed by:

$$\frac{(H_b)_{max}}{gT_{max}^2} = k \frac{D}{(H_b)_{max}} \qquad (5)$$

where $(H_b)_{max}$ is the maximum value of daily average breaker height during one interval between surveys, T_{max} is the average wave period of the day giving $(H_b)_{max}$, and k is a dimensionless coefficient. Figure 4 shows the demarcation of bar migrating directions (Table I). Kimura's (1976) data, obtained at Tatado Beach, Shimoda, Japan, are also plotted. Although some overlapping of the data points is seen, the demarcation can be reasonably described by the solid line:

$$\frac{(H_b)_{max}}{gT_{max}^2} = 20 \frac{D}{(H_b)_{max}} \qquad (6)$$

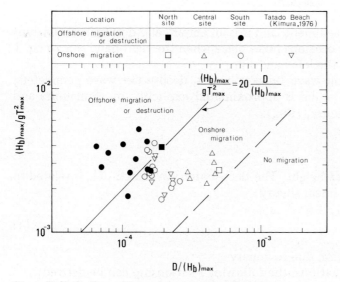

Fig.4. Delimitation of inner-bar migrating directions.

TABLE I

Data of bar migrating directions (notation explained in text)

Site	D (mm)	$(H_b)_{max}$ (m)	T_{max} (s)	Migrating direction	Remarks
North	0.76	1.53	7.6	Onshore	9/15—9/19/81
		4.02	10.2	Offshore	9/25—9/29/81
Central	0.66	2.24	9.3	Onshore	8/28—8/30/80
		2.16	9.6	Onshore	8/30—9/2/80
		1.63	8.7	Onshore	9/2—9/6/80
		1.49	6.5	Onshore	9/6—9/9/80
		1.44	6.9	Onshore	9/12—9/15/80
		1.42	7.5	Onshore	9/22—9/25/80
South	0.26	1.73	7.1	Onshore	9/7—9/10/80
		1.13	7.3	Onshore	9/13—9/16/80
		1.70	6.3	Offshore	9/16—9/20/80
		1.62	7.8	Offshore	9/20—9/23/80
		4.02	10.2	Offshore	9/26—10/4/80
		1.36	9.1	Onshore	10/8—10/11/80
		1.77	6.9	Onshore	10/11—10/14/80
		2.08	8.1	Offshore	10/14—10/15/80
		2.55	8.0	Offshore	10/23—10/28/80
		1.15	7.6	Onshore	11/7—11/12/80
		1.52	6.1	Onshore	11/12—11/17/80
		2.37	11.6	Offshore	2/22—2/28/81
		1.80	8.7	Onshore	5/3—5/11/81
		1.97	6.2	Offshore	5/11—5/17/81
		3.30	9.7	Offshore	5/17—5/24/81
		1.62	7.8	Offshore	6/1—6/7/81
		1.65	7.9	Onshore	6/7—6/15/81
		2.28	9.5	Offshore	6/15—6/21/81
		1.00	6.7	Onshore	7/9—7/14/81
		1.65	8.1	Onshore	9/11—9/16/81
		1.53	7.6	Onshore	9/16—9/20/81
		1.62	7.8	Onshore	10/13—10/18/81
		3.65	11.4	Offshore	10/18—10/24/81
Tatado Beach (Kimura, 1976)	0.23	1.34	6.3	Onshore	8/14—8/15/76
		1.40	8.0	Onshore	8/17—8/19/76
		1.15	8.0	Onshore	8/19—8/21/76
		1.01	6.8	Onshore	8/21—8/22/76
		1.03	6.4	Onshore	8/22—8/23/76
		1.36	6.4	Onshore	8/25—8/29/76

The dashed line would probably demarcate onshore migration and no migration of bars, although the data showing no migration have not been obtained. The area below the dashed line indicates lower-energy waves which are too small to move bars onshore effectively. The dashed line is written as:

$$\frac{(H_b)_{max}}{gT_{max}^2} = 5 \frac{D}{(H_b)_{max}} \tag{7}$$

Figure 4 indicates that the necessary condition for onshore bar-migration should be located in the area between the solid and the dashed lines.

Bar migration speed

Rates of onshore migration of bars have been measured at various beaches; for example, a rate of 1 inch 30 min^{-1} (1.2 m day^{-1}) has been reported from the coast of Lake Michigan by Evans (1939), 4 ft h^{-1} (29.3 m day^{-1}) from Nags Head, North Carolina by Sonu (1969), 9.23 cm h^{-1} (2.2 m day^{-1}) from Ajigaura, Japan by Hashimoto and Uda (1976), 13.7 cm h^{-1} (3.3 m day^{-1}) from the same location by Hashimoto and Uda (1977), 0.83—10 m day^{-1} from the coasts of Magdalen Islands, Quebec by Owens and Frobel (1977), 1—5 m day^{-1} from the Oregon coast by Fox and Davis (1978), and 4—5 m day^{-1} from Dai-nigori-zawa Beach, Japan by Sasaki (1982). Hayes (1972) stated that the rate of migration depends upon sediment grain size, nearshore gradient, wave conditions, length of wave duration, and tidal range. However, no studies have been performed to quantitatively relate the bar migration speed to the controlling factors.

When a bar is migrating onshore, wave breaking always occurs over the gently seaward-sloping surface of the bar. Waves after breaking form bores which advance over the bar. Such bores transport sediment onshore across the gently sloping surface, primarily in a bed-load manner, and eventually the material is moved to the steep landward edge of the bar where it is deposited on the slip-face of the bar, as is schematically illustrated in Fig. 5. This type of sediment transport and deposition on the slip-face causes onshore migration of bars.

Because bed-load is dominant for this type of sediment transport, Shields parameter would be useful:

$$\Psi = \frac{\tau_0}{(s-1)\rho g D} \tag{8}$$

where Ψ is Shields parameter, τ_0 is the bottom shear stress, $s = \rho_s/\rho$, ρ_s is the sediment density, and ρ is the fluid density. Shear stress, τ_0, can be expressed by:

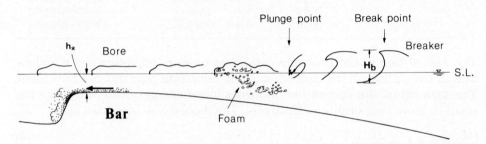

Fig.5. Schematic diagram showing wave transformation and sediment transport over a bar.

$$\tau_0 = \tfrac{1}{2}\rho f_w u_0^2 \tag{9}$$

where u_0 is the maximum bottom velocity on the bar crest (Fig.6) and f_w is the wave friction factor. Assume that the wave height at the bar crest, H_* (Fig.6), is linearly related to the breaker height, H_b, as:

$$H_* = BH_b \tag{10}$$

where B is a dimensionless coefficient ($B < 1$). Because H_* is a strongly depth-controlled quantity due to the shallow water depth, the following relationship (e.g., Komar, 1976, p.56) is employed:

$$\frac{H_*}{h_*} = 0.78 \tag{11}$$

where h_* is the water depth at the bar crest (Fig.6). Linear shallow-water wave approximation yields (see Appendix):

$$u_0 = \frac{H_*}{2}\sqrt{\frac{g}{h_*}} \tag{12}$$

Using eqs. 8 through 12, Shields parameter on the bar crest is related to the breaker height as:

$$\Psi = 0.098 \frac{B f_w H_b}{(s-1)D} \tag{13}$$

Part of the sediment transported onshore across the bar crest is deposited on the landward steep slope. The remaining part is transported alongshore by longshore currents which develop in the trough. Denoting q_* as the depositional rate on the steep slope and q_0 as the net onshore sediment transport rate on the bar crest (Fig. 7):

$$q_* = A q_0 \tag{14}$$

where A is a dimensionless coefficient ($A < 1$). This equation holds when two-dimensionality of sediment transport and of the resultant bar migration is maintained. It would be valid to assume that the depositional area, which is shown by the hatched area in Fig.7, can be expressed by the parallelogram for a short period of time. Then:

$$\Delta S = b \Delta l \tag{15}$$

Fig.6. Definition sketch.

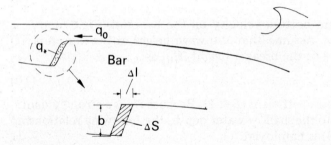

Fig.7. Definition sketch.

where ΔS is the cross-sectional area, Δl is the bar migration distance, and b is the bar height. Dividing both sides of this equation by the time interval, Δt, we have:

$$\frac{\Delta S}{\Delta t} = b \frac{\Delta l}{\Delta t} \tag{16}$$

The left-hand side of eq. 16, $\Delta S/\Delta t$, should be equal to q_* which is the depositional rate, and $\Delta l/\Delta t$ in the right-hand side is the bar migration speed. Namely:

$$q_* = bv \tag{17}$$

where v is the bar migration speed. Equations 14 and 17 lead to:

$$q_0 = \frac{bv}{A} \tag{18}$$

The dimensionless sediment transport rate, Φ, is described as:

$$\Phi = \frac{q_0}{wD} \tag{19}$$

where w is the fall velocity of sediment. According to the work of Madsen and Grant (1976a, b):

$$\Phi = c\Psi^3 \tag{20}$$

where c is a dimensionless coefficient. Using eqs. 13, 18, 19, and 20, the normalized bar migration speed, $v/(wD/b)$, is:

$$\frac{v}{wD/b} = Ac \left(\frac{0.098 B f_w}{s-1}\right)^3 \left(\frac{H_b}{D}\right)^3 \tag{21}$$

Assuming that all the coefficients involved in eq. 21, i.e., A, B, c, f_w, and s, are constants, we have:

$$\frac{v}{wD/b} = K' \left(\frac{H_b}{D}\right)^3 \tag{22}$$

where $K' = Ac[0.098 B f_w/(s-1)]^3$ which is an unknown constant. Equation 22 shows that the rate of onshore bar-migration is expressed as a function of the breaker height, the bar height, and the sediment properties (grain size and fall velocity). Because of the existence of temporal changes in wave climate, however, eq. 22 was replaced by:

$$\frac{\bar{v}}{wD/b} = K\left(\frac{\overline{H_b}}{D}\right)^3 \qquad (23)$$

where $\overline{H_b}$ is the time-averaged breaker height during one interval between surveys, $\bar{v}$ is the average speed of bar migration, and K is an unknown constant. The tidal effects have been neglected in the present study.

The speed of onshore bar-migration was examined using eq. 23. Due to assumptions involved in deriving eq. 23, only the data which satisfy the following two conditions were applied: (1) migrating bars have distinct two-dimensionality; and (2) the bar height does not significantly change on two successive beach profiles. The average bar-migration speed, $\bar{v}$, was obtained from the beach profile records. The bar height, b, was determined from the first of two successive surveys. The time-averaged, beach sediment grain size at each monitoring site was used for D, i.e., $D = 0.76$ mm for the North site, $D = 0.66$ mm for the Central site, and $D = 0.26$ mm for the South site. The fall velocity of sediment, w, was obtained from the $D - w$ relationship (e.g., U.S. Army Coastal Engineering Research Center, 1973, fig.4-31).

Selected data are listed in Table II, which includes Kimura's (1976) data obtained at Tatado Beach, Shimoda, Japan. Figure 8 gives a plot of these data. Although some scatter in the data points is found, the average speed of onshore migrating bars can be well described by:

$$\bar{v} = 2 \times 10^{-11} \frac{wD}{b} \left(\frac{\overline{H_b}}{D}\right)^3 \qquad (24)$$

Possible reasons for the data scatter are: (1) the sediment grain size, D, is treated as a constant, but this does vary slightly in time; and (2) the effect of tides upon the bar migration speed (Davis et al., 1972) is not considered.

Nomographs for bar migration speed

Rewriting of eqs. 6 and 7 on a daily basis gives the following relation which describes the condition for the onshore migration of inner bars:

$$5 \frac{D}{\tilde{H}_b} < \frac{\tilde{H}_b}{g\tilde{T}^2} < 20 \frac{D}{\tilde{H}_b} \qquad (25)$$

where $\tilde{H}_b$ and $\tilde{T}$ are the daily average breaker height and wave period, respectively. Similarly, eq. 24 can be rewritten as:

$$\tilde{v} = 2 \times 10^{-11} \frac{wD}{b} \left(\frac{\tilde{H}_b}{D}\right)^3 \qquad (26)$$

where $\tilde{v}$ is the daily average migration speed.

Using eqs. 25 and 26, nomographs giving the speed of onshore migrating bars were plotted in Figs.9a, b, and c for the beaches with a sediment grain

TABLE II

Data of onshore bar-migration speed (notation explained in text)

Site	D (mm)	$\bar{H}_b$ (m)	b (m)	$\bar{v}$ (m day^{-1})	Remarks
North	0.76	1.41	0.2	2.3	9/15—9/19/81
		1.41	0.2	2.5	9/15—9/19/81
Central	0.66	2.16	0.8	8.5	8/28—8/30/80
		2.16	0.7	11.5	8/28—8/30/80
		1.94	0.5	3.3	8/30—9/2/80
		1.14	0.2	3.3	9/12—9/15/80
		1.14	0.4	2.3	9/12—9/15/80
South	0.26	1.50	0.8	4.3	9/7—9/10/80
		1.06	0.3	5.0	9/13—9/16/80
		1.06	0.4	3.0	9/13—9/16/80
		1.06	0.4	3.3	9/13—9/16/80
		1.28	1.1	3.7	10/8—10/11/80
		1.28	0.5	3.0	10/8—10/11/80
		1.36	0.9	3.7	10/11—10/14/80
		1.36	0.8	2.7	10/11—10/14/80
		0.94	0.2	2.2	11/7—11/12/80
		0.94	0.2	2.4	11/7—11/12/80
		0.94	0.4	1.2	11/7—11/12/80
		1.29	0.3	2.2	11/12—11/17/80
		1.41	0.3	3.0	10/13—10/18/81
		1.41	0.4	2.6	10/13—10/18/81
Tatado Beach (Kimura, 1976)	0.23	1.34	0.4	3.5	8/14—8/15/76
		1.21	0.6	5.0	8/17—8/19/76
		1.03	0.6	5.0	8/19—8/21/76
		1.01	0.5	3.5	8/21—8/22/76
		1.03	0.5	3.0	8/22—8/23/76
		1.13	0.4	3.8	8/25—8/29/76

size of 0.2, 0.4, and 0.8 mm, respectively. These figures were constructed for the bars having a relative height of 0.5 and 1 m.

Assume, for example, that waves having a daily average breaker height of 2 m and a wave period of 10 s act on a beach composed of 0.4-mm sand. Use Fig.9b for this case. An intersecting point of two lines, i.e., $\widetilde{H}_b$ = 2 m and $\widetilde{T}$ = 10 s, is located near the line showing 5 m day^{-1} for a 1-m bar height case or 10 m day^{-1} for an 0.5-m case. Namely, under such wave and sediment conditions, the average speed of onshore bar-migration is estimated at 5 m day^{-1} if the bar height is 1 m, or 10 m day^{-1} if the bar height is 0.5 m.

CONCLUDING REMARKS

Demarcation of migrating directions of inner bars is shown in Fig.4. A necessary condition for landward bar-migration should be located in the area between the solid line (eq. 6) and the dashed line (eq. 7) in this figure. The

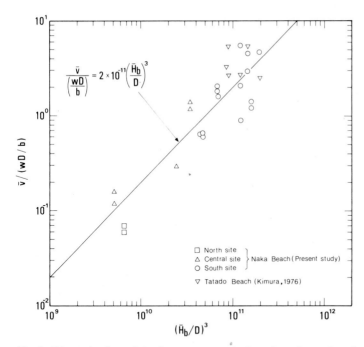

Fig.8. Dimensionless plot of average speed of onshore bar-migration.

average speed of onshore migrating bars is described by eq. 24, which is a function of the wave breaker height, the grain size and the fall velocity of the beach sediment, and the bar height. Nomographs for the migration speed are presented for the selected sediment grain size and bar height (Figs. 9a—c).

ACKNOWLEDGMENTS

Part of this study is financially supported through the Science Research Fund of the Ministry of Education (56020001, Principal investigator: Professor A. Ozaki, Hokkaido University) and the Kajima Foundation's Research Grant (Principal investigator: Associate Professor H. Nishimura, University of Tsukuba). Permission to use the wave data was given by the Sanpin Kowan Office of Ibaraki Prefecture. H. Tsujimoto, T. Mori, and T. Shimizu assisted in the field survey.

APPENDIX

Linear wave theory gives the maximum orbital velocity of the water particle near the bottom, u_m, as:

$$u_m = \frac{\pi H}{T \sinh(2\pi h/L)} \tag{A-1}$$

where H and L are the wave height and the wavelength at a water depth of h, respectively, and T is the wave period. Using shallow-water wave approximations such as:

$$\sinh \frac{2\pi h}{L} = \frac{2\pi h}{L}$$

and

$$\frac{L}{T} = \sqrt{gh} \ (= \text{wave velocity})$$

transformation of eq. A-1 leads to:

$$u_0 = \frac{H}{2}\sqrt{\frac{g}{h}} \qquad (A-2)$$

where u_0 is the maximum bottom velocity in a shallow water region.

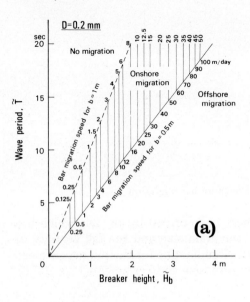

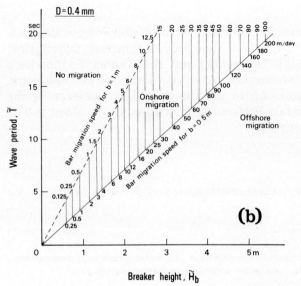

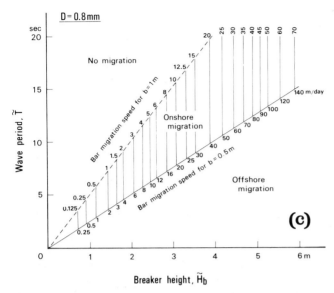

Fig.9. Nomographs for daily average speed of onshore migrating inner bars; $\tilde{H}_b$ and $\tilde{T}$ = daily average breaker height and wave period, respectively, D = sediment grain size, and b = bar height.

REFERENCES

Bascom, W., 1964. Waves and Beaches. Doubleday, Garden City, New York, 267 pp.
Dabrio, C.J. and Polo, M.D., 1981. Flow regime and bedforms in a ridge and runnel system, S. E. Spain. Sediment. Geol., 28: 97—110.
Davis, R.A. and Fox, W.T., 1972a. Coastal processes and nearshore sand bars. J. Sediment. Petrol., 42: 401—412.
Davis, R.A. and Fox, W.T., 1972b. Four-dimensional model for beach and inner nearshore sedimentation. J. Geol., 80: 484—493.
Davis, R.A. and Fox, W.T., 1975. Process-response patterns in beach and nearshore sedimentation: I. Mustang Island, Texas. J. Sediment. Petrol., 45: 852—865.
Davis, R.A., Fox, W.T., Hayes, M.O. and Boothroyd, J.C., 1972. Comparison of ridge and runnel systems in tidal and non-tidal environments. J. Sediment. Petrol., 42: 413—421.
Evans, O.F., 1939. Mass transportation of sediments on subaqueous terraces. J. Geol., 47: 325—334.
Fox, W.T. and Davis, R.A., 1978. Seasonal variation in beach erosion and sedimentation on the Oregon Coast. Geol. Soc. Am. Bull., 89: 1541—1549.
Greenwood, B. and Davidson-Arnott, R.G.D., 1975. Marine bars and nearshore sedimentary processes, Kouchibouguac Bay, New Brunswick. In: J. Hails and A. Carr (Editors), Nearshore Sediment Dynamics and Sedimentation. Wiley, London, pp.123—149.
Hashimoto, H. and Uda, T., 1976. Shore processes at Ajigaura Beach. Proc. 23rd Japan. Conf. Coastal Eng., pp.245—249 (in Japanese).
Hashimoto, H. and Uda, T., 1977. Shore processes at Ajigaura Beach (2). Proc. 24th Japan. Conf. Coastal Eng., pp.216—220 (in Japanese).
Hayes, M.O., 1972. Forms of sediment accumulation in the beach zone. In: R.E. Meyer (Editor), Waves on Beaches. Academic Press, New York, N.Y., pp.297—356.
Hine, A.C., 1979. Mechanisms of berm development and resulting beach growth along a barrier spit complex. Sedimentology, 26: 333—351.
Hoyt, J.H., 1962. High-angle beach stratification, Sapelo Island, Georgia. J. Sediment. Petrol., 32: 309—311.

Hoyt, J.H., 1971. Field Guide to Beaches. Houghton Mifflin, Boston, 44 pp.
Hunter, R.E., Clifton, H.E. and Phillips, R.L., 1979. Depositional processes, sedimentary structures, and predicted vertical sequences in barred nearshore systems, southern Oregon coast. J. Sediment. Petrol., 49: 711—726.
Kimura, R., 1976. Change of beach profile — in the case of Tatado Beach, Shimoda, Shizuoka Prefecture, Japan. Unpubl. BS Thesis, Dept. Geogr., Tokyo Kyoiku Univ., 53 pp. (in Japanese).
King, C.A.M. and Williams, W.W., 1949. The formation and movement of sand bars by wave action. Geogr. J., 112: 70—85.
Komar, P.D., 1976. Beach Processes and Sedimentation. Prentice-Hall, Englewood Cliffs, N.J., 429 pp.
Madsen, O.S. and Grant, W.D., 1976a. Sediment transport in the coastal environment. Tech. Rep., Dept. Civ. Eng., M.I.T., 209, 105 pp.
Madsen, O.S. and Grant, W.D., 1976b. Quantitative description of sediment transport by waves. Proc. 15th Int. Conf. Coastal Eng., pp.1093—1112.
Orford, J.D. and Wright, P., 1978. What's in a name? — Descriptive or genetic implications of "ridge and runnel" topography. Mar. Geol., 28: M1—M8.
Owens, E.H. and Frobel, D.H., 1977. Ridge and runnel systems in the Magdalen Islands, Quebec. J. Sediment. Petrol., 47: 191—198.
Sasaki, T., 1982. Three-dimensional topographic changes on the foreshore zone of sandy beaches. Unpubl. Doct. Thesis, Inst. Geoscience, Univ. Tsukuba, 172 pp.
Short, A.D., 1978. Wave power and beach-stages: a global model. Proc. 16th Int. Conf. Coastal Eng., pp.1145—1162.
Short, A.D., 1979. Three dimensional beach-stage model. J. Geol., 87: 553—571.
Sonu, C.J., 1969. Collective movement of sediment in littoral environment. Proc. 11th Int. Conf. Coastal Eng., pp.373—400.
Sonu, C.J., 1973. Three-dimensional beach changes. J. Geol., 81: 42—64.
Sunamura, T., 1980. Parameters for delimiting erosion and accretion of natural beaches. Annual Report, Inst. Geoscience, Univ. Tsukuba, 6: 51—54.
Sunamura, T., 1982. Determination of breaker height and depth in the field. Annual Report, Inst. Geoscience, Univ. Tsukuba, 8: 53—54.
Sunamura, T. and Horikawa, K., 1974. Two-dimensional beach transformation due to waves. Proc. 14th Int. Conf. Coastal Eng., pp.920—938.
Takeda, I., 1983. Beach changes by waves. Unpubl. Doct. Thesis, Inst. Geoscience, Univ. Tsukuba, 86 pp.
Tanaka, N., Ozasa, H. and Ogasawara, A., 1973. Note of the investigations on changes of shorelines in Japan, Part 1. Tech. Note, Port and Harbour Research Institute, Ministry of Transport, 163, 95 pp. (in Japanese).
U.S. Army Coastal Engineering Research Center, 1973. Shore Protection Manual, Vol. 1. U.S. Govt. Printing Office, Washington, D.C.

SEDIMENT FLUX AND EQUILIBRIUM SLOPES IN A BARRED NEARSHORE

BRIAN GREENWOOD and PETER R. MITTLER*

Departments of Geography and Geology, Scarborough Campus, University of Toronto, Scarborough, Ont. M1C 1A4 (Canada)
Department of Geography, Scarborough Campus, University of Toronto, Scarborough, Ont. M1C 1A4 (Canada)

(Received April 18, 1983; revised and accepted July 31, 1983)

ABSTRACT

Greenwood, B. and Mittler, P.R., 1984. Sediment flux and equilibrium slopes in a barred nearshore. In: B. Greenwood and R.A. Davis, Jr. (Editors), Hydrodynamics and Sedimentation in Wave-Dominated Coastal Environments. Mar. Geol., 60: 79—98.

Estimates of time-integrated values of total (ITVF) and net (INVF) sediment volume flux and the associated changes in bed elevation and local slope were determined for a crescentic outer nearshore bar in Kouchibouguac Bay, New Brunswick, Canada, for eight discrete storm events. A 100 × 150 m grid of *depth-of-activity rods* spaced at 10 m intervals was used to monitor sediment behaviour on the seaward slope, bar crest and landward slope during the storms, at which time winds, incident waves and near-bed oscillatory currents were measured. Comparisons between storm events and between these events and a longer-term synthetic wave climatology were facilitated using hindcast wave parameters. Strong positive correlations between storm-wave conditions (significant height and total cumulative energy) and total volume flux contrasted strongly with the zero correlation between storm-wave conditions and net volume flux. ITVF values ranged up to 1646 m^3 for the experimental grid and were found to have power function relations with significant wave height (exponent $\simeq$ 2) and cumulative wave energy (exponent $\simeq$ 0.4); values of INVF ranged from 0 up to 100 m^3 for the same grid indicating a balance of sediment volume in the bar form through time. Sediment reactivation increased linearly with decreasing depth across the seaward slope and bar crest reaching maxima of 20 cm for the two largest storms; bed elevation, and thus slope, changes were restricted to the bar crest and upper landward slope with near zero morphological change on the seaward slope. The latter represents a steady-state equilibrium with null net transport of sediment under shoaling waves. Measurements of the asymmetry of orbital velocities close to the bed show that the energetics approach to predicting beach slope of Inman and Bagnold (1963) is sound. Gradients predicted vary from 0.01 to 0.03 for a range of angles of internal friction appropriate to the local sediment (tan ϕ = 0.3—0.6). These compare favorably with the measured seaward slope of 0.015 formed under average maximum orbital velocities of 1.12 m s^{-1} (landward) and 1.09 m s^{-1} (seaward) recorded during the period of the largest storm waves.

---

*Present address: Dasco Data Products, 304-8495 Ontario Street, Vancouver, B.C. V5X 3E8, Canada.

0025-3227/84/$03.00 © 1984 Elsevier Science Publishers B.V.

INTRODUCTION

In many coastal environments dominated by waves the nearshore slope is characterized by one or more bars (Greenwood, 1982), which remain as stable bathymetric configurations throughout the annual cycle of wave climate (for example see Greenwood and Davidson-Arnott, 1975) and therefore appear in equilibrium with nearshore wave and current processes. However, the magnitude and frequency of occurrence of the sedimentation processes controlling bar growth and dynamics are still ill-defined and considerable debate continues concerning the mode and forcing of the sediment transport processes involved (Greenwood and Davidson-Arnott, 1979). Clarification of these issues requires knowledge of both fluid, sediment and morphological dynamics in the prototype over reasonably long periods to include especially periods of intense sediment transport. Furthermore, since the bar form consists of a very large number of slope facets each responding to local stresses, there is a need for knowledge of the spatial variability of process and form at this scale.

At the present time technological limitations prevent the necessary experiments to adequately answer the questions outlined above. However, recently it has been possible to obtain measures of both the sediment flux and bed elevation changes integrated over a storm cycle for a series of closely spaced locations within the nearshore zone (Greenwood and Hale, 1980; Greenwood et al., 1980). In this paper the sediment flux and morphological response in a nearshore bar is analysed for a series of storm-wave events of known frequency of occurrence and the data used to test one of the basic approaches to the prediction of local bed slope under wave activity.

EQUILIBRIUM NEARSHORE SLOPES

Currently no fully adequate theory of nearshore equilibrium exists although our general understanding of this equilibrium was well defined in the highly deductive and qualitative statement of Johnson as early as 1919:

> "At every point the slope is precisely of the steepness required to enable the amount of wave energy there developed to dispose of the volume of sediment there in transit."

Even earlier Cornaglia (1898) had proposed a *null point hypothesis* emphasizing the concept of the balance of forces (wave-generated currents, gravity, inertia) controlling the stability of single particles on nearshore slopes. This theory was quantified and tested experimentally by Ippen and Eagleson (1955), Eagleson and Dean (1961) and Eagleson et al. (1963), but met with only limited success. Qualitative models of the nearshore slope as a balance between wave-induced transport of sand landward and a seaward return via rip-current activity (Grant, 1943; Cook, 1970; Greenwood and Davidson-Arnott, 1979) have been supported by observation, but lack a quantitative form capable of being tested rigorously. Perhaps the most significant development in understanding coastal equilibrium under wave motion was that due to Inman and Bagnold (1963) based on Bagnold's (1963, 1966) energetics-based bedload transport model originally derived for unidirectional flow. In this formulation a local slope of zero net transport develops at equilibrium as a result of a balance of forces induced by asymmetry in the on-offshore velocities, the angle of internal friction of the sediment and the tangential component of gravity controlled by the slope itself. In essence the slope was defined by the ratio of offshore to onshore energy dissipation under oscillatory flow and the angle of internal friction of the sediment:

$$\tan \beta = \tan \phi \left(\frac{1-c}{1+c}\right) \tag{1}$$

where $\tan \beta$ = beach slope; $\tan \phi$ = coefficient of internal friction; and c is defined:

$$c = \frac{\text{offshore energy dissipation}}{\text{onshore energy dissipation}} \tag{2}$$

Thus, under shoaling waves, the increasing beach slope landward was a response to the increased total dissipation due to increased velocities at the bed and the relative difference in dissipation in the landward and seaward directions.

In 1979 the senior author proposed a conceptual model for equilibrium in nearshore bars whereby the seaward slope was maintained by the asymmetries in transport associated with the shoaling waves (Greenwood and Davidson-Arnott, 1979). If such is the case, then the Inman and Bagnold relationship should provide a prediction of the nearshore slope in this zone, if the energy dissipation ratio can be measured or predicted. Inman and Frautschy (1966), following Bagnold (1963), proposed that the energy dissipation ratio, c, was proportional to the third power of the ratio of the relevant onshore and offshore orbital velocities:

$$c = \left[\frac{U_m - \text{offshore}}{U_m - \text{onshore}}\right]^3 \tag{3}$$

where U_m is the maximum orbital velocity. This is in fact a measure of asymmetry in the orbital velocity field. Support for the basic mechanisms suggested by Bagnold (1963) and Inman and Bagnold (1963) can be seen in more recent sediment transport models which have adopted this approach. In 1980 Bowen proposed a quantitative model for predicting equilibrium beach slopes, which included a consideration of both suspended and bed load under oscillatory flows. An attempt to incorporate the effects of longshore currents is seen in the transport model of Bailard (1981, 1983) and Bailard and Inman (1981).

It is not possible in this study to provide a test of these later models, but rather a simple evaluation of the basic energetics approach is made by testing the predictor equation for slopes proposed in the original formulation.

LOCATION

The study was carried out in Kouchibouguac Bay, which is located at the western end of Northumberland Strait on the New Brunswick coast of Canada (Fig.1). Extensive detail on the form, structures, textures and bar dynamics, together with the general environmental constraints on the bar systems, have been documented previously (Greenwood and Davidson-Arnott, 1975, 1979; Davidson-Arnott and Greenwood, 1976; Greenwood and Hale, 1980) and only a brief synopsis need be presented here.

The area is a low-to-medium energy, micro-tidal (1.25 m maximum spring tide), storm-wave dominated environment, where sediment flux occurs as a highly discrete process during the passage of meteorological depressions (Greenwood and Hale, 1980). The nearshore bathymetry consists of a two-bar system with local slopes typified by those illustrated in Fig.2. The inner system is planimetrically variable with straight, oblique and crescentic forms cut through by well defined rip channels in places. In contrast the outer bar is generally continuous and crescentic in character (average wavelength = 500 m; average amplitude = 35 m) and ranges in height from 1.5 to 2.5 m. The outer bar, of particular relevance to this study is built in well-sorted, medium-to-fine sands (mean diameter 0.56—0.14 mm). Furthermore, it is a

Fig.1. Location of study area.

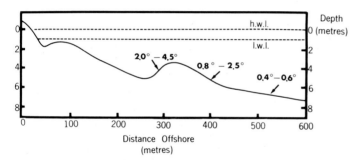

Fig. 2. Representative profile across the inner and outer bar systems of Kouchibouguac Bay.

highly stable feature in an environment of relatively high annual alongshore transport. Table I documents the sediment volumes and transports. It is clear that although large volumes of sediment exist in the bar system, collective transport alongshore due to bar migration contributes relatively little to the annual littoral drift.

A simple conceptual model of the equilibrium nature of these bars is based on a landward flux of sediments under asymmetric orbital motions due to waves, balanced by a seaward return of sediment in rip currents as part of a cellular nearshore circulation (Greenwood and Davidson-Arnott, 1979). Implicit to this hypothesis was that the seaward slope was essentially an equilibrium transport surface maintained by orbital asymmetries under the shoaling waves. Important questions therefore arise as to the mobility of the bar sediments (rates, spatial variability, relationship to wave energy, etc.), the morphological response to this mobility and the type of morphological equilibrium established (static, steady, dynamic, etc.). These will be addressed in this paper.

TABLE I

Sediment volumes and gross sediment flux Kouchibouguac Bay

Source	Amount	
Bar form: Total bar[1]	$\sim 2 \times 10^5$ m³	
: Per m length	$\sim 4 \times 10^2$ m³	
: Total bay[2]	$\sim 6 \times 10^6$ m³	
Net annual littoral drift[3]		$\sim 1 \times 10^5$ m³ yr^{-1}
Collective transport by bar migration[4]		$\sim 1 \times 10^2$ m³ yr^{-1}

[1]Based on single crescentic unit, 500 m long, 1.5—3 m high; [2]15 km of bar; [3]based on averaging dredging data, growth of inlet shoals and prediction based on CERC method 1973; [4]migration rate based on repeat surveys.

EXPERIMENTAL DESIGN

At the present state of technology no satisfactory instrument exists for monitoring simultaneously both suspended and bedload, and the morphological changes associated with a net sediment flux, across any large area in the nearshore. In this study a simple device is used which gives a measure of sediment transport (total and net) as well as local changes in elevation of the transport surface. This *depth-of-activity rod* (Greenwood and Hale, 1980; Greenwood et al., 1980) is a simple adaptation of the vigil network erosion pin and can be deployed in numbers sufficient to cover large areas. It consists of a round steel rod (0.5 cm diameter, 1—2 m in length) driven vertically into the sand by a Scuba diver until 0.45 m is left exposed; a loose fitting washer (0.6 cm internal diameter) is placed over the rod and allowed to fall to the bed. The rod is tagged with a fluorescent streamer tape and stamped with an identifier to assist re-location. Figure 3 illustrates the pre- and post-storm measurements made by a Scuba diver. These measurements are used to make estimates of the sediment flux taking place through discrete periods of time, i.e. the duration of the storm event. Assuming that the changing washer location indicates the maximum depth of reactivation (see Greenwood and Hale, 1980, for tests), then the minimum total volume of sediment in transport is also given. Since the rod measurements are time-integrated estimates of this transport rate (i.e., a volume) through the local control volume, this has been termed the *integrated total volume flux* (ITVF). In a similar manner the bed elevation changes measured at each rod represent the net volume transports integrated over the storm duration (again a volume measure): this has been termed the *integrated net volume flux* (INVF). The depth of activity and bed elevation measures record the response of the local slope to wave activity at the storm peak and the total sediment transport event.

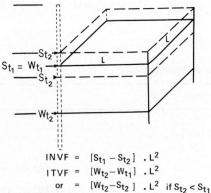

$INVF = [S_{t_1} - S_{t_2}] \cdot L^2$

$ITVF = [W_{t_2} - W_{t_1}] \cdot L^2$

or $= [W_{t_2} - S_{t_2}] \cdot L^2$ if $S_{t_2} < S_{t_1}$

S = surface elevation relative to top of rod
W = washer elevation relative to top of rod
t_2, t_2 = time intervals

Fig.3. Measurements made with the depth-of-activity rod and computations for sediment flux: *ITVF* = time-integrated total volume flux, *INVF* = time-integrated net volume flux. Both *ITVF* and *INVF* are volume measures.

To measure sediment flux and bed elevation change across the outer bar system, a reference grid (100 × 150 m) of *depth-of-activity rods* was established. Figure 4 illustrates the configuration of the grid, the primary rod coordinates and the relationship of the grid to a shore-normal profile. Rods were spaced at 10 m intervals and a subset was used to monitor the seaward slope. Control volumes for each grid were computed simply by extrapolating a line from the break-of-slope at the foot of the seaward slope to the trough and drawing vertical lines from the grid margins (Fig.5); a mean profile was assumed for the total grid to simplify calculations of these volumes potentially available for entrainment.

Fluid motions near the bed were recorded with an electromagnetic flowmeter (Marsh-McBirney Model 551) located in the centre of the seaward slope grid (Fig.4). The flowmeter sensor was mounted initially 0.2 m above the bed; however, depth of activity and bed elevation changes indicated possible variations in height of the sensor during storms between 0.15 and 0.38 m.

A continuous resistance wave gauge was deployed 700 m offshore in 7 m of water to record incident wave conditions, which provided a yardstick for storm magnitude. Winds, measured at the beach face with a Type 45 B Anemometer, were used to hindcast waves when direct measurements were unavailable. Hindcasting of waves using winds from the meteorological station at Chatham (Fig.1) allowed individual storms to be placed in the

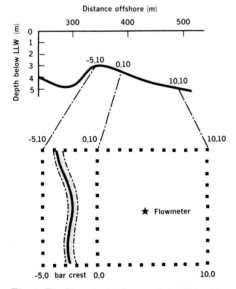

Fig.4. Deployment of experimental grids of depth of activity rods, Kouchibouguac Bay, 1977 and 1978. The 1977 grid was restricted to the seaward slope, row 0 to row 10 and this subset was used to monitor the seaward slope in both 1977 and 1978. Note the location of the flowmeter.

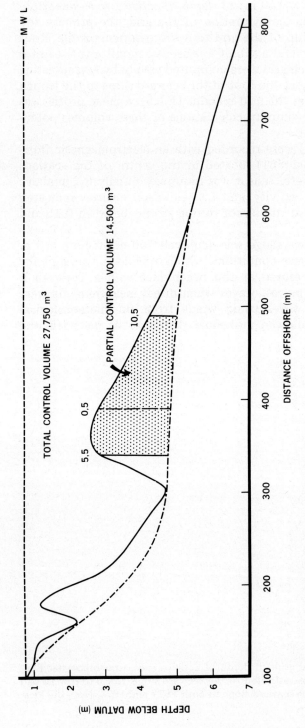

Fig.5. Control volume determinations for the experimental grids.

context of a synthetic wave climatology already established for the Bay (Hale and Greenwood, 1980). In this way the return periods and thus relative significance of the storms could be assessed.

Wave and current sensors were typically sampled at 2 Hz for periods of 5—7 min every hour during storms, with occasional record lengths up to 30 min. Wave spectra were computed using the University of California Biomedical Computer Package BMD02T (Dixon, 1971) to determine characteristic wave height (H_{mo} = 4 × standard deviation of the wave record) and peak period (T_{pk} = period corresponding to the frequency of the maximum energy density). Flowmeter records were resolved to produce true flow vectors (speed and orientation) and scatterplots used to describe the oscillatory motion at a particular period of time.

SEDIMENT FLUX AND EQUILIBRIUM SLOPES

Sediment flux

Table II summarizes data on the primary wave parameters and sediment flux for eight discrete storm events during 1977 and 1978, and reveals quite clearly the nature of the sediment balance. Although the percentage of the control volume mobilized by a single storm is not large it is significant, ranging from 1 to 12% with grid average maxima of 0.3 m³ per square metre of seaward slope in absolute terms. However, the time-integrated net volume flux (INVF) is extremely small, being less than one percent of the control volume in all cases. Small net additions of sediment and small net losses of sediment occur in approximately equal numbers.

To evaluate the response of the system to events of differing magnitude, which could then be placed within a correct time perspective, the time integrated total volume flux (ITVF) values were plotted against both measured

TABLE II

Sediment flux during storm events over the outer bar (seaward slope), Kouchibouguac Bay

Date	Duration (h)	H_{mo} (m)	T_{pk} (s)	ITVF (m³)	Percent* mobilized (%)	INVF (m³)	Percent change (%)
1977:07:06	8	1.30	6.4	661	5	110	—0.8
1977:07:23	11	1.17	6.0	319	2	36	—0.2
1977:08:24	15	1.00	5.5	208	1	20	+0.1
1977:08:30	17	1.06	5.3	182	1	61	—0.4
1978:05:23	5	No data		297	2	1	0
1978:06:01	10	No data		880	6	98	—0.7
1978:07:01	45	1.94	8.2	1734	12	29	—0.2
1978:07:24	17	1.61	7	1657	11	103	+0.7
				Average	5		—0.2

*Estimated control volume = 14,500 m³.

and hindcast wave data for each storm. Figure 6a, b and c illustrates double logarithmic plots of ITVF against both the maximum significant wave height ($H_{1/3}$) generated during the storms and the cumulative energy density (hindcast) for the duration of the storms. Both wave height and cumulative energy density are used here to express two characteristics of the storm: the former is a measure of the maximum energy input per unit time, while the latter is indicative of the total work done during a storm.

Strong positive correlations are evident in all cases with the percentage of explained variance ranging from a low of 45% in the case of hindcast

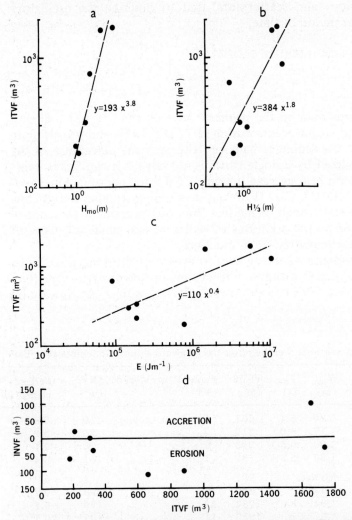

Fig. 6. Relationships between time-integrated total volume flux (ITVF), and (a) measured maximum significant wave height (H_{mo}); (b) hindcast maximum significant wave height ($H_{1/3}$); (c) hindcast cumulative energy density (E); (d) time-integrated net volume flux (INVF). Values for discrete storm events are plotted.

cumulative energy density to a high of 88% in the case of the measured significant wave height. Correlations with wave height are significant at the 0.05 level, while correlations with cumulative energy density are significant at the 0.10 level. The least squares regression lines suggest that a power function relationship exists between sediment response and wave energy. While it would be inappropriate to attach too great a significance to the absolute values of the exponents at this time it should be noted that sediment transport has always been considered as a power function of the forcing parameter. Indeed transport rates in the alongshore direction have been empirically established as being proportional to the square of the wave height (Galvin and Vitale, 1976). Certainly it is clear that the estimates of sediment reactivation are reasonably well recorded by the depth of activity rods and that maximum significant wave height is the better of the two wave parameters for predicting total sediment flux.

Of greater importance in the present context are the near-zero values for the INVF (Table II), even though the ITVF values range over an order of magnitude (Fig.6d). It would appear, therefore, that, at least in terms of sediment volume, the seaward slope exists in a state of steady equilibrium with the prevailing wave climate. If it can be established that the bar form remains stable both throughout storms and from storm to storm, then a steady morphodynamic equilibrium would be established.

Spatial variability of sediment flux and bar morphodynamics

Figure 7 illustrates the depths of activity and bed elevation changes for the two largest storms monitored (1978:01 and 1978:24). Table III documents the general storm characteristics. These two events had recurrence intervals close to 1.5 yr based upon hindcast wave parameters and might be expected to indicate bar response to storms equal to that of the most probable annual maximum. Although differing in intensity to some degree it is evident that the sedimentary response was similar. In both cases maximum reactivation occurred on the bar crest with similar absolute values of 28 and 32 cm; the average values for reactivation were also comparable (14 and 13 cm). A general tendency for decreasing depth of activity with increasing water depth is also evident as might have been expected, but the seaward slope was subjected to considerable sediment motion even at its outer margin. It should be noted, however, that the data are somewhat "noisy" due to sampling variability, experimental error and variations in the bedforms generated. Bed elevation changes are somewhat less "noisy" overall and illustrate again a comparable response of the bar slope to the two storms. Large areas of the grid in both cases exhibit less than ±1 cm change, which is close to the limit of measurement. Changes that do occur are more prevalent on the landward side of the bar crest and the landward slope; the seaward slope in contrast remained essentially stable in both cases even though large volumes of sediment were in motion. In order to generalize the shore-normal variability in the two measured parameters, grid row averages were calculated:

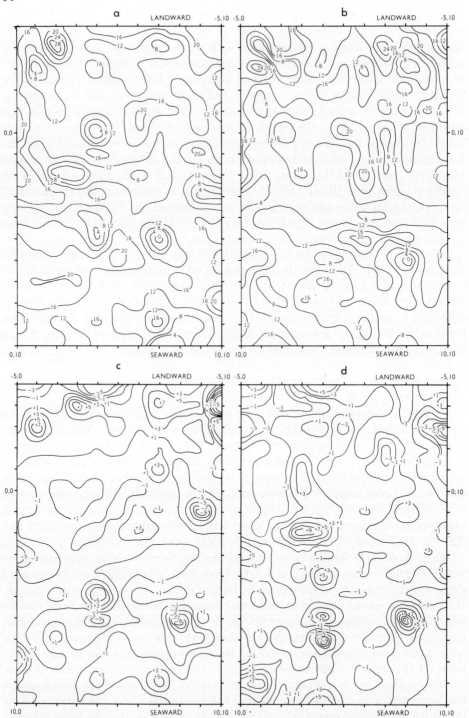

Fig.7. Spatial distribution of depth of activity (a and b) and bed elevation change (c and d) over the outer bar during the storms of 1978:07:01 (a and c) and 1978:07:24 (b and d). Contours are at 4 cm intervals (a and b) and 2 cm intervals (c and d). The grid coordinates are check marked along the edges.

TABLE III

Peak storm characteristics Kouchibouguac Bay, 1978:07:01 and 1978:07:24

Storm characteristics	Storm event			
	1978:07:01		1978:07:24	
	Measured	Predicted	Measured	Predicted
Maximum wind speed	58 kph	32 kph	48 kph	26 kph
Wind direction	E	NE	NE	N
Wind duration	45 h	37 h	17 h	23 h
Characteristic wave height (H_{mo})	1.94 m	0.72 m	1.64 m	0.62 m
Characteristic wave period (T_{pk})	8 s	4 s	7 s	3.2 s
Maximum orbital velocity*	1.26 m s^{-1}	—	1.20 m s^{-1}	—
Average orbital velocity*	0.42 m s^{-1}	—	0.39 m s^{-1}	—
Recurrence interval**: $H_{1/3}$	1.45 yr	(0.14 yr)	1.3 yr	(0.10 yr)
: ΣE	1.1 yr	(0.40 yr)	1 yr	(0.20 yr)

*Determined from resolved flowmeter vectors; sample period 5 min.
**Figures determined from both the annual maximum series and, in parentheses, the partial duration series of storm wave predictions.

this gave an overall view of the sedimentary responses along a profile by reducing much of the "noise" in the data. Figure 8 illustrates these data and confirms that:

(a) As waves shoal and orbital velocities near the bed increase so there is an increase in the rates of sediment motion. An almost linear increase in the depth of activity is suggested here which follows the almost linear seaward slope. Although there is a slightly greater variation associated with the larger of the two storms there appears to be no difference in the average response of sediment transport. This is reasonable in that although measured wave heights were greater in the July 1 storm, the near-bed currents measured on the seaward slope were almost the same (Table III).

(b) With the reduction in orbital velocities as a result of height loss through breaking on the bar crest and with the increase in water depth, the lower landward slope and trough experienced reduced rates of sediment motion. Again a linear decrease in the depth of activity is suggested down the landward slope.

(c) In general bed elevation and thus slope changes on the seaward slope are extremely minor even under intense sediment reactivation by the shoaling waves. Most of the variability in the averaged values is less than ±1 cm.

(d) The bar crest and landward slope experience the greatest redistribution of sediment and morphological change. This is to be expected since these areas are subject to wave breaking and wave reformation with high orbital velocities and strong flow asymmetries. Intense turbulence and interactions between waves and secondary flows, such as seaward flowing rip currents, are common. In the July 1 storm it appears that on average the bar was shifted seaward with erosion of the landward slope and accretion on the bar

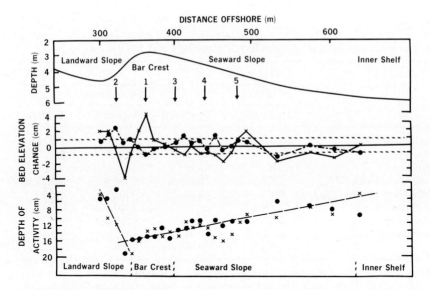

Fig.8. Average depth of activity and bed elevation change for the storms of 1978:07:01 (crosses) and 1978:07:24 (dots) across the outer bar. The profile is that taken along the centre line of the grid on 1978:07:14. The numbered arrows indicate locations of box cores.

crest: in the following storm of July 24, with only slightly lower wave heights, accretion on the landward slope and thus landward migration of the bar was the average condition. It should be stressed that these are averages, since along a 100 m length of bar crest re-orientation of the crest location could produce landward migration in one section and seaward migration in another. This has been observed frequently on these bars (Greenwood and Davidson-Arnott, 1975) with the crest oscillating about some mean position in association with the slow alongshore migration of the crescentic form and the transient positions of seaward flowing rip currents.

Two important aspects of the sediment flux patterns during these storms cannot, however, be addressed directly with the depth of activity rod data. The paths of net transport associated with bed elevation changes are not expressed, for example, nor is it possible to say whether the morphologically stable seaward slope is maintained by high but uniform rates of transport or whether a true oscillating equilibrium exists in the sediment transport. Examination of sedimentary structures preserved in the reactivated layer can provide some indication of transport paths (Greenwood and Davidson-Arnott, 1975; Davidson-Arnott and Greenwood, 1976; Greenwood and Mittler, 1979; Greenwood and Hale, 1980) as in the case of the 1978:07:24 storm. Epoxy peels of box cores from the bar crest and upper landward slope along the central profile (locations 1 and 2 in Fig.8) of the experimental grid all exhibit strong landward dipping cross-stratification indicative of landward migrating lunate megaripples (Fig.9a and b). This would suggest that the accretion of the landward slope indicated by the depth of activity rods results from sediment transferred landward from the crest where the rods suggest some slight degradation (Fig.8). Cores from the bar crest for a distance of 90 and 120 m either side of the grid all reveal similar structures supporting both a general transfer of sediment and a displacement of the bar in a landward direction (Mittler, 1981). Structural indices from the reactivation layer of the morphologically stable seaward slope (locations 3 and 4 in Fig.8) reveal a preponderance of planar stratification (Fig.9c and d). Thus the high sediment transport indicated by the depth of activity and the near-zero net sediment flux indicated by the bed elevation change were associated primarily with sediment motion under flat bed conditions: no directional information is provided by these structures but continuity considerations would suggest that during the phase of high transport rates the sediment was likely in a state of near oscillating equilibrium rather than one of high uniform net transport.

The most striking aspect of these data is, however, the morphological stability of the seaward slope with high rates of sediment motion. This clearly supports the thesis that this unit is a surface of null *net* transport and thus in a state of steady equilibrium with the orbital velocity field associated with the shoaling waves.

Seaward slope equilibrium

From the preceding analysis it would seem that a balance of forces involving wave energy dissipation and the morphological gradient in the manner suggested by Inman and Bagnold (1963) would be appropriate to explain the steady state of the seaward slope of the bars in Kouchibouguac Bay. To test this theory, measured orbital velocities on the seaward slope were analyzed. In order to satisfy constraints imposed by the model, the flow field must be: (1) solely the result of the onshore-offshore motion of waves approaching along the shore-normal (line of maximum topographic gradient); (2) be representative of the lower boundary layer; and (3) great enough not only to produce motion but to ensure transport under flat bed conditions since, as was pointed out by Inman and Bowen (1962), transport over rippled beds under oscillatory flow produces a complex phase-dependent process involving suspension associated with vortex generation. Indeed, as the basic model requires bedload transport, a flat bed sheet flow model would seem most appropriate.

Current records were obtained from a number of the storms (Mittler, 1981), but in only one case did the velocity field satisfy the above mentioned criteria. Figure 10 illustrates this velocity field, recorded on July 1 at the peak of the largest storm monitored. The ellipses represent the limits of the distribution of current vectors determined from the electromagnetic flowmeter records. The principal axes of the ellipses, determined by least squares, illustrate the shore-normal approach of the waves at this time, the strength of the wave oscillations, and, most importantly, the landward asymmetry of the maximum orbital velocities. During other storms, either the wave

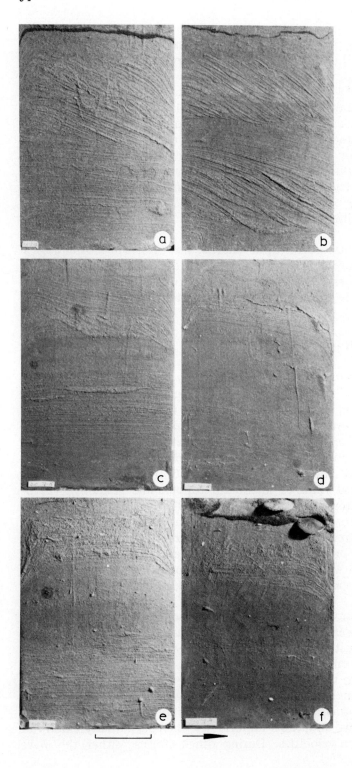

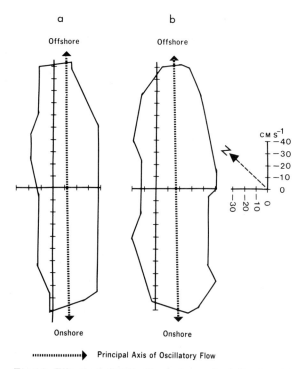

Fig.10. Elliptical distribution of near bed flow vectors on the seaward slope of the outer bar, Kouchibouguac Bay, during the peak of the storm, 1978:07:01; (a) 1520 h, (b) 1627 h.

approach (and thus the primary orbital velocity vector) was not normal to shore, or the measured maximum offshore orbital velocity was greater than its onshore equivalent. Even with the very strong shore-normal currents of the July 1 storm, a slight displacement of the centroid of the ellipse suggests a superimposed shore-parallel flow (Fig.10). Detailed analysis of similar offsets in a large number of discrete flowmeter records shows such flows to be of tidal origin (Mittler, 1981). For sand of the size common on the seaward slope (mean diameter ~0.18 mm), even the mean orbital velocities (Table III) at this time would theoretically produce flat bed. Using the criterion of Dingler (1974) as expressed by Clifton (1976) for the transition velocity between rippled and flat bed this grain size would require a velocity of $\simeq 0.26$ m s^{-1}. This is far exceeded by the mean velocities measured and would indicate flat bed sediment transport at this time. Corroboration for this is provided by post-storm box cores from the seaward slope (locations 4 and 5

Fig.9. Epoxy peels of box cores and the associated depths of activity (in parentheses) after the storms of 1978:07:24 (a, b, c, d) and 1978:07:01 (e, f). The horizontal bar is 0.10 m; the arrow indicates the direction of the shoreline. (a) bar crest (0.15 m); (b) upper landward slope (0.15 m); (c) upper seaward slope (0.13 m); (d) mid-seaward slope (0.14 m); (e) mid-seaward slope (0.13 m); (f) lower seaward slope (0.13 m).

in Fig.8) which illustrate dominance of planar stratification in the most recent active layer (Fig.9e, f).

Table IV documents the maximum onshore and offshore velocities, the ratio of offshore to onshore energy dissipation as defined by Inman and Frautschy (1966) and the predicted equilibrium beach slopes for a range of values of the angle of internal friction appropriate to the sand materials present. Also given is the average slope along the profile measured after the storm from a position 10 m landward to a position 10 m seaward of the flowmeter location. From the earlier analysis of the depth of activity it is evident that differences in bed elevation would not have exceeded 2 cm even at the storm peak over this section of slope and this could have no effect on the slope value used in this comparison. Predicted slope values are given for both the maximum instantaneous velocity recorded, and the average maximum velocity based on three discrete sample records spaced 1 h apart during the storm peak. The correspondence between measured and predicted values is good, particularly, if the average maximum velocity for the storm peak is used. The latter is a more reasonable figure to use for sediment transport in any case, since transport is a time integrated phenomenon; further, the instantaneous maxima in a record could be the result of single waves.

CONCLUSIONS

Time-integrated values of total sediment flux over a large outer crescentic bar in Kouchibouguac Bay, Canada, were found to be positively correlated with measured and predicted wave energy parameters: power function relationships with exponents of ~2 to ~4 for significant wave height and ~0.4 for cumulative energy values provide statistically significant explanation. Time-integrated values of net sediment flux, in contrast, were close to zero and bear no relationship to the total reactivation of sediment. Local bed elevation changes were only significant over the bar crest and landward slope, with the seaward slope existing in a state of steady equilibrium with the wave processes. Continuity considerations suggest a state of oscillating equilibrium for the large

TABLE IV

Measured maximum orbital velocities, dissipation coefficients, predicted and measured nearshore slopes, Kouchibouguac Bay, 1978:07:01

Maximum orbital velocity (u_m)		Dissipation coefficient (c)	Nearshore slope tan β			
Onshore (cm s^{-1})	Offshore		Predicted tan ϕ[4] = 0.30　0.45　0.60			Measured
128[1]	123[1]	0.89	0.018	0.026	0.035	0.015[3]
112[2]	109[2]	0.92	0.012	0.019	0.025	

[1]Maximum instantaneous velocity; [2]average maximum during storm peak; [3]average slope from post-storm survey; [4]tan ϕ = angle of internal friction.

volumes of sediment set in motion on this slope by storm waves. Morphological changes on the bar crest and landward slope result from the number and variability of processes occurring here: asymmetric oscillatory currents associated with propagating surface gravity waves are present, but also turbulence associated with wave breaking and interactions between these and secondary longshore and rip currents.

Results of this study support both the basic energetics concept inherent in the Inman and Bagnold (1963) model for equilibrium nearshore slopes and the predictor equation based on the asymmetry in the maximum orbital velocities (Inman and Frautschy, 1966). Predicted values for the seaward slope of 0.012—0.025 compare favourably with the measured value of 0.015 using measured average maximum orbital velocities of 1.12 m s^{-1} (landward) and 1.09 m s^{-1} (seaward), and a range for the appropriate angle of internal friction of 0.3—0.6.

REFERENCES

Bagnold, R.A., 1963. Mechanics of marine sedimentation. In: M.N. Hill (Editor), The Sea, Vol. 3. Wiley-Interscience, New York, N.Y.
Bagnold, R.A., 1966. An approach to the sediment transport problem from general physics. Prof. Pap. 422-1, U.S. Geol. Surv., Washington, D.C., 37 pp.
Bailard, J.A., 1981. An energetics total load sediment transport model for a plane sloping beach. J. Geophys. Res., 86: 10,938—10,954.
Bailard, J.A., 1983. Modeling on-offshore transport in the surfzone. Proc. 18th Coastal Engineering Conf., Cape Town, pp.1419—1438.
Bailard, J.A. and Inman, D.L., 1981. An energetics bedload model for a plane sloping beach: local transport. J. Geophys. Res., 86: 2035—2043.
Bowen, A.J., 1980. Simple models of nearshore sedimentation; beach profiles and longshore bars. In: S.B. McCann (Editor), The Coastline of Canada. Geol. Surv. Can., Pap. 80-10, pp.1—11.
Clifton, H.E., 1976. Wave-formed sedimentary structures: a conceptual model. In: R.A. Davis, Jr. and R.L. Ethington (Editors), Beach and Nearshore Sedimentation. Soc. Econ. Paleontol. Mineral., Spec. Publ., 24: 126—148.
Cook, D.O., 1970. Models for nearshore sand transport. Proc. 13th Conf. on Great Lakes Research, Buffalo, N.Y., pp.210—216.
Cornaglia, P., 1898. On beaches. Accadamia Nazionale dei Lincei Atti, Classe di Scienze Fisiche, Matematiche e Naturali, Mem. 5, Ser. 4, pp.284—304. Reproduced in translation in: J.S. Fisher and R. Dolan (Editors), Beach Processes and Coastal Hydrodynamics. Benchmark Pap. in Geol., Stroudsberg, Pa., 39: 11—26.
Davidson-Arnott, R.G.D. and Greenwood, B., 1976. Facies relationships on a barred coast, Kouchibouguac Bay, New Brunswick, Canada. In: R.A. Davis, Jr. and R.L. Ethington (Editors), Beach and Nearshore Sedimentation. Soc. Econ. Paleontol. Mineral., Spec. Publ., 24: 149—168.
Dingler, J.R., 1974. Wave-formed ripples in nearshore sands. Unpubl. Ph.D. Diss., University of California, San Diego, Calif., 136 pp.
Dixon, W.J., 1971. BMD: Biomedical Computer Programs. University of California Press, Berkeley, Calif., 600 pp.
Eagleson, P.S. and Dean, R.G., 1961. Wave-induced motion of bottom sediment particles. Trans. Am. Soc. Civ. Eng., 126(1): 1161—1189.
Eagleson, P.S., Glenne, B. and Dracup, J.A., 1963. Equilibrium characteristics of sand beaches. J. Hydraul. Div. Am. Soc. Civ. Eng., 89(HY1): 37—57.

Galvin, C.J. and Vitale, P., 1976. Longshore transport prediction — SPM 1973 Equation. Proc. 15th Coastal Engineering Conf., Honolulu, Hawaii, pp.1113—1148.
Grant, U.S., 1943. Waves as a sand transporting agent. Am. J. Sci., 241: 117—123.
Greenwood, B., 1982. Bars. In: M.L. Schwartz (Editor), Encyclopaedia of Beaches and Coastal Environments. Dowden, Hutchinson and Ross, Stroudsberg, Pa., pp.135—139.
Greenwood, B. and Davidson-Arnott, R.G.D., 1975. Marine bars and nearshore sedimentary processes, Kouchibouguac Bay, New Brunswick, Canada. In: J. Hails and A. Carr (Editors), Nearshore Sediment Dynamics and Sedimentation. Wiley, New York, N.Y., pp.123—150.
Greenwood, B. and Davidson-Arnott, R.G.D., 1979. Sedimentation and equilibrium in wave-formed bars: a review and case study. Can. J. Earth Sci., 16: 312—332.
Greenwood, B. and Hale, P.B., 1980. Depth of activity, sediment flux and morphological change in a barred nearshore environment. In: S.B. McCann (Editor), The Coastline of Canada: Littoral Processes and Shore Morphology. Geol. Surv. Can., Pap. 80-10, pp.89—109.
Greenwood, B. and Mittler, P.R., 1979. Structural indices of sediment transport in a straight wave-formed nearshore bar. Mar. Geol., 32: 191—203.
Greenwood, B., Hale, P.B. and Mittler, P.R., 1980. Sediment flux determination in the nearshore zone: prototype measurements. In: Workshop on Instrumentation for Currents and Sediments in the Nearshore Zone, National Research Council of Canada, Associate Committee for Research on Shoreline Erosion and Sedimentation, Ottawa, Ont., pp. 99—120.
Hale, P.B. and Greenwood, B., 1980. Storm wave climatology: a study of the magnitude and frequency of geomorphic process. In: S.B. McCann (Editor), The Coastline of Canada: Littoral Processes and Shore Morphology. Geol. Surv. Can., Pap. 80-10, pp.70—88.
Inman, D.L. and Bagnold, R.A., 1963. Littoral processes. In: M.N. Hill (Editor), The Sea, Vol. 3. Wiley-Interscience, New York, N.Y., pp.529—533.
Inman, D.L. and Bowen, A.J., 1962. Flume experiments on sand transport by waves and currents. Proc. 8th Coastal Engineering Conf., Mexico City, pp.137—150.
Inman, D.L. and Frautschy, J.D., 1966. Littoral processes and the development of shorelines. Proc. Coastal Engineering Speciality Conf., Santa Barbara, Calif., pp.511—536.
Ippen, A.T. and Eagleson, P.S., 1955. A study of sediment sorting by waves shoaling on a plane beach. Beach Erosion Board, U.S. Army Corps of Engineers, Tech. Mem., 63, 83 pp.
Johnson, D.W., 1919. Shore processes and shoreline development. Columbia University, Facsimile Ed., Hafner, New York, N.Y., 1972, 584 pp.
Mittler, P.R., 1981. Storm related sediment flux and equilibrium in a barred nearshore, Kouchibouguac Bay, New Brunswick, Canada. Unpubl. Ph.D. Diss., University of Toronto, Scarborough, Ont., 419 pp.

SEDIMENT TRANSPORT AND MORPHOLOGY AT THE SURF ZONE OF PRESQUE ISLE, LAKE ERIE, PENNSYLVANIA

DAG NUMMEDAL[1], DAVID L. SONNENFELD[1]* and KENT TAYLOR[2]

[1]*Department of Geology, Louisiana State University, Baton Rouge, LA 70803 (U.S.A.)*
[2]*Department of Geology, Mercihurst College, Erie, PA 16546 (U.S.A.)*

(Received June 1, 1983; revised and accepted September 1, 1983)

ABSTRACT

Nummedal, D., Sonnenfeld, D.L. and Taylor, K., 1984. Sediment transport and morphology at the surf zone of Presque Isle, Lake Erie, Pennsylvania. In: B. Greenwood and R.A. Davis, Jr. (Editors), Hydrodynamics and Sedimentation in Wave-Dominated Coastal Environments. Mar. Geol., 60: 99—122.

A four-year investigation of surf zone sedimentation at Presque Isle, Pennsylvania, was undertaken in preparation for the design of a segmented breakwater system. Sediment transport calculations were based on hind-cast annual wave power statistics and "calibrated" by known accretion rates at the downdrift spit terminus. 30,000 m³ of sediment reaches the peninsula annually from updrift beaches. The transport volume increases downdrift due to shoreface erosion and retreat of the peninsular neck. At the most exposed point on Presque Isle (the lighthouse) the annual transport is 209,000 m³. East of the lighthouse is a zone of net shoreface accretion as the longshore transport rate progressively decreases.

The downdrift variation in sediment supply, combined with increasing refraction and attenuation of the dominant westerly storm waves produce a systematic change in prevailing surf zone morphology. Storms produce a major longshore bar and trough along the exposed peninsular neck. The wave energy during non-storm periods is too low to significantly alter the bar which consequently becomes a permanent feature. The broad shoreface and reduced wave energy level east of the lighthouse produce a morphology characterized by large crescentic outer bars, transverse bars, and megacusps along the beach. At the sheltered and rapidly prograding eastern spit terminus the prevalent beach morphology is that of a ridge and runnel system in front of a megacuspate shore.

The morphodynamic surf zone model developed for oceanic beaches in Australia is used as a basis for interpretation of shoreface morphologic variability at Presque Isle. In spite of interference by major shoreline stabilization structures, and differences between oceanic and lake wave spectra, the nearshore bar field at Presque Isle does closely correspond to the Australian model.

INTRODUCTION

Presque Isle is a 10.5 km long, compound recurved spit off Erie, Pennsylvania. Originally a natural eastward-migrating "flying" spit, Presque

---

*Present address: Louisiana Geological Survey, Louisiana State University, Baton Rouge, LA 70803, U.S.A.

0025-3227/84/$03.00 © 1984 Elsevier Science Publishers B.V.

Isle has over the last 150 years been subjected to increasingly intensified stabilization efforts. Presque Isle's dual role as a state park with the best beaches along the U.S. shores of Lake Erie and as a shelter for the industrial Erie harbor provides more than adequate economic incentives for the increasingly expensive stabilization efforts. The investigations summarized in the paper were undertaken in direct response to the need for better information about the patterns of nearshore sediment movement prior to the design and construction of a new large-scale lake shore protection system planned for the 1980's (U.S. Army, Corps of Engineers, 1979).

This phase of design-related studies consisted of field monitoring of changes in the Presque Isle beach profiles and nearshore bathymetry (Fig.1). The field surveys were supplemented by vertical aerial mapping photography. Both ground surveys and aerial photography were done on a seasonal basis. Field surveys lasted from the spring of 1978 through the fall of 1980. Annual reports to the U.S. Army Corps of Engineers (Nummedal, 1979, 1980, 1981) present all relevant survey data as well as preliminary analyses. The emphasis in this paper is on the patterns and mechanisms of sediment transport as deduced from the surveys.

The two main topics of this paper are:

(1) *The sediment budget for Presque Isle.* The budget is based on the premise that the *relative* transport rates along different segments of Presque Isle can be calculated based on the longshore wave power distribution during dominant southwesterly wave approach. Absolute values for longshore transport are determined by "calibrating" the computed relative rates by

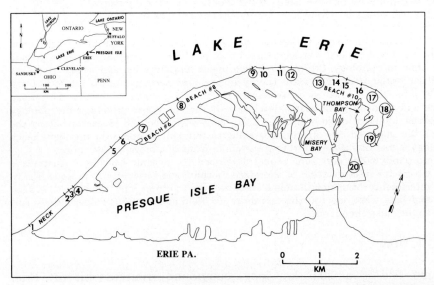

Fig.1. Location map of Presque Isle, Pa. Numbered sites designate beach profile locations reoccupied seasonally during the 3-year shoreline monitoring program (1978—1980). Circled numbers designate sites where both beach and nearshore bathymetric profiles were obtained. These profiles are designated with their "PI numbers" throughout the text.

using known rates for maintenance dredging at the harbor entrance and historical rates of accretion at Gull Point.

(2) *Bar morphology along the lake shore of Presque Isle.* Observed seasonal changes and longshore variability in the bar field are documented. The morphologic variability is explained in terms of the derived sediment budget and the inferred longshore changes in absolute wave power and nearshore current regime.

LONGSHORE VARIABILITY IN SEDIMENT TRANSPORT RATES

Wave climate

No long-term wave gauge has been operating in eastern Lake Erie. Therefore, data on the lake shore wave climate are based on Summaries of Synoptic Meteorological Observations (SSMO-data from the U.S. Naval Weather Service Command, 1975); hindcast wave data from meteorological reports (Saville, 1953; Resio and Vincent, 1976); and occasional periods of LEO observations.

The most severe weather disturbances affecting the Great Lakes region are extratropical cyclones. Analysis of a large number of synoptic charts have demonstrated that common cyclone paths take the center of the low-pressure system on an eastward course passing to the north of Lake Erie. The strongest associated winds, therefore, commonly blow out of the west-southwest (Nummedal et al., 1976). Because this corresponds to the direction of maximum fetch with respect to the shore of Presque Isle one would expect to find that the dominant waves and the maximum wave power reach Presque Isle from the west. Saville's (1953) wave hindcast (Fig.2) and the derived wave power distribution diagram (Fig.3) demonstrate that such is indeed the case. Observations summarized in the SSMO data files are generally consistent with those hindcast by Saville (1953).

Resio and Vincent (1976) conducted a more recent hindcast study of wave conditions on Lake Erie. Resio and Vincent (1976) applied a numerical hindcasting technique developed by Pierson and Moskowitz (1964) which is better suited to the short and variable fetches encountered at Lake Erie, than is the Sverdrup-Munk-Bretschneider approach (Coastal Engineering Research Center, 1973) as used by Saville. Comparison of the results indicate that the SMB-technique tends to overpredict the Great Lakes wave heights by a factor of 10—20%. Although significant in design wave considerations, this level of discrepancy does not appear to seriously impair the applicability of Saville's (1953) results in studies of nearshore sediment transport.

Lake Erie has a seasonal wave climate. Hindcasting demonstrates that winter is the stormy season (Fig.4). However, Lake Erie is often frozen between late December and March. Thus, most of the wave energy effective in sediment transport hits the Presque Isle beaches in the months of April, May and November. The total average hindcast, deep water, wave power at Erie, Pennsylvania, is 566 W m^{-1}.

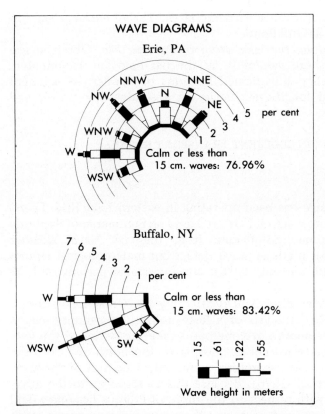

Fig.2. Summary diagram of the annual wave height distribution in eastern Lake Erie based on data obtained from hindcasting performed by Saville (1953). The hindcast is for deep water conditions for the whole year, including those months which some years have a lake-wide ice cover.

Long periods of calm interrupted by short-lived high-energy events characterize a storm wave environment such as Lake Erie. As will be demonstrated below, bar dynamics and beach stability at Presque Isle correspond closely to what has been documented on micro-tidal, storm-dominated oceanic shores.

Longshore distribution of wave power

The longshore sediment transport rate can be calculated for the lake shore of Presque Isle based on the wave power distribution in Fig.3. The assumptions used in the calculations are: (a) the waves responsible for the bulk of the annual transport have a period of 4 s and they break in 1 m of water (see Nummedal et al., 1976, for field data supporting this assumption); and (b) the bathymetry consists of simple shore-parallel contours. This permits analytical determination of the refracted wave angle at breaking from Snell's law (Komar, 1976).

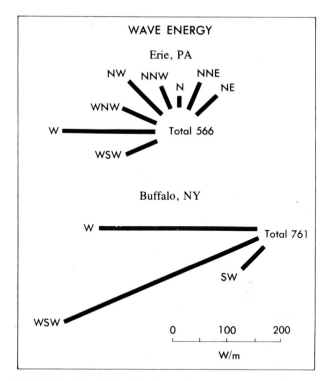

Fig.3. Directional distribution of mean annual wave power in eastern Lake Erie. The distribution has been calculated from hindcast data of Saville (1953). The listed total (e.g. 566 W m^{-1} at Erie) designates the mean annual wave power at that station irrespective of wave approach direction. Deep-water data.

The deep-water wave power from a given direction is designated $P_0(m)$. Its longshore component can be shown to be $P_1(m)$, where:

$$P_1(m) = P_0(m) \sin a_b \cos a_0 \tag{1}$$

This equation is derived in Walton (1973). Equation 1 properly accounts for refraction and shoaling but assumes no friction or bottom percolation. Angle a_0 is determined by the local shoreline orientation and a_b is calculated using Snell's law.

The results of the computations are summarized in Table I. The peninsula was divided into five straight shoreline segments (Fig.5). There are distinct breaks in shoreline trend at groin 11, the big groin by the lifeguard station, the lighthouse, the eastern end of Beach 10, and at Gull Point. These points were used as segment boundaries.

Gross longshore wave power (sum of components to the right and to the left along a shoreline) is a fairly good measure of the total sediment transport potential along a shoreline segment. As seen in Table I, the gross longshore power is essentially invariant between the neck of the peninsula and Beach 10. The gross longshore power is, as expected, much less in the Gull

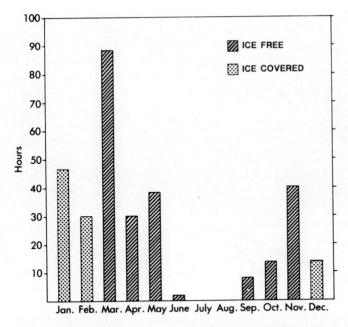

Fig.4. Annual variation in wave characteristics. Maximum storm activity occurs in March, generally associated with the break-up of lake ice. May and November have about the same storm frequency. Short-lived summer thunderstorms, which do generate high waves, are not adequately represented in the hindcast statistics. From Nummedal et al. (1976).

TABLE I

Wave power calculations for individual segments of Presque Isle. See Fig.5 for location of segments, and Nummedal (1983) for a more detailed account of the computations

Az[a]	P_0[b]	P_1I[c]	P_1II[c]	P_1III[c]	P_1IV[c]	P_1V[c]
WSW	61.5	15.1	14.3	9.2	offshore	offshore
W	172.0	29.7	37.3	42.0	23.4	offshore
WNW	69.0	0	4.9	12.7	16.6	1.5
NW	87.9	−15.1	−10.0	1.5	17.8	16.4
NNW	45.6	−10.8	−10.4	−7.1	2.1	10.9
N	22.5	−3.6	−4.5	−5.1	−2.9	3.3
NNE	55.1	0	−3.8	−9.8	−12.7	1.1
NE	52.4	offshore	offshore	−0.86	−10.5	−9.6
Net power	—	15.2	27.8	43.0	33.8	21.2
Gross power	566	74.3	85.2	88.9	86.0	42.8

[a]Direction of incident waves; [b]deep-water wave power in W m⁻¹ (from Fig.3); [c]longshore component of wave power within segments I, II, III, IV, V. Negative sign indicates movement to the left (westward) along the beach.

Point area because the shoreline here faces away from the dominant incoming westerly waves.

Net longshore wave power is uniformly directed toward the east along the entire shore but the rates vary greatly. There is a continuous downdrift increase in longshore power from the neck all the way to the lighthouse, then a reduction from there eastward. This pattern is consistent with net erosion along the entire peninsular neck and net deposition to the east of the lighthouse for the shoreface as a whole. Individual beaches may erode or accrete within both of these areas in response to local, in part man-made, shoreline perturbations. The erosion along the neck will be most severe where there is a sudden increase in the net longshore wave power. This occurs where there are abrupt changes in shoreline trend, as for example, in the area between groin 9 and Beach 6. This has, in recent years, been an area of rapid beach erosion.

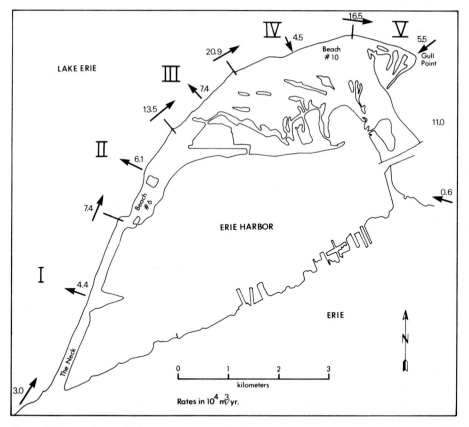

Fig.5. Calculated rates of sediment transport, erosion and accretion along the Lake Erie shore of Presque Isle. The budget was derived from the distribution of wave power, historical accretion rates at Gull Point and Erie harbor dredging records.

Sediment transport rates

The wave power values in Table I are used only to determine the relative rates of longshore sediment transport along the Presque Isle beaches. The assumption is made that the volume rate of sediment transport is linearly related to wave power (e.g. eq. 4-40 in the Shore Protection Manual, CERC, 1973). Sediment delivery rate to the updrift end of Presque Isle has been estimated from bluff retreat rates measured by Carter (1976) by the U.S. Army, Corps of Engineers (1979). Data on the downdrift accretion were derived from historical records of shoreline and bathymetric change, and Erie harbor dredging records, in two independent budget estimates (Nummedal, 1979, 1983; U.S. Army Corps of Engineers, App. C, 1979). The results of these budget estimates, for average conditions over the 1955—1978 time period (which covers periods of heavy artificial nourishment), are summarized in Table IIA.

Knowing sediment input and downdrift accretion rates, as well as the relative magnitude of the wave power along the five segments of Presque Isle (Table I), one can write a set of simple linear equations to solve for the actual transport rates between the individual shoreline segments. Along shorelines where there is an increase in the longshore transport rate the supplied material comes from erosion of the shoreface, where there is a diminishing transport rate the shoreface accretes. The results of these computations are summarized in Table IIB and Fig.5.

The computations document shoreface erosion along the neck of Presque Isle; the rate of erosion being the highest in segment III. Segments IV and V are seen to accrete. This is consistent with the observed shallow water depths and multiple bars in segment IV, and net spit progradation in segment V.

TABLE II

Input data and results of calculations of sediment transport along the lake shore of Presque Isle. All rates in $m^3 \ yr^{-1}$

A. Input data	
Net accretion at Gull Point (seg. V):	55,000
Delivery of littoral material from Presque Isle beaches to Erie harbor entrance:	110,000
Supply of littoral material from beaches to the west of Presque Isle:	30,000
B. Results of computations	
Transport between segments I and II:	74,000
Net loss from segment I:	44,000
Transport between segments II and III:	135,000
Net loss from segment II:	61,000
Transport between segments III and IV:	209,000
Net loss from segment III:	74,000
Transport between segments IV and V:	165,000
Net gain in segment IV:	45,000

Beach nourishment and shoreline erosion east of the lighthouse

The beach east of the lighthouse and east of Beach 10 is characterized by large-scale shoreline megacusps. The megacusps typically migrate yet maintain their identifiable form over periods ranging from months to years. The area immediately east of the lighthouse was closely monitored during the time period Oct. 1977 through Sept. 1980 (Fig.6). Because the road was threatened by erosion, artificial nourishment was initiated in the spring of 1977. By fall of that year some of the nourished sand had moved eastward and begun building a megacusp between profiles PI 11 and PI 12 (Fig. 6). Continued nourishment at profile PI 10, combined with sediment bypassing of the lighthouse groin from beaches nourished farther updrift, led to dramatic widening of the beach at PI 10 by 1980. The associated megacusp migrated to the east side of profile PI 12 (Fig.6C) and the intervening beach, centered at PI 11, became subject to intense erosion. An ironic situation had developed: after filling of about 124,000 tons of beach material east of the lighthouse groin between 1977 and 1980 (Nummedal, 1981), the beach at PI 11, 600 m downdrift, was eroding faster than before the nourishment.

Sediment dispersing eastward from the lighthouse fill area migrates along

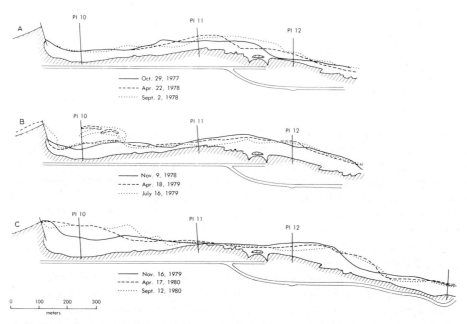

Fig.6. Shoreline changes downdrift of the lighthouse groin between October 1977 and September 1980. Incipient beach fill led to rapid megacusp development and beach accretion between profiles PI 11 and PI 12 (panel A). Continued nourishment caused downdrift migration of the megacusp and the beginning of an erosion phase at PI 11 (panel B). By 1980 erosion at PI 11 had become severe; the bulk of the megacusp had now migrated to the east of PI 12, a total distance of 500 m downdrift from the site of its incipient formation in 1978 (panel C). See Fig.1 for location.

an arcuate bar system. Some of this re-enters the beach at the bar-attachment point on the PI 12 megacusp, some sediment continues farther eastward. The sole source of sediment to the megacusp is the bar; there is a totally insignificant amount moving along the beach. As a consequence of this bar bypassing the beach at PI 11 did erode. Eastward extension of the artificial beach fill in 1981 alleviated the erosion problem at PI 11.

NEARSHORE BARS

A detailed echo-sounder survey of 87 bathymetric profiles conducted in the summer of 1979 by the U.S. Army Corps of Engineers, Buffalo District (drawings no. 79S-PIP-2/4.7 sheets), combined with vertical aerial photography obtained on July 16th, 1979, a day with excellent water clarity, form the basis for a set of figures depicting the three-dimensional geometry of the entire Presque Isle bar field (Fig.7A—F). This is probably the most extensive continuous bar field ever mapped in such detail. The maps form the basis for the following description of the component bars.

Bar conditions at Presque Isle are a function of location and wave conditions preceding the observation period. The spit neck is characterized by a dual bar system: a transient, highly variable inner bar which typically crests in 0.5—1.5 m of water about 50 m offshore, and a permanent outer bar with a crest in 3—4 m of water about 150—200 m offshore. East of the lighthouse, the bar field is more complex with strongly developed shore-normal (transverse) bar components. In segment V (Gull Point) the only bars present are small and close to shore.

Bars along the peninsular neck

Inner bars. The inner bars in the groin field (segment I) are linear or crescentic in plan form. The geometries seen in Fig.8 are most commonly encountered. When present, the crescents vary in degree of concavity. The most arcuate ones extend about 100 m offshore and have well developed complete crescents. Other bars have a rather flat longshore form with broad landward-pointing horns and poorly developed rhythmicity. Bar relief and cross-sectional profile varies widely. At some profiles, the inner bar is fairly symmetrical with a relief of more than a meter (Fig.7B, profile 14); at other profiles, the bar is expressed as a gentle shoulder.

Selected bathymetric profiles were measured on a seasonal basis over three years (April 1978—November 1980). Because of the long time separation between successive surveys it was impossible to tie changes in bar relief to specific wave conditions. However, the repetitive pattern each year was characterized by maximum bar relief at the spring survey, a moderate or non-existent bar in the summer and only a gentle shoulder in the fall. This suggests that the high-energy events which are most common in early winter and spring are directly responsible for the growth of the inner bar.

In areas devoid of regularly spaced large groins, the inner bar system is

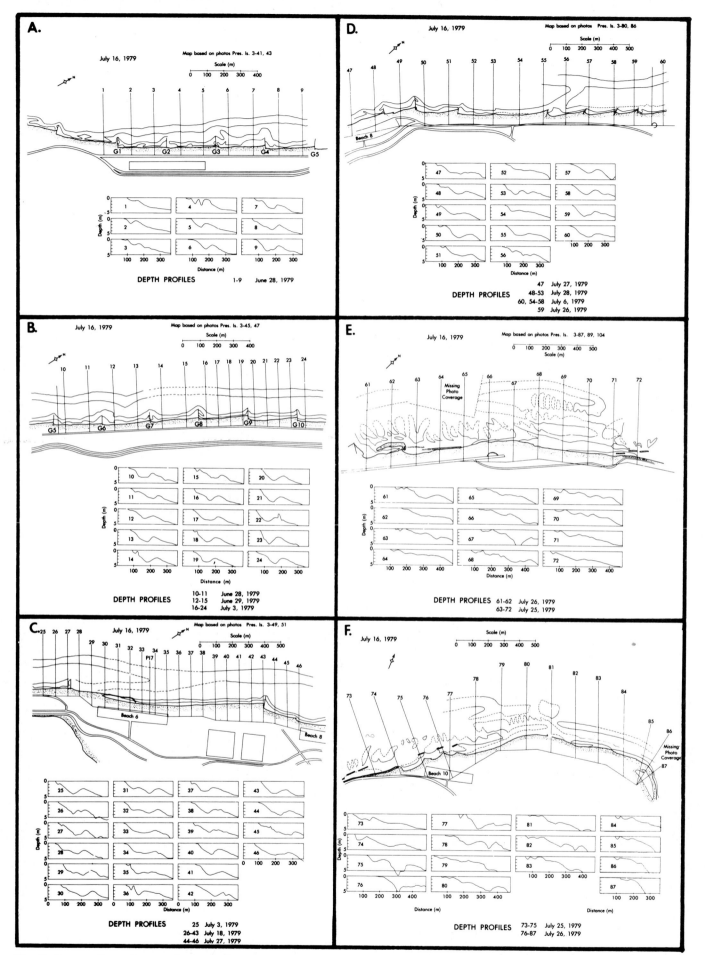

Fig. 7. Bar configurations on Presque Isle. Map is based on air photos and Army Corps of Engineers surveys during June and July, 1979. Panels A through F sequentially cover the whole lake shore of Presque Isle.

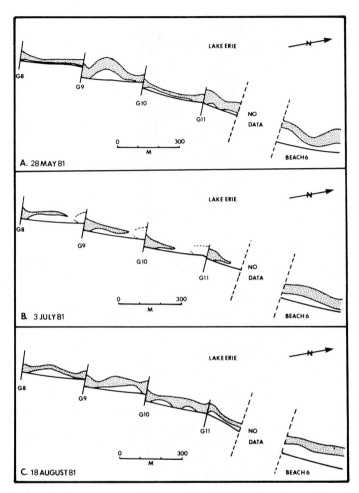

Fig.8. Morphology of the inner bars in the groin field along the neck of Presque Isle during the summer of 1981. (A) Bars after more than one month of essentially calm lake conditions. (B) Bars one week after a summer storm with 1 m high primary breakers. (C) Bars after a long period of essential calm interspersed with small summer storms. From Sonnenfeld (1983).

generally straight. These bars dominate within segments II and III. The bar crest typically lies less than 50 m offshore in a water depth of 1—1.5 m. Near all groins the bar is invariably deflected lakeward. Welding of the bar to the beach has never been directly observed, but the seasonal reduction in bar relief generally coincides with periods of beach accretion suggesting that the bar does supply some of the sediment.

The beach along the peninsular neck is generally straight or cuspate without any large-scale rhythmic features.

Outer bar. Segments I, II, and III are characterized by one continuous outer bar. This bar disappears as a single continuous feature to the east of the lighthouse (between bathymetric profiles no. 61 and 63, Fig.7E). The outer bar merges with the inner one at the updrift attachment point of Presque Isle (Fig.7A). Along the adjacent Lake Erie mainland shore there is only one bar. Downdrift from the point of merger the outer bar gradually moves offshore until an apparent "equilibrium" distance of between 150 and 200 m is attained in the vicinity of bathymetric profile no. 7 (Fig.7A). The bar crest typically lies in 3—4 m of water; its associated landward trough attains depths of between 4 and 5 m. The transverse bar profile is quite variable. It may be distinctly asymmetric with a steep flank toward land (Fig.7A, profiles 5, 7 and 8), or symmetrical (Fig.7A and B, profiles 6, 12 and 14). In plan view the outer bar is gently crescentic with a "wavelength" on the order of 400—1000 m. This bar rhythm displays no relationship to any obvious shoreline features.

A longshore structure is apparent in the crest depth and the relief of the outer bar (Fig.9). The water depth at the bar crest rapidly increases as the bar moves farther offshore downdrift from its point of merger with the inner bar. Bar relief, measured as the ratio between trough and crest depth, has an average value of 1.305 for all the 60 profiles. This relief is low compared to earlier published bar relief (Komar, 1976, p.301).

The relief exceeds the average along the first 33 profiles, it is less than the average along the rest of the peninsula. The high-relief section roughly corresponds to the major groin field (segment I).

Bathymetric profiles across the outer bar at PI 4, PI 7 and PI 9 demonstrate seasonal changes in bar relief (Fig. 10). The most distinctive pattern is one of maximum relief a few days after major storm events. Storm magnitude is measured by the associated water level set-up recorded on the NOS gage in Erie harbor (Fig.11). The largest storm during the 3-year monitoring period occurred on April 6, 1979. The shoreface profiles along the neck of Presque Isle recorded four weeks after this event displayed the greatest bar relief of all the nine surveys conducted. PI 7 (Fig. 10) is a typical example. The ratio between the trough depth and the crest depth was 1.38 compared to an average for all nine surveys at this location of 1.19. The profile with the second highest relief ratio (1.37) was recorded on

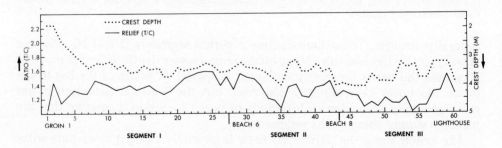

Fig.9. Crest depth and relief of the outer bar along the neck of Presque Isle. Relief is defined as the ratio between the trough and crest water depth. Numbers along the x-axis (1—60) refer to Army Corps of Engineers' bathymetric profiles shown in Fig.7.

Nov. 3, 1980, only a few days after a storm on October 25th (Fig.11). All other shoreface profiles were recorded after long periods of fair weather. During fair weather the bar relief is reduced through combined scour of the crest and infill of the associated landward trough (Fig.10, 1979 set of profiles). The storm of October 25th, 1980, caused other profile changes as well. The inner bar was reformed from the pre-existing shoulder which prevailed through the summer and early fall and a third bar of low relief appeared in deeper water more than 250 m offshore. Evidence for this third bar can also be seen in the profile of May 3rd, 1979.

These seasonal bathymetric profiles demonstrate that onshore sediment transport with attendant shoreface shoaling is the rule during the summer and fall. Much of this shoaling is a direct consequence of onshore bar migration; in each of the three survey years the bar crest moved about 50 m landward between the spring and fall surveys.

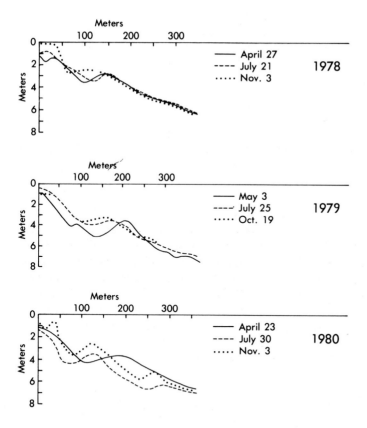

Fig.10. Seasonal changes in shoreface profile PI 7 during the three survey years. Note the generally higher bar relief at the spring surveys. The biggest storm of the study period occurred about 4 weeks prior to the recording of the May 1979 profile. See Fig.1 for profile location.

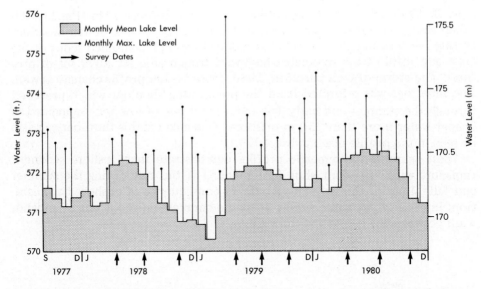

Fig.11. Water levels at the Erie, NOS gage during the three year study period. Arrows designate survey dates. Note the 175.6 m one-day water level for the April 6th, 1979 storm; 4 weeks prior to that year's spring survey.

Bars east of the lighthouse

As demonstrated in Fig.7E and F these bars have a very complex morphology. A shore-parallel component is clearly present. In profiles 68—70 one can trace four longshore bars; farther west or east the pattern is more confused. This segment of Presque Isle's shoreface (Fig.5, segment IV) has a significantly gentler slope than the others (Table III). This is consistent with the budget-prediction of shoreface accretion within this segment. As a consequence of this high rate of sediment supply, a multibarred shoreface is developed. There is a clear relationship between the nearshore slope and the prevailing bar types. Where the slope is very steep, as at Gull Point (east end of segment V), the shore is characterized by one, or at most two, simple longshore bars close to shore. Slightly gentler slopes characterize the area immediately west of Gull Point and the whole peninsular neck. The slope of these is about 1° (Table III). This shore is fronted by a transient inner bar, a permanent outer bar and a third (discontinuous?) deep-water bar. Finally, the gentle shoreface in segment IV has multiple parallel longshore bars superimposed on transverse bar components.

TABLE III

Nearshore slopes along the lake shore of Presque Isle. The listed slope is the average between the water line and 5 m depth contour. Slope in degrees

Segment	I	II	III	IV	V
Mean slope	1.05	0.97	0.97	0.67	0.91
Range	0.82—1.3	0.86—1.14	0.82—1.19	0.53—0.77	0.71—1.5

DISCUSSION

Morphodynamics

Nearshore bars, in general, respond to storms by a decrease in relief, offshore crest migration, and subsequent rapid recovery. The Presque Isle shoreface bars follow this same pattern. Short (1979) and Wright et al. (1979) documented the detailed morphological changes occurring in an oceanic surf zone bar system during storm and post-storm recovery cycles. In this progressive change one can recognize a set of six morphodynamic stages characterized by distinct morphology, wave forcing mechanisms and current structure. According to Wright et al. (1982), the morphologic changes go from low-amplitude straight bars and troughs at the storm peak, through crescentic and transverse bar stages to a non-barred straight beach with a well-developed berm at the end of the completed recovery cycle (Table IV). During storms the surf zone generally is dissipative (incident waves dissipate their energy through the process of repetitive breaking; Guza and Inman, 1975), and the energy spectrum, at least in oceanic settings, is dominated by long-period infragravity oscillations. The surf zone velocity field is stratified with offshore residual currents near the bed. Morphodynamic stages 2—5 (Table IV), associated with the progressive post-storm recovery of the surf zone bed, generally have dissipative conditions at the bars and reflective conditions at the beach face. Rather than being vertically stratified, the surf zone current field is horizontally segregated into a series of rip current circulation cells. On the fully accreted beach the wave processes are dominated by reflection of the incident and subharmonic components of the wave spectrum.

The completion of a recovery cycle of an oceanic beach may take many weeks; therefore, a cycle is rarely brought to completion before the advent of a new storm. Observed over long time, any beach is likely to occupy the whole range of possible morphodynamic stages. The most frequently encountered stage is referred to as the "modal beach stage" (Short, 1979). Years of observation at Presque Isle have demonstrated that the configurations described earlier in this paper are the modal stages for this system. The modal stages progressively change along the Presque Isle beaches in response to differences in prevailing surf zone processes.

The objective here is to explain this observed variability in the Presque Isle bar field in terms of the inferred modes of surf zone motion. The selection of such modes is a function of the deep water wave spectrum, and the shallow water refraction and energy attenuation. These factors, in turn, are controlled by the lake wave climate, large-scale bathymetry and sediment supply.

TABLE IV

Characteristics of the surf zone morphodynamic stages in the model used in this paper. Data from Wright et al. (1982) and other sources

Stage	Morphology	Dynamics	Dominant spectral band	Currents
1	Low-amplitude straight bars and troughs	Dissipative	Long-period infragravity oscillations (T: 1—3 min)	Vertically stratified. Offshore flow near bed
2	Straight/crescentic well-developed bars. Straight beach[1]	Dissipative across bar. Reflective at beach	Broad band. Low-period infragravity (T: 30—60 s) and subharmonic oscillations	Weak rips
3	Highly crescentic bar. Megacusps on beach	Dissipative at bar. Reflective at beach[2]	Same as 2	Moderate rip currents
4	Transverse bars and rip channels	Dissipative at bars. Dissipative or reflective in rip channels	Same as 2	Strong rip currents
5	Ridge and runnel or low-tide terrace	Dissipative at low water level. Reflective at high water level.	Same as 2	Small rip currents
6	Well-developed berm on straight beach	Reflective	Incident wave frequency and first subharmonic	No horizontal or vertical segregation

[1]Small cusps may be present in the swash zone for all stages 2—6.
[2]Dissipative conditions across transverse bars induce water level set up, driving rip current circulation.

Lake Erie mainland shore

Because of refraction and attenuation of the prevailing westerly waves before they reach this mainland shore, the wave power is relatively low. Persistent longshore currents to the east effectively remove sand and gravel supplied by the bluffs preventing the building of a wide shoreface. Consequently, this segment of shore has only one transient nearshore bar. After storms the bar quickly attains the welded and transverse bar stages (Table IV, stages 3 and 4). Continued fair-weather bar migration forms a straight (or cuspate) berm after a few weeks.

The neck of Presque Isle

Storm waves commonly arrive nearly perpendicular to this section of the Presque Isle shore. A large proportion of their energy is transferred into storm water-level set-up against the beach; the surf zone current pattern becomes stratified with net lakeward flow near the bed as documented by fluorescent tracer studies (Sonnenfeld, 1983; Taylor, 1983). Sand moving lakeward from the beach and inner bar system under such conditions appear to feed the outer bar, which typically is moved lakeward to an "equilibrium" distance of 200 m during storms.

Historical data are also consistent with the interpretation that the adjacent beach is the primary source of bar sediment. The outer bar along the Presque Isle neck attained its maximum recorded relief shortly after completion of a major beach nourishment project in 1955—1956. The average outer bar relief (trough depth/crest depth) in a 1957 survey was 1.74 (Nummedal, 1981).

The outer bar remains in morphodynamic stage 2 (or 3) (Table IV) during fair-weather periods throughout the year. Because of the absence of long-period swell conditions in Lake Erie, the storm-generated bar remains as a "relic" form rather than migrating onshore during "intermediate" energy conditions. This is probably the explanation for the permanence of the outer bar along many exposed Great Lakes shorelines (Short, 1979).

The inner bar along the neck of Presque Isle is directly affected by the groins. Standing waves between the shore-normal groins may interfere with components of the incident wave spectrum to produce a crescentic pattern of bottom sediment convergence (Bowen and Inman, 1971).

After long periods of low summer waves the inner bar evolves through the "low-tide" terrace stage (stage 5, Table IV) and welds onto the beach as a (frequently cuspate) berm.

Observations show that the inner bar is removed during major storms. Tracer dispersal patterns and drogue studies of the currents suggest that sediment is moving lakeward both by bottom return flow and groin-associated rip currents (Sonnenfeld, 1983).

The essential difference between the modal bar configuration in segments II and III and that in segment I is that the inner bar in the former segments generally has no crescentic components. This suggests that the current field lacks longshore structure. Such structure is generally induced by groins.

Bars east of the lighthouse

As documented in the transport section, sediment for this shoreface region is largely derived from sources updrift rather than the local beach. Sediment abundance has built a permanently multi-barred, dissipative surf zone. Because the storm-wave energy level is lower than farther west this surf zone maintains a morphodynamic modal stage different from the peninsular neck. Crescentic bars and megacusps (stage 3, Table IV) are the norm. Megacusps slowly migrate eastward and beach erosion is tied to rip current embayments between the megacusps. In spite of abundant sediment on the shoreface, strong rip circulation keeps these beaches erosion-prone. This explains why beach nourishment immediately east of the lighthouse groin did not stop downdrift erosion (Fig.6). The nourished sand moved lakeward by rip current action onto the crescentic bar.

The complex bar morphology in this area may be explained as follows. The outer bar becomes a post-storm "relic" feature in stage 2 (or 3) (analogous to the outer bar farther west). The multiple inner bars are progressively altered through lower (more reflective) morphodynamic stages during the post-storm recovery phase thus generating distinct transverse bar components. The entire bar-rip channel-megacusp system remains strongly skewed to the east because of strongly eastward-directed incident waves.

Gull Point area

This beach (segment V, Fig.5) receives the lowest annual wave power along the entire outer shore of Presque Isle. The modal morphodynamic stages are 4 and 5 (Table IV). Because of strongly eastward-directed incident waves the rip channels and transverse bars are skewed to become nearly shore-parallel (Fig.7F). As such, they are hard to differentiate from ridge and runnel systems. Associated megacusps maintain a typical spacing of 700—800 m. The surf zone is narrow; the shoreface profile is the steepest one along the peninsula (Table III). This profile is maintained by the prevailing onshore transport. Storms with incident waves from the northeast are the only ones which could generate sufficient energy for offshore transport and shoreface broadening. Such storms are infrequent and moderate in strength. Gull Point has a high rate of sediment influx from updrift and is therefore rapidly accreting. The wave climate is such that this sediment will be kept close to shore maintaining a steep prograding shoreface profile.

CONCLUSIONS

Volume rates of sediment transport for the Lake Erie shore of Presque Isle were determined from hindcast wave power distributions and historical records of shoreline change and dredging volumes at the downdrift end of the transport system. The calculations show that: (1) the maximum net longshore transport occurs in the vicinity of the lighthouse where the annual volume rate is 209,000 m^3; (2) due to downdrift changes in sediment transport rate, the shoreface along the neck of Presque Isle is eroding whereas the shoreface to the east of the lighthouse is accreting; and (3) the erosional or accretionary state of the shoreface is not, by itself, adequate information to determine the stability of the adjacent beach. Beach stability depends on the morphodynamic stage of the shoreface. Specifically, the beaches east of the lighthouse experience erosion in mega-cusp embayments in spite of large-scale beach nourishment immediately updrift. Only at Gull Point do the beaches experience sustained accretion.

The morphodynamic surf zone model developed for oceanic beaches by Short (1979) and Wright et al. (1979) is used as a basis for discussion of shoreface morphologic variability at Presque Isle. In spite of interference by major shoreline stabilization structures and expected differences between oceanic and lake wave climates the nearshore bar field at Presque Isle does closely correspond to that model. The Presque Isle lakeshore can be divided into three sections with distinctly different morphodynamics. These are:

(1) *The neck*. A permanent outer bar is maintained by storm waves in stages 2 and 3 (Table IV). The inner bar may occupy any morphodynamic stage in response to varying wave conditions. It is commonly crescentic because of waves induced by the groin field.

(2) *Shore east of the lighthouse*. Abundant sediment supply has produced a multi-barred shoreface of gentle slope. Strongly skewed transverse bar components maintained by rip current circulation are generally present.

(3) *Gull Point*. Prevailing low incident waves maintain this shore in the ridge and runnel morphodynamic stage.

ACKNOWLEDGEMENTS

This work has been supported by the U.S. Army Corps of Engineers, Buffalo District, through contracts DACW49-78-C-0020, DACW49-79-C-0055 and DACW49-80-C-0047. The authors acknowledge valuable discussions with Joan Pope. Thorough reading by two anonymous reviewers greatly helped improve the manuscript.

REFERENCES

Bowen, A.J. and D.L. Inman, 1971. Edge waves and crescentic bars. J. Geophys. Res., 76: 8662—8671.
Carter, C.H., 1976. Lake Erie shore erosion, Lake County, Ohio: setting, processes and recession rates from 1876 to 1973. Ohio Geol. Surv., Rep. of Invest. 99, 105 pp.
Coastal Engineering Research Center, 1973. Shore Protection Manual, 3 vols.
Guza, R.T. and D.L. Inman, 1975. Edge waves and beach cusps. J. Geophys. Res., 80: 2997—3012.
Holman, R.A., 1981. Infragravity energy in the surf zone. J. Geophys. Res., 86: 6442—6450.
Komar, P.D., 1976. Beach Processes and Sedimentation. Prentice-Hall, Englewood Cliffs, New Jersey, 429 pp.
Messinger, D.J., 1977. Form and change of a recurved sand spit, Presque Isle, PA. M.Sc. thesis, State University of New York, Fredonia, N.Y., 125 pp.
Nummedal, D., 1979. Monitoring of shoreline change, Presque Isle, Pa. Annual Report to Buffalo District, U.S. Army Corps of Engineers, for contract no. DACW49-78-C-0200, 52 pp.
Nummedal, D., 1980. Monitoring of shoreline change, Presque Isle, Pa. Annual Report to Buffalo District, U.S. Army Corps of Engineers, for contract no. DACW49-79-C-0055, 72 pp.
Nummedal, D., 1981. Monitoring of shoreline change, Presque Isle, Pa. Annual Report to Buffalo District, U.S. Army Corps of Engineers, for contract no. DACW49-80-C-0047, 62 pp.

Nummedal, D., 1983. Sediment transport and surf zone morphodynamics of Presque Isle, Pennsylvania. Coastal Zone '83, Am. Soc. Civ. Eng., III: 2612—2629.

Nummedal, D., Hayes, M.O. and Fahnestock, R.K., 1976. Littoral processes and sedimentation in the Cattaraugus Embayment, N.Y., Appendix E. In: General Design Memorandum, Sunset Cattaraugus Harbor, N.Y., 246 pp.

Pierson, W.J. and Moskowitz, L., 1964. A proposed spectral form for fully developed wind seas based on the similarity theory of S.A. Kitaigorodskii. J. Geophys. Res., 69: 5180—5190.

Resio, D.T. and C.L. Vincent, 1976. Design wave information for the Great Lakes; report 1, Lake Erie, U.S. Army, Waterways Experiment Station, Tech. Rep. H-76-1, 54 pp.

Saville, T., 1953. Wave and lake level statistics for Lake Erie. Beach Erosion Board, Tech. Memo. 37, 24 pp.

Seabergh, W.T., 1983. Design for prevention of beach erosion of Presque Isle beaches, Erie, Pa. — Hydraulic model investigation. U.S. Army, Waterways Experiment Station, Vicksburg, MS, Draft Report to Buffalo District, U.S. Army Corps of Engineers.

Short, A.D., 1979. Wave power and beach stages: a global model. Proc. 16th Int. Conf. Coastal Engineering, ASCE, II: 1145—1162.

Sonnenfeld, D.L., 1983. Inner bar sediment dynamics, Presque Isle, Pa. M.Sc. thesis, Louisiana State University, Baton Rouge, La.

Sonu, C.J., 1973. Three-dimensional beach changes. J. Geol., 81: 42—64.

Taylor, K.B., 1983. Sand dispersal patterns on the outer bar, Presque Isle, Pa. M.Sc. thesis, State University of New York, Fredonia, N.Y.

U.S. Army Corps of Engineers, 1953. Presque Isle Peninsula, Erie, Pa., Beach erosion control study, House Document no. 231, 83rd Congress.

U.S. Army Corps of Engineers, 1979. Presque Isle Pennsylvania, Erie, Pa., Draft Phase I, General Design Memorandum, Dept. of the Army, Buffalo District.

U.S. Naval Weather Service Command, 1975. Summary of synoptic meteorological observations for Great Lakes area. National Climatic Center, Asheville, N.C., 201 pp.

Walton, T.L., 1973. Littoral drift computations along the coast of Florida by means of ship wave observations. Coastal and Oceanographic Engineering Laboratory, Univ. of Fla., Tech. Rept. 15, 96 pp.

Wright, L.D., Thom, B.G. and Chappell, J., 1979. Morphodynamic variability of high energy beaches. Proc. 16th Int. Conf. Coastal Engineering, ASCE, II: 1180—1194.

Wright, L.D., Short, A.D. and Nielsen, P., 1982. Morphodynamics of high energy beaches and surf zones: a brief synthesis. Tech. Rept. 82/5, Coastal Studies Unit, Univ. of Sydney, 64 pp.

SEDIMENTOLOGY AND MORPHODYNAMICS OF A MACROTIDAL BEACH, PENDINE SANDS, SW WALES

C.F. JAGO and J. HARDISTY

Department of Physical Oceanography, Marine Science Laboratories, U.C.N.W., Menai Bridge, Gwynedd, LL59 5EY (U.K.)
Department of Geology, University of Bristol, Bristol (U.K.)

(Received March 14, 1983; revised and accepted January 4, 1984)

ABSTRACT

Jago, C.F. and Hardisty, J., 1984. Sedimentology and morphodynamics of a macrotidal beach, Pendine Sands, SW Wales. In: B. Greenwood and R.A. Davis, Jr. (Editors), Hydrodynamics and Sedimentation in Wave-Dominated Coastal Environments. Mar. Geol., 60: 123—154.

The foreshore of Pendine Sands forms the seaward part of an extensive, sandy coastal barrier in a shallow Carmarthen Bay, SW Wales. The sedimentological features of the macrotidal foreshore reflect a tide-induced modification of nearshore wave characteristics. As the tide ebbs, the breaker height may decrease, the surf zone widens and becomes increasingly dissipative, and swash/backwash velocities diminish. A concomitant change from plunging to spilling breakers and increasingly symmetrical swash zone flows are associated with a decreasing beach gradient.

A zero net transport model demonstrates that the beach profile is self-stabilising in the short-term, and periodic levelling has shown that the beach is in long-term equilibrium with prevailing conditions, though this does not preclude a significant dynamic response to changing tides and waves.

The flow regimes of wave-generated currents decline as the tide ebbs, and normal beach processes do not usually affect the lower foreshore. Accordingly, there is an overall seaward-fining of the primary framework component of the sands. In more detail, this framework component displays a slight seaward-coarsening across an upper foreshore dominated by high water swash and surf; a rapid seaward-fining across the mid-foreshore in response to the ebb-attenuating swash zone flow velocities; and a slight seaward-fining across the lower foreshore under the action of nearshore shoaling waves. Bedforms vary from a swash/backwash emplaced flat bed across the upper foreshore to the small ripples of nearshore asymmetric oscillatory flows across the lower foreshore.

The surface sediment veneer is not representative of the subsurface sediments which form in response partly to fairweather conditions, partly to storms. The upper foreshore is characterised by swash/backwash emplaced plane bedding in fine sands frequently disrupted by bubble cavities. The mid-foreshore is composed of coarser-grained shelly traction clogs arranged as landward- and seaward-dipping large-scale cross bedding and/or plane bedding; these are probably storm breaker/surf deposits. The lower foreshore, though partially and sometimes totally bioturbated, shows landward-dipping small-scale cross bedding in very fine sands sorted by nearshore shoaling waves.

Tide- and storm-induced modification of the nearshore flow regimes therefore produces a distinctive shore-normal array of sedimentary facies. Each facies is characterised by diagnostic textural and structural signatures. A prograding sequence of such macrotidal deposits would be similar to, but more extensive than, a comparable microtidal sequence.

0025-3227/84/$03.00 © 1984 Elsevier Science Publishers B.V.

INTRODUCTION

The nearshore environment can be envisaged as a shoreward progression of distinct dynamic zones — shoaling waves, breaker, surf, swash — each characterised by a particular mode and intensity of sediment mobility (see, for example, Komar, 1976, for summary). Such a progression of wave effects develops a shore-normal sequence of bedforms and syndepositional sedimentary structures arranged in more-or-less linear bands parallel to the strandline (Clifton et al., 1971; Davidson-Arnott and Greenwood, 1976).

This arrangement of sedimentary features must respond to fluctuating nearshore dynamics and water levels. Thus, on macrotidal shorelines, while wind-generated surface waves remain the principal, if variable, source of energy, dissipation of this energy may be controlled by tidal processes. As a consequence of the tide, the nearshore dynamic and sedimentary zones sweep the foreshore to a degree that depends on the tidal range. This must create a certain migration of sedimentary facies with each tidal cycle; Clifton et al. (1971) briefly described just such a migration across a mesotidal foreshore.

The aim of the study outlined below is firstly to examine the shore-normal intertidal textural gradient and array of sedimentary structures, and thence to derive a simple dynamic model that incorporates grain-size, bedforms, beach slope and wave characteristics on a depositional macrotidal shoreline.

PENDINE SANDS

The beach at Pendine Sands forms the seaward part of an embayed Quaternary beach/dune barrier which extends some 10 km from a rocky headland in the west to the confluence of three small estuaries in the east (Fig.1). Carmarthen Bay is rather shallow, a mere 10 m deep some 6 km seaward of the dunes at low spring tide. A reconnaissance study of the bay has established that it is floored with abundant offshore sand waves and with nearshore bars and intertidal sandbanks (particularly off the eastern end of the barrier). Historical evidence, field observations and aerial photographs emphasise the mobility of the nearshore features in response to the rapidly changeable dynamics of the bay. The predominant sedimentological trend is depositional in the northern and eastern nearshore areas (Fig.1) coupled with a progressive and rapid movement of marine sand into the estuaries (Jago, 1980). The area is therefore of considerable geological interest as a model of coastal sedimentation.

Carmarthen Bay has a southwesterly aspect and so is exposed to an oceanic swell with a fetch of over 5000 km. Draper (1972) estimated that the highest 50 year storm wave in the Celtic Sea should be 30 m. Analysis of waves logged during 1968 by the St. Gowan Lightvessel at the western entrance of the bay gives: $\bar{H}$ = 1.2 m, $H_{1/3}$ = 2.0 m, $T_{1/10}$ = 6 s; while Darbyshire's (1963) analysis of data during 1960/61 from the "Helwick Lightvessel" at the eastern entrance of the bay gave a modal wave height of

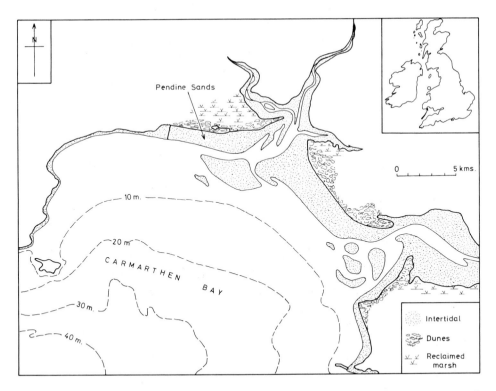

Fig.1. Carmarthen Bay and Pendine Sands, showing working transect at the western end of the beach.

3.0 m. However, our observations at Pendine Sands suggest that the energy loss as waves sweep shoreward across the shallow bay is such that few large waves reach the shoreline. It may be predicted (Hydraulics Research Station, 1978) that, for the prevailing waves that approach from 203°, 4 s waves will be reduced in height by 50% and 10 s waves by 70% as they shoal from the 40 to the 5 m isobaths off the eastern end of the barrier. Constriction of the shelf tidal wave in the Bristol Channel produces semi-diurnal tides in Carmarthen Bay of mean spring range 7.5 m and mean neap range 3.7 m. Extreme springs are of 10.0 m range. Surface currents reach 1.0 m s^{-1} in the middle of the bay (R.J. Uncles, pers. commun., 1980). With such a tide, and a gentle foreshore/shoreface slope (<1°), the intertidal zone becomes very wide at low water springs — 1500 m or more at the widest part of the barrier.

FIELD AND ANALYTICAL PROCEDURES

Seven shore-normal transects, spaced at approximately 1.4 km intervals along the foreshore, were regularly levelled from the dune to the low water mark of a spring tide at 3 monthly intervals during 1968—1970. The shoreface was profiled once by echo sounder along seaward extensions of the beach transects using a small boat positioned by theodolites from the shore.

Surficial sediment samples were collected along the transects at low tide, each sample consisting of a composite of four closely-spaced sub-samples of the upper 3 cm of the beach sediment. These composites were sieved at 0.25 phi unit intervals. It was established that the between-sample station variation was significantly greater than any sampling/sieving errors which could therefore be ignored. Textural parameters (M_z = mean grain size, σ_I = graphic standard deviation) were calculated after Folk and Ward (1957). Senckenberg box cores were taken at intervals across the transects at low tide and impregnated with either nitrocellulose lacquer or epoxy resin (after Bouma, 1969). Additional grain-size analyses were made of individual laminae within selected box cores.

The results discussed below are mostly limited to the transect at the western, non-barred end of the barrier (Fig.1) where we have additional annual surveys during 1980—1982. Concurrent with the sedimentological procedures, certain breaker, surf and swash characteristics were monitored along this transect at 15 min intervals during several tidal cycles. These measurements included breaker height (using a hand-held staff), surf and swash zone widths (with measuring tape), times of uprush (of surf and swash), swash/backwash velocities (using hand-held Ott and Braystoke flowmeters), and swash/backwash bedload transports (using a simple sediment trap).

MORPHODYNAMICS OF THE FORESHORE

Beach morphology

All the beach profiles at the western end of the barrier have a uniform concave-upward shape with a marked absence of either longshore features, such as ridges and runnels, or rhythmic structures such as cusps, crescentic bars, etc. The foreshore is backed by established foredunes (ca. 5 m high) and a backshore of variable width (0—15 m depending on the tide). Except at the eastern end of the barrier, where offshore shoals are important, the foreshore profile continues smoothly seaward at a diminishing gradient (Fig.2). The foreshore can be divided into three zones on morphological and sedimentological grounds: the upper foreshore, from highest spring to mean neap high tide marks, has the steepest gradient (though still, on average, less than 1.5°); the middle foreshore, between mean neap high and low tide marks, is of intermediate gradient ($\simeq 0.7°$); and the lower foreshore, below the mean neap low tide mark, has the lowest gradient ($\simeq 0.4°$). The profile continues offshore with a gradient of $<0.1°$. Obviously the frequency of exposure of the intertidal zone during a lunar half cycle diminishes from upper to lower foreshores.

Tidal variation of wave characteristics

Carmarthen Bay is up to 10 m shallower at low water than it is at high water. Because of the concave-up intertidal/subtidal profile, the shoreface

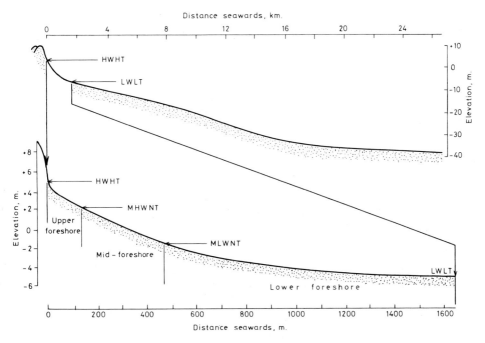

Fig.2. Nearshore and foreshore morphology. *HWHT* = high water highest tide; *LWLT* = low water lowest tide; *MHWNT* and *MLWNT* = mean high water and mean low water neap tide, respectively.

(i.e., that part of the seafloor affected by waves) may be twice as wide at low water as at high water (Fig.3). Since the frictional dampening of wave energy is a function of water depth and shoreface width, the progressive shoreward modification of shoaling waves must vary according to the stage of the tide. Wave attenuation will increase, and effective nearshore wave energy decrease, as the tide ebbs.

The morphodynamic variables are reflected in the changing character of the nearshore waves during the tidal cycle. Thus, for example, the breaker coefficient ($B_0 = H_b/gT^2 \tan \alpha$, where H_b = breaker height, $\tan \alpha$ = beach

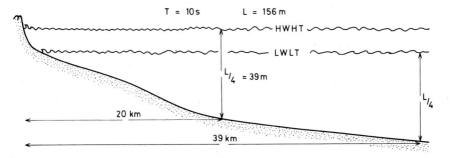

Fig.3. Variable shoaling modification of a 10 s wave. Significant shoaling begins at L/4. HWHT and LWLT = high water highest tide and low water lowest tide, respectively.

slope, T = wave period) after Galvin (1968): plunging breakers, with $B_0 < 10^{-1}$, over the upper foreshore; spilling breakers, with $B_0 > 10^{-1}$, over the middle and lower foreshores (Fig.4). The breaker height may diminish as the tide ebbs (Fig.4).

There are concomitant changes in the surf zone as the profile slope changes. The surf scaling parameter, $\epsilon = 4\pi^2 B_0/\tan\alpha$, after Guza and Inman (1975) and Wright et al. (1979), is always high and increases during the ebb (Fig.4). Hence the flat lower foreshore is highly dissipative under any conditions; the uppermost foreshore could be very moderately reflective with respect to exceptionally low waves, but such conditions were not observed. The phase difference ($P = t_b/T$, where t_b = duration of surf and swash flows) of Kemp (1961) increases with the ebb as surf and swash zones widen and swash/backwash currents decline (Fig.5).

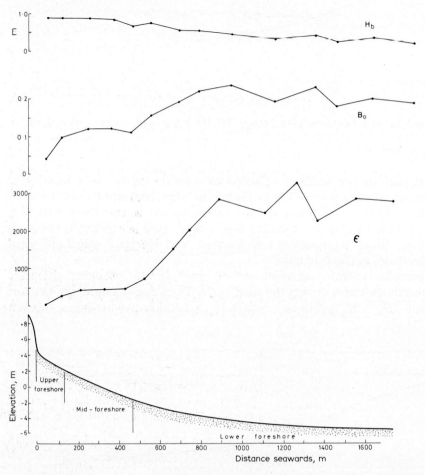

Fig. 4. Variation, on an ebbing tide, of breaker height (H_b), breaker coefficient (B_0) and surf scaling parameter (ϵ). ϵ is displaced landward since surf is landward of breaker.

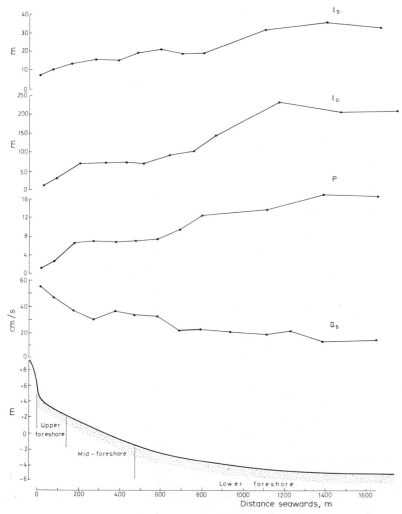

Fig.5. Variation, on an ebbing tide, of swash width (l_s), surf width (l_b), phase difference (P) and mean swash/backwash velocity ($\bar{u}_s$).

Beach stability

Taken as a whole, the seven surveyed transects showed a mean vertical erosion of 0.5 m during the period 1968—1970 (this seems very little but nevertheless represents a total volume of 0.46×10^6 m^3). However, much of this erosion occurred in the vicinity of the estuary due to movement of estuarine channels. The western end of the beach was remarkably stable (Fig.6). The greatest change between surveys of this westernmost transect was a mean erosion of 0.32 m (amounting to 96 m^3 m^{-1} width of profile) in response to a westerly gale. After nine surveys, the standard deviation of volumetric variation of the beach prism (above mean spring tide low water

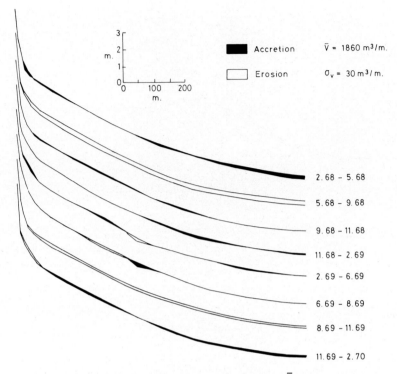

Fig.6. Periodic beach profile changes, 1968—1970. $\bar{V}$ = mean beach prism volume above mean low water spring tide mark. σ_V = standard deviation of V.

mark) was 30 m³ m⁻¹ (cf. Short, 1980). The stability of the profile suggests that the foreshore was in long-term equilibrium with the prevailing conditions over the period of study. The beach is similar, in this respect, to microtidal dissipative beaches which rarely experience substantial profile changes because of the low sediment exchange between nearshore, surf and subaerial zones (Short, 1980).

SEDIMENTOLOGY

Interpretation of grain-size distributions

Like most grain size frequency distributions, the Pendine Sands sediments plot on probability paper not as straight lines but as S-shaped curves. There is little doubt that these curves represent composite distributions and that they can be interpreted as mixtures of three or more subpopulations. These subpopulations are the result of bed-building and depositional processes (Moss, 1972) or of hydraulic and transport processes (Middleton, 1976) or of both.

Many of the Pendine sands can be dissected into three components using a simple graphical technique (Cassie, 1954). A typical example is shown in Fig.7A. These will be referred to as contact, framework and

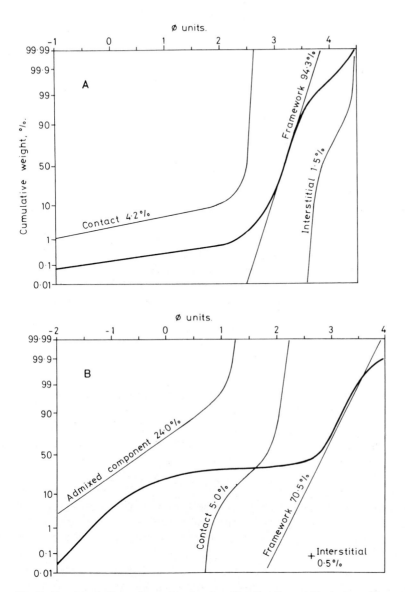

Fig.7. Graphical dissection of grain size distributions, (A) single sedimentation unit; (B) dual sedimentation unit.

interstitial, after Moss (1972), while acknowledging that they probably relate also to different modes of transport, after Middleton (1976) (although it should be noted that both Moss and Middleton were concerned with steady, unidirectional flows and not with rapidly reversing or oscillatory flows).

As shown in Fig.7, the framework component is lognormal and very well sorted. This indicates the degree of selectivity of the bed-building process and/or the efficiency of the intermittent suspension transport mechanism.

It is not clear whether the remaining components form log-normal or truncated populations since in many cases the data permit either solution. The precision of the method (sieving plus graphical presentation) is too low to resolve this question but, on balance, we favour truncation as this is most compatible with Moss's (1972) field observations.

Not all samples resolve into three components. Some contain a sizeable fourth component (Fig.7B). It would be expected that any number of components could be present if the sample intersects several depositional horizons and this seems to be the case here. The mixing reflected in the grain-size distribution is therefore a sampling 'error', and reflects the difficulty of recognising and sampling discrete sedimentation units in the field. This difficulty was resolved by sampling the Senckenberg box cores used to examine sedimentary structures (see below). After removal of the impregnated face, the sand remaining in the box retains a perfect record of the structure and can be sampled at any points of interest. Our intention was to sample individual laminae but this proved impossible in the fine-grained, often bioturbated, sands. Where feasible, samples of visually homogeneous units of several laminae were taken at about 3 cm intervals down the core. After this more precise sampling we find that the grain-size distributions of the sands can always be resolved into three components.

Grain-size characteristics

The surficial sediments of the Pendine foreshore are mostly fine and very fine sands ($M_z > 2.50$ phi units) and usually very well sorted ($\sigma_I < 0.35$ phi units). While textural variations within the 30 cm vertical sections of the box cores are generally small, there are pronounced shore-normal variations, so that the upper, mid, and lower foreshore sands are distinctive.

Upper foreshore

On the upper foreshore, the subsurface sands are virtually identical to the surficial sands (Fig.8). The framework component (with $\sigma_F = 0.27$ phi units) is dominant; the contact component contributes less than 3% and gives the grain size distribution a short tail; the interstitial component is variable, 0.1—11.0%, and greatest at high water mark where the grain size curve can show a marked gradient change toward its fine end. Where well-developed, the interstitial component consists mostly of heavy minerals deposited at the upper limit of swash action.

Mid-foreshore

Over the mid-foreshore, the subsurface sands are again frequently uniform but differ from the surface layer (Fig.8). The subsurface sands contain three components: a framework component ($\sigma_F = 0.40$ phi units); a large contact component, up to 70%, which is poorly sorted ($\sigma_c = 0.70$—1.35) and whose mean is about 3.0 phi units coarser than the framework; a negligible interstitial component (<0.1%). These sands make up almost the entire mid-

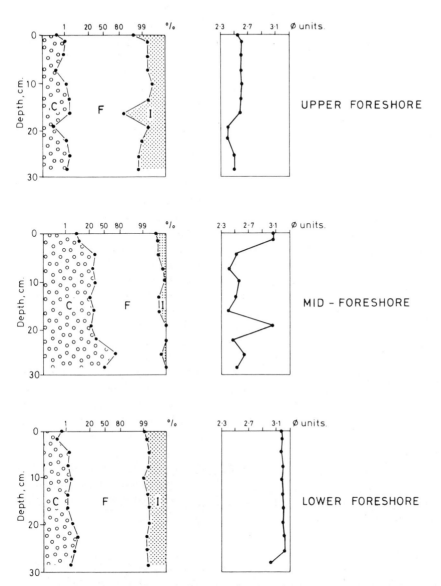

Fig.8. Vertical variation of textural parameters in box cores. C, F and I refer to the contact, framework and interstitial component, respectively.

foreshore section. They are distinctive because of their large contact component which consists of ill-shaped shell debris, whole shells and some pebbles. This material is found only in the mid-foreshore deposits. We have yet to observe any swash zone characteristics that would account for such a mid-tide phenomenon. A likely explanation is that it represents a storm lag produced under the breaker/surf at high water still-stand. If we take the breaking criterion as $H_b/h_b = 0.78$ (where H_b = breaker height and h_b = water depth at break point), then a 2 m breaker will break in 2.6 m of water.

Corresponding to the limits of mean spring and mean neap high tides, breakers of this size would produce a sweep zone which matches quite closely the observed range across the mid-foreshore of this coarse grained component (Fig.9). This implies that coarse-grained material may concentrate close to the breaker during storms. That such a mechanism can occur has been demonstrated by Ingle (1966) who showed, using fluorescent tracers, that coarse grains move into the breaker from the surf zone under non-equilibrium conditions. It is likely that, on a macrotidal beach, the breaker and surf zones are too mobile for a significant deposit to form except at high water (and perhaps at low water). The resulting deposits resemble traction clogs (Moss, 1972) which may form when accumulating grains of the contact component control bed-building processes. The sands are often cross-bedded (see below) which suggests that some of the coarse material may accumulate as lags in the troughs of large bedforms (B. Greenwood, pers. commun., 1983). Micro-layering of alternate coarser and finer sands occurs within the major cross-bedded units and this may result from the strongly asymmetrical oscillatory flows of the breaker zone. The traction clogs are missing on the eastern half of the Pendine barrier where the nearshore shoals apparently protect the beach from storm breakers.

The surface layers of the mid-foreshore differ in that they contain a framework component that is finer grained than the subsurface frameworks; a smaller, but still important (up to 10%), contact component; and an interstitial component that is insignificant (<0.5%), though greater than in the subsurface. As a result, the mean grain size is finer in the surface layers than in the subsurface (Fig.8). The surface layers must be derived, at least in part, from reworking of the uppermost portion of the subsurface. Blackley and Heathershaw (1982) have shown that selective transport of different grain sizes may take place in both alongshore and shore-normal directions on a comparable macrotidal beach. It is likely that reworking by swash and/or surf of the framework component, and a little of the contact component, of the storm deposits creates the finer-grained framework and contact components of the surface layers.

Lower foreshore

The lower foreshore sands are identical throughout the vertical section (Fig.8). This is partly the result of bioturbation, and the subsurface bioturbated sands are texturally indistinguishable from the surficial cross bedded sands. The framework (σ_F = 0.20 phi units) is again the major component; the contact component, though <5%, is poorly sorted so that the parent grain size curve has a long coarse tail (Fig.13); and the interstitial component makes up to 2% of the curve.

Dual sedimentation units

Sampling methods which cut through bundles of laminae will clearly be sampling several sedimentation units. Samples may represent a range of transport processes and depositional events created by the tidal sweep of

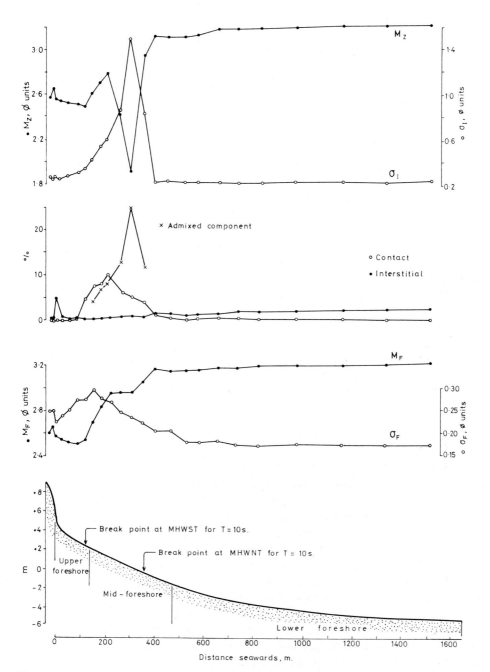

Fig.9. Shore-normal variations of mean grain size (M_z) and sorting (σ_I); population frequencies; framework mean grain size (M_F) and sorting (σ_F).

the foreshore. It is noteworthy that the surface veneer over the mid-foreshore differs from, and is consequently unrepresentative of, most of the mid-foreshore deposits. Comparable units have been observed on microtidal (Otvos, 1965) and mesotidal beaches (Williams, 1971). Graphical dissection of the grain-size distributions of the composite samples from the Pendine mid-foreshore often produces four components (Fig.7B). These are dual sedimentation units consisting of: (1) the usual three components (surface layers); and (2) an admixed component (subsurface). The relative proportion of the two units depends solely on the depth of sampling relative to the depth of reworking of the storm deposits by fairweather swash and surf. King (1951) found that on a Carmarthen Bay beach of similar grain size and slope, the depth of disturbance was about 1 cm for every 30 cm of wave height. Wave conditions prior to sediment sampling at Pendine should have disturbed the upper and mid-foreshores down to 2—3 cm and the lower foreshore to about 1 cm or less. With a sampling depth of 3 cm, the mid-foreshore samples contain up to 24% of sediment undisturbed by reworking (Fig.7B).

These dual sedimentation units appear only on the mid-foreshore. The laminae of the upper foreshore sands, while clearly deposited at different times and subsequently reworked, are always exposed to the same kind of process — i.e. swash and backwash. The laminae are accordingly similar in texture. The lower foreshore sands are frequently bioturbated and suffer minimal reworking by waves (see below). Little selective sorting of grains seems to occur (once deposited), so the sands are remarkably similar in texture in a vertical section (Fig.8).

Shore-normal textural variations

From the foregoing, it is obvious that the grain-size distributions of the sands will be influenced by the choice of sampling procedure. Hence, composite samples, which cut through more than one depositional unit, show a pronounced coarsening and become poorly sorted over the mid-foreshore (Fig.9). Dissection of the grain-size curves shows that this is largely due to the presence of a coarse admixed component — from the storm-emplaced subsurface layers. This component can therefore be isolated by analysis of the curves of composite samples or by sampling of discrete laminae (or bundles of laminae) in box cores. Both methods have been used here.

Considering first the surface sands: these are dominated by a very well sorted framework component which becomes finer grained down the foreshore. This seaward-fining gradient parallels the seaward-attenuating swash zone velocities. In more detail, the shore-normal framework gradient has three components (Fig.9): (1) a slight seaward-coarsening across the upper foreshore; this must be established at high water when surf energy and swash zone velocities decline from the break point to the landward limit of the uprush (cf. Evans, 1939); (2) a rapid seaward-fining across the mid-foreshore, matching the progressive ebb-attenuation of surf energy and swash zone velocities; and (3) a slight seaward-fining across the lower foreshore; swash zone flows are too subdued to entrain these sands during most conditions and the textural gradient here is probably generated by nearshore shoaling waves (cf. Komar, 1976; Jago and Barusseau, 1981). The lower foreshore framework components are the best sorted — perhaps because rates of deposition are slower under shoaling waves than in the swash zone. Repeated sampling of the transect has shown that minor temporal variations can occur as tides and waves change, but the above trends are invariably present.

The interstitial component is barely represented in these sediments since fine grained material is removed by surf zone turbulence. It is sometimes present at the upper limit of swash action (where it consists mostly of heavy minerals) and is marginally more abundant on the low energy lower foreshore than elsewhere (Fig.9).

The contact component also is unimportant except over the mid-foreshore where it makes up to 10% of the surface sands (Fig.9). Its increased frequency here is presumably because it can draw from material in the underlying storm deposits — the result, therefore, of a 'source' control.

The shore-normal trends — in the surface sands — therefore reflect the passage of the fairweather tide across the foreshore: sorting by swash/surf action across the upper and mid-foreshores, by shoaling waves across the lower foreshore.

The subsurface sands show the added influence of storms over the mid-foreshore. The shore-normal textural gradients in the subsurface are therefore disrupted by an abrupt coarsening across the mid-foreshore. It appears therefore that the upper foreshore sands are emplaced by swash and surf action, the mid-foreshore by storm breakers, and the lower foreshore by shoaling waves. Since these subsurface sands are the accumulating beach *deposits*, this subsurface trend is obviously of the most geological significance. The grain-size curves are distinctive but hardly individually diagnostic of the beach environment. But the shore-normal, ultimately vertical, sequence of textures may be a useful indicator of ancient beaches (Fig. 13).

Bedforms and sedimentary structures

The foreshore surface changes in appearance depending on conditions. The shore-normal sequence commonly begins with a flat surface and/or antidunes (length $\simeq$ 0.5 m, height <0.02 m) on the upper part of the profile, followed further down by small near-symmetrical ripples, then small asymmetric landward-facing ripples across the lower part (length $\simeq$ 0.05 m, height <0.05 m). After storms the sequence consists almost entirely of flat beds with small asymmetric ripples near low water mark. Both sequences indicate a seaward-diminishing flow regime. Larger-scale bedforms, like those reported by Clifton et al. (1971) on Oregon beaches and by Hawley (1982) at Rhosilli beach (also in Carmarthen Bay), have not been observed on the exposed foreshore under any conditions.

Upper foreshore

Subsurface structures vary consistently across the foreshore. Plane bedding is ubiquitous on the highest part of the upper foreshore (Fig.10A). Whole shells of subtidal organisms are sometimes embedded in the sands but do not much disturb the laminations. Postdepositional cavity or bubble structures are as commonly developed as the plane beds they disrupt (Fig.10B). To seaward, the upper foreshore displays a greater variety of structures. Plane bedding always characterises the near-surface layers, but both small and large-scale cross bedding may occur below. The small-scale sets are <0.02 m in thickness, festoon-shaped (Fig.10D and F) and frequently obscured by bubble cavities (Fig.10C and D). The large-scale sets ($\simeq 0.05$ m thick) are both planar and trough bedded. Both scales of cross beds always have a landward-dipping component (Fig.10C and E). While the small-scale sets are usually parallel to the strand, the large-scale sets have longshore components with dips to both east and west (Fig.10D). Bioturbation sometimes destroys the primary bedding after long calm spells but is confined to the subsurface layers (Fig.10D, E and F). These upper foreshore structures are found shoreward of the high water breaker zone and must therefore form in the surf or swash zones. The plane beds are clearly swash zone products and are typical of swash/backwash flows (Clifton, 1969). The cross beds probably form in the surf zone and the large-scale sets probably indicate surf-generated longshore currents. However, the megaripples which generate the large-scale structures are never exposed by the ebbing tide and must be washed out by swash and backwash.

Mid-foreshore

This is invariably dominated by shelly traction clogs arranged as either plane beds or large-scale cross beds (Fig.11). The cross beds, usually planar, and up to 0.08 m in thickness, have both landward- and seaward-facing sets (Fig.11A, C and E) and longshore dips to both east and west (Fig.11B and D). Sharp erosional contacts are characteristic. These are the presumed storm breaker/surf deposits mentioned above and are analogous to the structures formed at the break-point of microtidal beaches. Whether the bimodal cross sets represent variable surf-zone current velocities or superimposed reversing tidal currents or both is not yet known. The large-scale bedforms that must be responsible for these structures are not to be seen on the exposed foreshore. Reworking of the surface layers on the tide ebbs can give rise to the plane beds of the swash zone (Fig.11A). Or the surface layers can be remoulded to landward-dipping small-scale cross beds (0.01—0.02 m thick) during low flow regime phases (Fig.11C, note preserved ripple form). The latter do not appear to be swash or surf zone features but instead form under the shoaling waves seaward of the breaker and then retain their identity despite the ebbing surf and swash. Bioturbation modifies, though generally does not totally obliterate, the primary structures during prolonged calms (Fig.11, all cores).

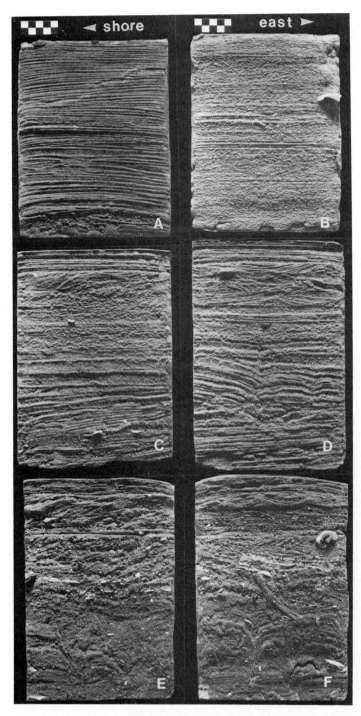

Fig.10. Box cores from the upper foreshore. Left-hand column, shore-normal. Right-hand column, shore-parallel. Scale in cm.

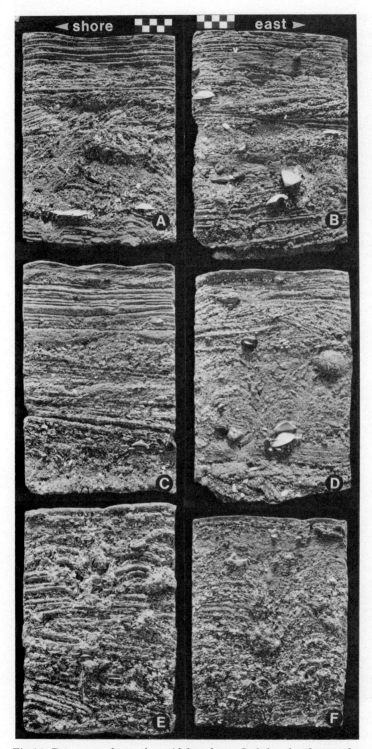

Fig.11. Box cores from the mid-foreshore. Left-hand column, shore-normal. Right-hand column, shore-parallel. Scale in cm.

Lower foreshore

The lower foreshore sands almost always exhibit small-scale, trough cross beds (usually <0.02 m thick) at and near the surface (Fig.12A and B). They are landward-facing structures, usually parallel to the strand. Climbing ripples are not uncommon and foresets sometimes develop offshoots which pass the troughs and reach adjacent flanks (Fig.12A and C). These structures apparently develop under the asymmetric oscillatory flows seaward of the breaker; similar structures are to be found at the mid-foreshore surface during calm periods (see above). The asymmetric ripples are sometimes just modified by the ebbing backwash to give a form-discordant internal structure (Fig.12A). Rather larger cross sets of thickness $\simeq$ 0.04 m and with a pronounced longshore dip to either east or west (Fig.12B and D) can occur, and this must be an indication of longshore currents and/or tidal flows. Rarely, the upper part of the lower foreshore is free of small ripples and then the upper layers are plane bedded (Fig.12F). The couplets of small ripple cross beds and planar truncation surfaces near the top of Fig.12F may be either storm- or tide-induced. Few primary structures are preserved more than 10 cm below the beach surface across the lower foreshore. Benthos burrows and small shells can be distinguished and frequently the sands are completely bioturbated (Fig.12, especially 12E). Both the physical and the biogenic structures of the lower foreshore would suggest a dominance of subtidal rather than intertidal processes.

Shore-normal sequence of structures

The shore-normal sequences of bedforms and sedimentary structures show that, while swash and surf/breaker processes control the upper and midforeshore, respectively, subtidal shoaling waves dominate the lower foreshore (which is therefore better described as a low tide terrace rather than as a beach). The structural suite contains elements of day-to-day activity (confined mostly to the surface layers) but is dominated by extremes of activity (plane bedding, large-scale cross bedding) and inactivity (bioturbation). A prograding sequence (Fig.13) should produce a succession that commences with very fine grained sands mostly bioturbated but with some small-scale cross bedding, passes up through coarser grained, cross-bedded sands also partially bioturbated, and ends with fine-grained sands either plane bedded or structureless (from bubble cavities). Small-scale ripple bedding may have low preservation potential since it is destroyed both by hydraulic processes during storms and by biological processes during calms.

Bedforms and flow regime

Under steady, unidirectional flows the bedforms of a sandy bed undergo a sequential transformation as the fluid power increases (Simons et al., 1965). For fine sands the sequence is: *no movement — small ripples — dunes — plane bed*. The bedforms arranged on a mesotidal foreshore were

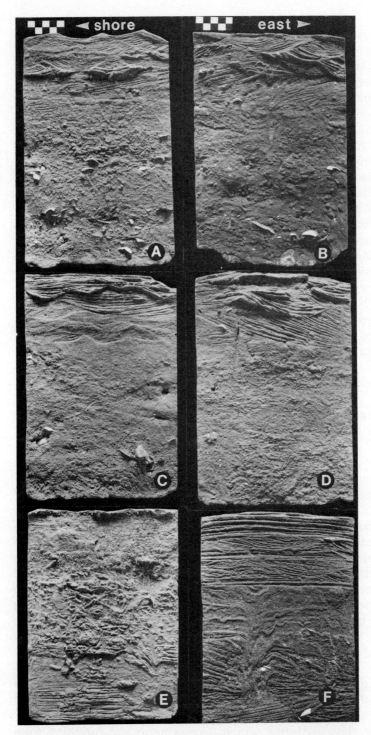

Fig.12. Box cores from the lower foreshore. Left-hand column, shore-normal. Right-hand column, shore-parallel. Scale in cm.

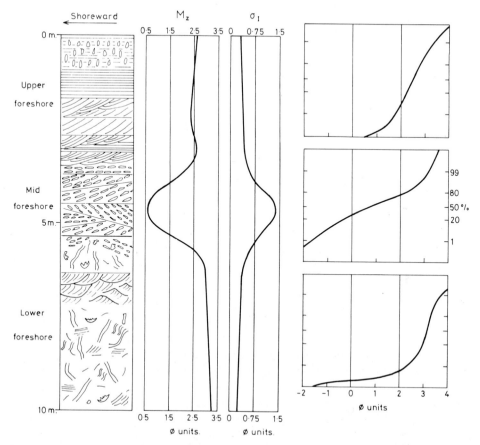

Fig.13. Sedimentary facies of a prograding macrotidal foreshore.

accordingly qualitatively assessed by Clifton et al. (1971) in terms of a flow regime model. More recent flume studies (Southard, 1975) suggest that, for very fine sands, the dune stage is left out. In Fig.14 we have plotted some swash and backwash flow data from Pendine Sands on the depth-velocity diagram given by Southard (1975). Note that we have used maximum instantaneous velocities and not the mean velocities of the original flume data. Clearly there is some accord between the Pendine and the flume bedform sequences: diminishing swash zone velocities and depths as the tide ebbs leave plane beds and backwash antidunes on the upper foreshore and small ripples on the lower; movement ceases altogether near the low water mark.

Obviously, a macrotidal beach and a flume differ in many major respects. The nearshore zone is subjected to interacting flows of several kinds: symmetrical oscillatory flow, asymmetrical oscillatory flow and unidirectional flow (Clifton, 1976; Davidson-Arnott and Greenwood, 1976); a macrotidal shoreline is also subjected to reversing tidal flow. Furthermore, on a macrotidal foreshore, a sequence may not start from a flat bed stage with no

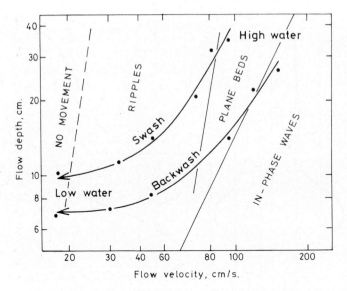

Fig.14. Depth-velocity bedform diagram (after Southard, 1975) with Pendine data superimposed.

movement. As the tide sweeps the foreshore, a dynamic zone will be bequeathed a bed configuration from the preceding zone and hence the flow regime and bedform will be temporarily out of phase; the bed phase will in turn impose a rhythmic roughness and create a flow separation. Thus the unidirectional flow model correctly predicts ripples over much of the mid and lower foreshores but in fact the ripples are formed not by the swash/backwash but by shoaling waves. While the swash zone flows may be competent to form ripples, they do not do so because the ripples are already there; instead the backwash may superficially modify the existing bedforms to give form-discordant internal structures. Such a bedform disequilibrium must be common during storms when megaripples/dunes create the large-scale cross beds in the coarser sands of the mid-foreshore.

Moreover, such is the speed of lateral migration of the tide across the foreshore (an average 4 cm s^{-1} on a mean spring) that it seems likely that the growth of bedforms may lag so far behind flow conditions that the appropriate bedforms may not form at all. A particular flow regime sequence may therefore be incomplete.

There is very little information on disequilibrium bedforms (except Allen, 1973, 1974; Lofquist, 1978), and the flow regime beach model of Clifton et al. (1971) may not be appropriate on a macrotidal shoreline. But there seems little doubt that the upper foreshore plane beds represent Clifton's *inner planar facies* which is typical of swash zone activity (Clifton, 1969; Davidson-Arnott and Greenwood, 1974). The landward-facing cross beds which sometimes contribute to the upper foreshore have something in common with structures occasionally observed by Clifton in his *inner rough facies*. We have still to see the larger storm-generated bedforms which produce the mid-foreshore cross beds, but tentatively assign these to the *outer rough facies* as the structures are similar to those described by both Clifton et al. (1971) and Davidson-Arnott and Greenwood (1976) in the breaker zone. The lower foreshore small-scale cross bedding is clearly the *asymmetric ripple facies*. The plane beds that can occur from time to time on any part of the foreshore may be the result of symmetrical or asymmetrical oscillatory flow, or unidirectional flow, or disequilibrium effects.

The sedimentary facies of Pendine Sands are therefore recognisably similar to those elsewhere on microtidal and mesotidal beaches and are apparently established more by waves than by tidal currents. However, without further observation it would be imprudent to attempt a more detailed flow regime interpretation, especially if some of the structures are generated by disequilibrium bedforms.

MORPHODYNAMIC MODEL

A more complete understanding of these particular sedimentological associations and variations on the macrotidal profile depends upon a comprehension of the processes which lead to this sedimentary accumulation and also of the relationships between these processes and the local tide and wave regimes.

The levelling data have highlighted the longer-term morphological equilibrium of the beach and yet the bedform survey has revealed quite considerable disturbance on a shorter time scale. A simple, and very general, morphodynamic model will now be developed and tested against the field data in an attempt to reconcile these observations and to clarify the important processes.

Definition of the flow ratios

Three ratios may be employed to characterise the dynamics of the beach system. Consider that, at a point on the bed beneath a single wave, the onshore flow has an effective mean velocity u_{in} which persists for a time t_{in} and transports sediment at a rate j_{in} producing a net shoreward sediment mass transport of J_{in}. Define the corresponding parameters for the subsequent offshore flow as u_{ex}, t_{ex}, j_{ex} and J_{ex}, respectively. The ratio of the flow velocities is the 'velocity ratio', V_r:

Velocity ratio = $V_r = u_{in}/u_{ex}$ (1)

This is analogous to Kemp's (1975) 'velocity magnitude asymmetry'. A large velocity ratio indicates that the onshore flow exceeds the offshore flow, a value of unity indicates that the two are symmetrical, and a value of less than one indicates that the offshore flow exceeds the onshore. The flow durations are characterised by the 'duration ratio', D_r:

Duration ratio = $D_r = t_{in}/t_{ex}$ (2)

This is analogous to Kemp's (1975) 'velocity time asymmetry'. The two ratios may be usefully combined by neglecting the effect of residual shore-normal flow components so that continuity of water allows the equation of the onshore and offshore discharges. That is $u_{in}t_{in} = u_{ex}t_{ex}$ so that:

$$D_r = 1/V_r \tag{3}$$

The ratio of the net sediment mass transports is defined as the 'transport ratio' J_r:

$$\text{Transport ratio} = J_r = J_{in}/J_{ex} \tag{4}$$

This ratio will now be used to develop a simple beach equilibrium concept which will be tested with Pendine Beach data.

Equilibrium sediment dynamics

A state of morphological equilibrium can result from either of two general conditions, known as 'zero transport' and 'zero net transport' (Hardisty, 1981). The former state, wherein flows remain below threshold values, so that no sediment is moving, has been applied to beaches in the 'null point hypothesis' (for example, Johnson and Eagleson, 1966). The latter state, wherein equal amounts of sediment enter and leave the system, was applied theoretically to beaches by Bagnold (1963), Inman and Frautschy (1966), Bowen (1980) and Bailard and Inman (1981). This more promising approach to the problem can be employed with the flow ratios defined above to present a simple general beach model.

Zero net transport occurs on the beach when the quantity of sediment carried shoreward by the wave, J_{in}, is precisely balanced by the quantity carried seaward by the returning flow, J_{ex}. The transport ratio (eq. 4) in this equilibrium condition is then equal to unity:

$$J_r = J_{in}/J_{ex} = 1 \tag{5}$$

Swash zone measurements were collected throughout a number of tidal cycles on the profile at Pendine to test this equilibrium concept. A differential bedload trap (Fig.15) was buried in the beach and the top was smoothed level with the sediment surface. After submergence by the rising tide the lid was removed and, after the passage of a single wave, the trap was removed. The net onshore sediment transport and the net offshore sediment transport were thus collected separately and later dried and weighed. These data (Fig. 16) show that the beach was close to the equilibrium condition defined by the transport ratio model. The scatter is probably due to the fluctuating nature of the flows from one wave to another, highlighting the need for longer-term measurements.

These experimental results show that the foreshore was in short term equilibrium during the period of study. Our levelling data (discussed above) suggest that the foreshore (at this western end of the barrier) is also in long-term equilibrium with the nearshore flows. However, the sedimentary

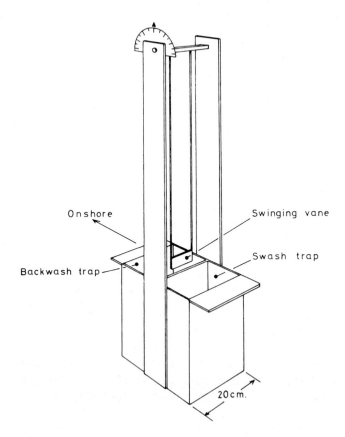

Fig.15. Bedload trap.

structures of the upper and mid-foreshore show that this is indeed a 'zero net transport' dynamic equilibrium and that considerable disturbance of the beach sand does occur under certain conditions.

Equilibrium beach slope

The transport ratio equilibrium concept outlined and tested above can be extended to relate the surface slope of the beach sediments to the measure of the flow asymmetry provided by the velocity ratio. For this a relationship between the flow speeds and the sediment transport rate is required. Bagnold's (1963, 1966) equation has proven useful (Langhorne, 1982) and more accurate than others in the marine environment (Heathershaw, 1981). The equation correlates the bedload transport rate with the cube of the flow velocity (Hardisty, 1983; Greenwood and Mittler, 1984, this volume).

The onshore, upslope transport rate (Bagnold, 1963) is:

$$j_{in} = k\, u_{in}^3 / (\tan\phi + \tan\alpha)$$

where ϕ and α are the angle of internal friction and the beach gradient,

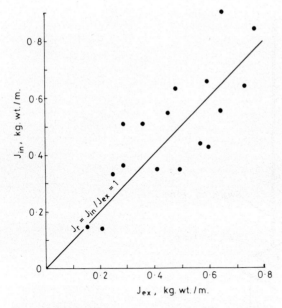

Fig.16. Measured bedload transport J_{in} and J_{ex}.

respectively, and k is an empirical constant. The $(\tan \phi + \tan \alpha)$ term sensibly reduces the transport rate with increasing bed slope to the limiting value of $\alpha = \phi$, at which the bed fails. The net onshore mass transport, $J_{in} = j_{in}t_{in}$ is therefore:

$$J_{in} = k\, u_{in}^3 t_{in}/(\tan \phi + \tan \alpha)$$

Similarly the offshore, downslope transport rate (Bagnold, 1963) is:

$$j_{ex} = k\, u_{ex}^3/(\tan \phi - \tan \alpha)$$

which sensibly increases the rate with increasing downslope gradient to the limiting value of $\alpha = \phi$ at which the bed fails and offshore transport continues regardless of the flow. The net offshore transport is therefore:

$$J_{ex} = k\, u_{ex}^3 t_{ex}/(\tan \phi - \tan \alpha)$$

These two net transport equations are potentially very useful for beach investigations but an assessment of the calibration coefficient k is presently difficult due to the paucity of published beach transport data. The limited measurements reported here do, however, yield a value for the coefficient of about 10 kg m^{-4} s^{-2} when the appropriate units are used and u_{in} and u_{ex} are taken as the *mean* flow velocities.

Combining these two formulations the transport ratio (eq. 4) becomes:

$$J_r = \frac{k\, u_{in}^3 t_{in}\, (\tan \phi - \tan \alpha)}{k\, u_{ex}^3 t_{ex}\, (\tan \phi + \tan \alpha)}$$

Substituting for V_r (eq. 1) and D_r (eqs. 2 and 3) yields:

$$J_r = (\tan \phi - \tan \alpha) V_r^2/(\tan \phi + \tan \alpha)$$

The zero net transport equilibrium slope, $\tan \alpha_e$, which was discussed above as occurring when $J_r = 1$, is given by solving this equation for $\tan \alpha$ as:

Equilibrium slope, $\tan \alpha_e = \dfrac{\tan \phi \, (V_r^2 - 1)}{(V_r^2 + 1)}$ (6)

This relationship is plotted in Fig.17 for a range of values of the velocity ratio showing that the more asymmetric the flow then the steeper the equilibrium gradient that is required to maintain a balance between the onshore and the offshore sediment transports.

Furthermore the diagram appears to correctly display the self-stabilising nature of such systems. Two unstable situations may be identified. Firstly the region above the equilibrium line where the bed gradient is too steep and the velocity ratio is too low. Here the higher offshore velocity combines with the steeper slope to increase the offshore sediment transport, moving material down the beach and thus flattening the profile until equilibrium is re-established. Alternatively within the region below the equilibrium line, the higher onshore velocities and flatter slopes combine to increase the onshore transport thus steepening the profile gradient until again equilibrium is re-established. In morphodynamic terms the beach profile is therefore self-stabilising and any perturbation from the equilibrium induces a response which opposes the change and returns the beach toward the equilibrium condition.

The onshore and offshore flow velocities were monitored at various positions along the profile to test these slope predictions. The flow velocities were used to calculate the velocity ratio (eq. 1) and thence the theoretical equilibrium slope (eq. 6). These theoretical values are compared with the actual beach slopes from the levelling data as shown in Fig.18. There is clearly considerable agreement between the theoretical and observed values for this limited data set.

The model suggests that the beach gradient steepens across the profile

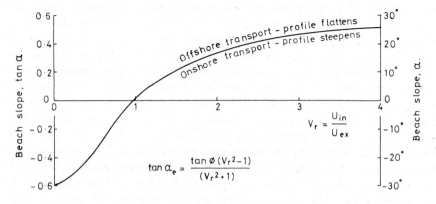

Fig.17. Equilibrium slope as a function of the velocity ratio.

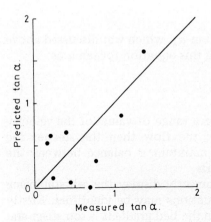

Fig.18. Theoretical and measured equilibrium beach slopes.

from low to high water because the flow velocities become more asymmetric. This is in line with an earlier argument (Hardisty, 1981) where the inherent asymmetry of the Stokes Wave Theory was preferred to the symmetry of the Airy Wave Theory and was related to the occurrence of different breaker types. At Pendine the rising tide submerges steeper sections of the profile; this produces a gradation from dissipative spilling breakers, which have relatively symmetrical onshore and offshore flow components and hence low sediment surface gradients, over the lower foreshore, up to the relatively asymmetric flows and hence steeper beach gradients associated with the narrow surf zones and plunging breakers over the upper foreshore. The breaker variations caused by the macrotidal range at Pendine result in increasingly asymmetric flows and equilibrium beach gradients toward high water. These in turn produce the shore-normal changes in sedimentological characteristics described earlier.

CONCLUSIONS

The foreshore of Pendine Sands is exposed to wind-generated waves and swell and is subject to a large tidal range. The sedimentological characteristics of this macrotidal beach are comparable to those of mesotidal and microtidal beaches but there are important differences. These differences arise partly because the daily to-and-fro sweep of the tide across the foreshore causes variations in breaker, surf and swash zone characteristics and energy dissipation. Furthermore, the large tidal range creates a variable shoaling modification of unbroken nearshore waves such that the breaker height can change significantly during the tidal cycle. The result is that, as the tide ebbs, the breaker height decreases, swash and surf zones widen, the surf zone becomes increasingly dissipative and swash zone velocities diminish. These changes are related to breaker type, beach slope, and swash/backwash velocity asymmetry. The plunging breakers and narrow surf zone at high water generate relatively asymmetric swash zone flows and are associated with a steeper beach gradient while the spilling breakers and dissipative surf at low water produce more symmetrical swash zone flows and hence a gentle beach gradient.

The important sedimentological consequence of these tide-induced modifications of wave-induced currents is that the hydraulic flow regimes of the nearshore circulation diminish as the tide ebbs. This is reflected in the overall seaward-fining textural gradient of the primary framework population of the surficial sands. In more detail, the shore-normal framework gradient has three components: slight seaward-coarsening (upper foreshore); rapid seaward-fining (mid-foreshore); slight seaward-fining (lower foreshore). These gradients are established by the high water swash/surf, the ebbing swash/surf, and nearshore shoaling waves, respectively. There is a corresponding change down the beach profile from plane beds to small ripples. The plane beds are a product of unidirectional swash zone flows, the small ripples of asymmetric oscillatory flows in the wave build-up zone.

Subsurface sedimentary structures reflect this tidal variation with plane bedding characteristic of the upper foreshore and landward-dipping small-scale cross bedding of the lower foreshore. The upper foreshore structures are frequently modified by bubble cavities, the lower foreshore structures destroyed by bioturbation. Variations in wave energy result in cross-bedding, both large- and small-scale, associated with longshore currents, across the lower part of the upper foreshore. The mid-foreshore subsurface is distinguished by coarse shelly and lithic traction clogs arranged as plane beds and both seaward- and shoreward-inclined large-scale cross beds presumably deposited under storm breakers and highly dissipative surf at high water still-stand. The storm deposits become bioturbated during calms. These sequences of sedimentary structures suggest that swash zone processes establish the upper foreshore facies, storm breakers and surf the mid-foreshore facies, and shoaling waves the lower foreshore facies. Whereas the texture of the surface veneer of sediments displays a short-term equilibrium with foreshore dynamics, the subsurface deposits reflect long-term responses to both storm and fairweather conditions. Consequently, the upper, mid and lower foreshore facies have distinctive, and diagnostic, textural and structural signatures. A composite vertical section through a regressive sequence would be, from bottom to top: shoreward-dipping small-scale cross bedding in very fine sands, much bioturbated; seaward- and shoreward-dipping, large-scale cross bedding in shelly coarse and medium sands; plane bedded fine sands; aeolian dune cross bedding. Such a sequence would be similar to that of a microtidal shoreline but should be laterally and vertically more extensive.

Periodic levelling of beach profiles has shown that the foreshore is in long-term equilibrium with prevailing meteorological and dynamical conditions. A zero net transport model for the beach shows that, in morphodynamic terms, the beach profile is self-stabilising in the short-term. The sedimentary structures, however, indicate a depth of disturbance of at least 30 cm over much of the foreshore. A long-term dynamic equilibrium is therefore not incompatible with a significant short-term mobility.

ACKNOWLEDGEMENTS

The bulk of the beach profiling programme was carried out by C.F.J. during the tenure of a research studentship, from the Natural Environmental Research Council, in the Department of Geology, Imperial College, London. Dr. Graham Evans initiated the Carmarthen Bay programme, advised during the early part of this study, and critically reviewed an early draft of the paper. John Malcolm, Peter Bush, and especially the late Peter Gee, provided invaluable assistance in the field, as did many undergraduates too numerous to name. Mr. F. Dewes drafted the figures and Mr. W. Rowntree photographed the box cores. Everyone's help is most gratefully acknowledged.

NOTATION

B_0: breaker coefficient
H: deep-water wave height
$H_{1/3}$: mean height of highest one-third of the waves
H_b: breaker height
h_b: water depth at break point
J: net bedload transport
j: bedload transport rate
L: deep-water wavelength
l_b: width of surf zone
l_s: width of swash zone
P: phase difference
T: wave period
$T_{1/10}$: mean period of longest one-tenth of the waves

t: flow duration
t_b: flow duration, surf and swash
u: mean flow velocity
u_s: mean swash/backwash velocity
ϵ: surf scaling factor
α: beach slope, degrees
ϕ: angle of repose of sand
$(\ )_C$: denotes contact component
$(\ )_F$: denotes framework component
$(\ )_I$: denotes interstitial component
$(\ )_{in}$: denotes swash
$(\ )_{ex}$: denotes backwash
$(\ )_e$: denotes equilibrium

REFERENCES

Allen, J.R.L., 1973. Phase differences between bed configurations and flow in natural environments, and their geological significance. Sedimentology, 20: 323—329.

Allen, J.R.L., 1974. Reaction, relaxation and lag in natural sedimentary systems: general principles, examples and lessons. Earth-Sci. Rev., 10: 263—342.

Bagnold, R.A., 1963. Mechanics of marine sedimentation. In: M.N. Hill (Editor), The Sea, Vol. 3. Wiley, New York, N.Y., pp.507—582.

Bagnold, R.A., 1966. An approach to the sediment transport problem from general physics. U.S. Geol. Surv., Prof. Pap., 422-I.

Bailard, J.A. and Inman, D.L., 1981. An energetics bedload model for a plane sloping beach: local transport. J. Geophys. Res., 86: 2035—2043.

Blackley, M.W.L. and Heathershaw, A.D., 1982. Wave and tidal-current sorting of sand on a wide surf-zone beach. Mar. Geol., 49: 345—356.

Bouma, A.H., 1969. Methods for the Study of Sedimentary Structures. Wiley, New York, N.Y., 458 pp.

Bowen, A.J., 1980. Simple models of nearshore sedimentation: beach profiles and offshore bars. Proc. Conf. on Coastlines of Canada, Halifax, 1978, Can. Geol. Surv., Pap. 80-10: 1—11.

Cassie, R.M., 1954. Some uses of probability paper in the analysis of size frequency distributions. Aust. J. Mar. Freshwater Res., 5: 513—522.

Clifton, H.E., 1969. Beach lamination — nature and origin. Mar. Geol., 7: 553—559.

Clifton, H.E., 1976. Wave formed sedimentary structures — a conceptual model. In: R.A. Davis and R.L. Ethington (Editors), Beach and Nearshore Sedimentation. Soc. Econ. Paleontol. Mineral., Spec. Publ., 24: 126—148.

Clifton, H.E., Hunter, R.E. and Phillips, R.L., 1971. Depositional structures and processes in the non-barred high energy nearshore. J. Sediment. Petrol., 41: 651—670.

Darbyshire, M., 1963. Wave measurements made by the National Institute of Oceanography. In: R.C. Vetter (Editor), Ocean Wave Spectra. Prentice-Hall, Englewood Cliffs, N.J., pp. 258—291.

Davidson-Arnott, R.G.D. and Greenwood, B., 1974. Bedforms and structures associated with bar topography in the shallow-water wave environment, Kouchibouguac Bay, New Brunswick, Canada. J. Sediment. Petrol., 44: 698—704.

Davidson-Arnott, R.G.D. and Greenwood, B., 1976. Facies relationships on a barred coast, Kouchibouguac Bay, New Brunswick. In: R.A. Davis and R.L. Ethington (Editors), Beach and Nearshore Sedimentation. Soc. Econ. Paleontol. Mineral., Spec. Publ., 24: 149—168.

Draper, L., 1972. Extreme wave conditions in British and adjacent waters. Proc. 13th Conf. Coastal Engineering, Vancouver, B.C., 1: 157—165.

Evans, O.F., 1939. Sorting and transportation of material in the swash and backwash. J. Sediment. Petrol., 9: 28—31.

Folk, R.L. and Ward, W.C., 1957. Brazos river bar: A study in the significance of grain size parameters. J. Sediment. Petrol., 27: 3—26.

Galvin, C.J., 1968. Breaker type classification on three laboratory beaches. J. Geophys. Res., 73: 3651—3659.

Greenwood, B. and Mittler, P.R., 1984. Sediment flux and equilibrium slopes in a barred nearshore. In: B. Greenwood and R.A. Davis, Jr. (Editors), Hydrodynamics and Sedimentation in Wave-Dominated Coastal Environments. Mar. Geol., 60: 79—98 (this volume).

Guza, R.T. and Inman, D.L., 1975. Edge waves and beach cusps. J. Geophys. Res., 80: 2997—3012.

Hardisty, J., 1981. Sediment dynamics on the intertidal profile. Unpubl. Ph.D. thesis, University of Hull, Hull, 194 pp.

Hardisty, J., 1983. An assessment and calibration of formulations for Bagnold's bedload equation. J. Sediment. Petrol., 53: 1007—1010.

Hawley, N., 1982. Intertidal sedimentary structures on macrotidal beaches. J. Sediment. Petrol., 52: 785—795.

Heathershaw, A.D., 1981. Comparison of measured and predicted sediment transport rates in tidal currents. Mar. Geol., 42: 75—104.

Hydraulics Research Station, 1978. Coast erosion at Pendine: a wave refraction study. Rep. No. EX 809, Wallingford, 22 pp.

Ingle, J.C., 1966. The Movement of Beach Sand. Elsevier, Amsterdam, 211 pp.

Inman, D.L. and Frautschy, J.D., 1966. Littoral processes and the development of shorelines. Proc. Spec. Conf. Coastal Engineering, Am. Soc. Civ. Eng., Santa Barbara, Calif., pp. 511—536.

Jago, C.F., 1980. Contemporary accumulation of marine sand in a macrotidal estuary, Southwest Wales. In: A.H. Bouma, D.S. Gorsline, C. Monty and G.P. Allen (Editors), Shallow Marine Processes and Products. Sediment. Geol., 26: 21—49.

Jago, C.F. and Barusseau, J.P., 1981. Sediment entrainment on a wave graded shelf, Roussillon, France. In: C.A. Nittrouer (Editor), Sedimentary Dynamics of Continental Shelves. Mar. Geol., 42: 279—299.

Johnson, J.W. and Eagleson, P.S., 1966. Coastal processes. In: A.T. Ippen (Editor), Estuary and Coastline Hydrodynamics. McGraw-Hill, New York, N.Y., pp.405—493.

Kemp, P.H., 1961. The relationship between wave action and beach profile characteristics. Proc. 7th Conf. Coastal Engineering, Am. Soc. Civ. Eng., pp.262—277.

Kemp, P.H., 1975. Wave asymmetry in the nearshore zone and breaker area. In: J. Hails and A. Carr (Editors), Nearshore Sediment Dynamics and Sedimentation. Wiley, London, pp.47—68.

King, C.A.M., 1951. Depth of disturbance of sand on beaches by waves. J. Sediment. Petrol., 21: 131—140.

Komar, P.D., 1976. Beach Processes and Sedimentation. Prentice-Hall, Englewood Cliffs, N.J., 429 pp.

Langhorne, D.N., 1982. A study of the dynamics of a marine sandwave. Sedimentology, 29: 571—594.

Lofquist, K.E.B., 1978. Sand ripple growth in an oscillatory-flow water tunnel. C.E.R.C., Tech. Pap. 78-5.

Middleton, G.V., 1976. Hydraulic interpretation of sand size distributions. J. Geol., 84: 405—426.

Moss, A.J., 1972. Bed load sediments. Sedimentology, 18: 159—219.

Otvos, E.G., 1965. Sedimentation—erosion cycles of single tidal periods on Long Island Sound beaches. J. Sediment. Petrol., 35: 604—609.

Short, A.D., 1980. Beach response to variations in breaker height. Proc. 17th Int. Conf. Coastal Engineering, Am. Soc. Civ. Eng., Sydney, pp. 1016—1035.

Simons, D.B., Richardson, E.V. and Nordin, C.F., 1965. Sedimentary structures generated by flow in alluvial channels. In: G.V. Middleton (Editor), Primary Sedimentary Structures and their Hydrodynamic Interpretation. Soc. Econ. Paleontol. Mineral., Spec. Publ., 12: 34—52.

Southard, J.B., 1975. Bed configurations. In: Depositional Environments as Interpreted from Primary Sedimentary Structures and Stratification Sequences. Soc. Econ. Paleontol. Mineral., Short Course No. 2, Dallas, Texas, pp.5—44.

Williams, A.T., 1971. An analysis of some factors involved in the depth of disturbance of beach sand by waves. Mar. Geol., 11: 145—158.

Wright, L.D., Chappell, J., Thom, B.G., Bradshaw, M.P. and Cowell, P., 1979. Morphodynamics of reflective and dissipative beach and inshore systems: Southeastern Australia. Mar. Geol., 32: 105—140.

HIGH-FREQUENCY SEDIMENT-LEVEL OSCILLATIONS IN THE SWASH ZONE

ASBURY H. SALLENGER, Jr. and BRUCE M. RICHMOND

U.S. Geological Survey, 345 Middlefield Road, Menlo Park, CA 94025 (U.S.A.)
(Received June 14, 1983, revised and accepted January 16, 1984)

ABSTRACT

Sallenger Jr., A.H. and Richmond, B.M., 1984. High-frequency sediment-level oscillations in the swash zone. In: B. Greenwood and R.A. Davis, Jr. (Editors), Hydrodynamics and Sedimentation in Wave-Dominated Coastal Environments. Mar. Geol., 60: 155—164.

Sediment-level oscillations with heights of about 6 cm and shore-normal lengths of order 10 m have been measured in the swash zone of a high-energy, coarse-sand beach. Crests of oscillations were shore parallel and continuous alongshore. The oscillations were of such low steepness (height-to-length ratio approximately 0.006) that they were difficult to detect visually. The period of oscillation ranged between 6 and 15 min and decreased landward across the swash zone. The sediment-level oscillations were progressive landward with an average migration rate in the middle to upper swash zone of 0.8 m min^{-1}. Migration was caused mostly by erosion on the seaward flank of the crest of an oscillation during a period of net seaward sediment transport. Thus, the observed migration was a form migration landward rather than a migration involving net landward sediment transport. The observed sediment-level oscillations were different than sand waves or other swash-zone bedforms previously described.

INTRODUCTION

Most previous studies on beach-profile changes have focussed on changes measured at intervals of weeks, days, or hours. We know of only one previously published study (Waddell, 1973) that focussed on high-frequency changes, changes measured at intervals approaching the swash period.

On a medium-sand low-energy beach, Waddell (1973) measured sediment level at two locations in the upper half of the swash zone after the backwash of each wave swash. He found significant oscillations of sediment level with periods of 40 s and longer and presented evidence that the sediment-level oscillations had some characteristics of sand waves. Discussing the same data, Waddell (1976) hypothesized that the bed oscillations were caused by ground-water oscillations of the same frequency. The rising and falling ground water would cause areas to erode and accrete similar to the ground-water control over tidal cycle sedimentation discussed by Duncan (1964). When the water table is high, swash infiltration into the beach is relatively low and erosive backwash is enhanced. When the water table is low, backwash is

0025-3227/84/$03.00   © 1984 Elsevier Science Publishers B.V.

diminished due to swash infiltration into the beach and accretion results. The low frequency ground-water oscillations were thought to be caused by waves of the same frequency in the surf zone.

In the present study, we measured changes in sediment-level at numerous locations across the swash zone of a coarse-sand high-energy beach. We will show that sediment level changed in a surprisingly rapid and well-organized manner. We found sediment-level oscillations which were low-amplitude, landward-progressive, and had a unique mode of migration. We will show that these oscillations had characteristics different than sand waves or the several types of swash-zone bedforms previously described in the literature. The processes of formation are presently unclear, although we point out that ground-water oscillations could not explain a critical characteristic of the observed sediment-level oscillations.

EXPERIMENT DESCRIPTION

Our experiment was conducted during January 1981 at Fort Ord, California. Fort Ord is located on the shore of southern Monterey Bay about 150 km south of San Francisco. Average foreshore slope during our study was 7.5°. The foreshore was composed of coarse sand with a median diameter of 0.8 mm. During the experiment, waves were of normal incidence, breaker heights were 3.0—4.0 m, and the surf zone width was approximately 100 m. Wind speeds were low during the experiment although local strong winds associated with squalls occurred both before and after the experiment. Low-energy sea waves associated with the passing squalls were present, but incident waves were dominantly of the swell type with a period of about 16 s.

Stakes, 1 cm in diameter and 2 m long, were driven into the swash zone at the locations shown in Fig.1. The shore-normal array of seven stakes had 3-m spacings between stakes and the shore-parallel array of five stakes had 4-m spacings. Stake numbers for the shore-normal array refer to the distance from the landward stake (for example, the landward stake is called 0 and the stake 15 m seaward is called 15). Stakes in the shore-parallel array are referenced by the letters indicated on Fig.1.

Stake heights above the bed, initially about 0.5 m, were measured using modified meter sticks. A circular base plate (15-cm diameter) was attached to one end of a meter stick to inhibit settling into the bed. A movable pointer, free to slide over the length of the meter stick, was used to determine stake heights. Repeated measurements of a stake not exposed to the swash showed that the accuracy of the technique was 1—2 mm. In the upper swash zone, the accuracy of our measurements approached that of our test case. In the lower swash zone, where measurements needed to be made rapidly and the base plate tended to sink into the bed more than in the upper swash zone, accuracy was about 5 mm.

During our experiment, stake heights were measured after a backwash when the swash zone was subaerial. Heights were read to the closest millimeter. Three persons measured stake heights and one person recorded the

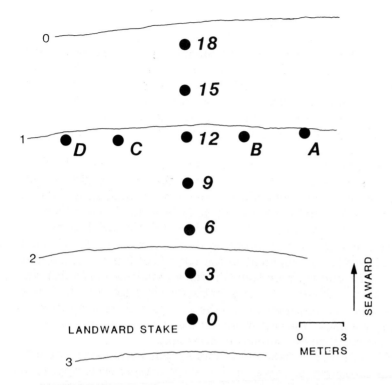

Fig.1. Stake locations and foreshore contours. Contour interval is 1 m; the vertical datum is arbitrary.

stake-height data and time. Persons measuring stakes were careful to stand as far as possible in a longshore direction away from the stakes. Scour holes caused by persons standing in the swash did not extend to the stakes. The stakes were of sufficiently small diameter, 1 cm, to prevent significant scour around themselves. Stakes in the lower swash zone were measured on the average every 40—50 s or about every third swash. Stakes in the extreme upper swash zone were measured only after a swash reached their location; they were not reached by every wave swash. Stakes were measured for a period of about 70 minutes midway through a flooding tide. Although the diurnal tide range for the area is 1.6 m, the range from low to high tide on the day of our experiment was 0.6 m. Tide range during our experiment was 0.09 m which caused a 0.7 m landward shift of the swash zone.

The mean swash position was between stakes 12 and 15. Due to the powerful swash, it would have been very difficult to obtain data lower in the swash zone than stake 18. In fact, the person who measured stake 18 was frequently knocked down by the upwash. The maximum landward excursion of upwash during our experiment was 1—2 m landward of the landwardmost stake, stake 0. The seaward excursion of backwash was typically 3—6 m seaward of the seawardmost stake, stake 18, although some backwashes reached as much as 9 m seaward of stake 18.

RESULTS

Significant sediment-level oscillations are apparent in the time series from the shore-normal array (Fig.2). These oscillations showed a shift in frequency across the swash zone. At the lower stakes (15 and 18) which were below the mean swash position, the period of oscillation was roughly 6 min, whereas at the middle stakes (9 and 12) period increased to 10—15 min. Oscillations disappeared, or were not as apparent, in the extreme upper swash zone.

A measure of the height of sediment-level oscillations is twice the standard deviation of a sediment-level time series with the linear trend removed. For stakes in the seaward end and middle of the shore-normal array, heights were uniformly about 5.5 cm (Fig.3 and Table I). Heights decreased significantly to about 1.5 cm for stakes in the landward end of the swash zone.

Significant negative and positive elevation trends, indicating net changes to the swash zone profile, are apparent in the records of Fig.2. The amount of net vertical change due to these trends (over a period of 67.7 min) was plotted versus distance offshore in Fig.3 (tabulated in Table I). The lower swash zone underwent net erosion whereas the upper swash zone underwent net accretion. The volumes, however, were not equal; the amount of erosion was significantly greater than the amount of accretion.

Sediment-level data for each stake in the shore-parallel array are plotted in Fig.4. Oscillations measured at stakes in the shore-parallel array appear to be in phase. Heights and net sediment-level changes were computed for these records as before. Heights ranged from 4.7 to 6.0 cm, but did not vary systematically alongshore (Table I). The in-phase relationships among records and the lack of systematic change in heights suggest the oscillations were

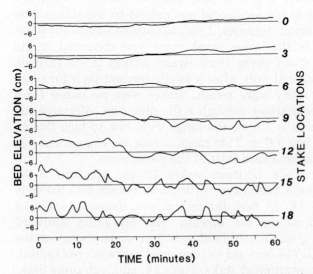

Fig.2. Time-series of sediment-level for the shore-normal array.

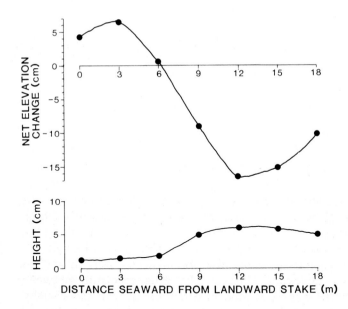

Fig.3. In the upper plot, net elevation change (over 67.7 min) is plotted versus location for the shore-normal array. Net changes were computed from the linear slope of each record. In the lower plot, heights of sediment-level oscillations are plotted for the shore-normal array.

TABLE I

Statistics on sediment elevation records

Stake	Net change* (cm)	Heights** (cm)
0	4.5	1.3
3	6.5	1.4
6	0.5	1.7
9	9.0	5.0
12	16.0	6.0
15	15.0	5.8
18	10.2	5.2
12A	17.1	5.6
12B	17.4	5.2
12	16.0	5.0
12C	10.2	4.7
12D	8.5	5.5

*Computed over 67.7 min; **this is twice the standard deviation of the record with the trend and mean removed.

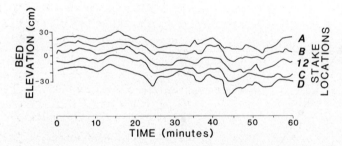

Fig.4. Time-series of sediment-level for the shore-parallel array.

shore parallel and continuous alongshore, at least over the 16 m length of the shore-parallel array. However, net changes varied systematically alongshore. The amount of erosion decreased significantly to the south along the shoreline (Table I). The reason for this pattern of net change is unknown.

The 6-min oscillations apparent in records obtained at stakes 15 and 18 appear to be migrating landward (Fig.2). The cross-spectrum between records at stakes 15 and 18 was calculated in order to confirm this visually apparent landward migration. Unfortunately, in order to resolve 6-min peaks we had to use such a narrow bandwidth to smooth raw spectral estimates that little confidence could be placed in the resolved peaks. The landward migration of longer period oscillations measured more landward in the swash zone is confirmed below in another manner.

Figure 5 shows a time sequence of shore-normal profiles of bed elevations. These profiles were plotted at one minute intervals using records with the mean and linear trend removed. In essence, an individual profile shows the difference in bed elevation from a mean profile at a given time; the mean profile is horizontal and is defined by the means of all shore-normal records. By plotting the profiles in a vertical time sequence, the shore-normal movements of areas of erosion and accretion are made visible. Clearly, the plot shows the crest of an oscillation that progressed landward across the swash zone. The migration rate slowed in a landward direction; the mean rate of migration was about 0.8 m min^{-1}. These landward migrating features were associated with the 10- to 15-min oscillations of Fig.2; the higher frequency oscillations lower in the swash zone were not well represented in Fig.4.

Figure 6 shows how a sediment-level oscillation migrates across the swash zone. At 11 min, a distinct accretional area had developed at the seaward end of the swash zone (Fig.6A). Within 4 min, a crest of an oscillation was apparent in the midst of our shore-normal array. Excluding the initial landward migration (between 15 and 18 min), the crest progressed landward mostly by erosion of its seaward flank (Fig.6B). Note that the eroded sediment was apparently transported seaward. Thus, the migration involved more of a form migration landward than a net landward sediment transport.

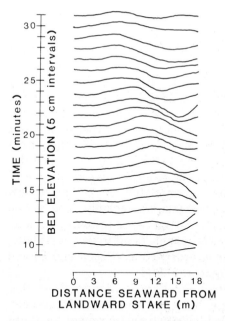

Fig.5. Time sequence of shore-normal profiles of sediment-level elevations. An individual profile shows the difference in bed elevation from a mean profile at a given time; the mean profile is horizontal and is defined by the means of all shore-normal records. Linear trends were removed from records prior to constructing these profiles.

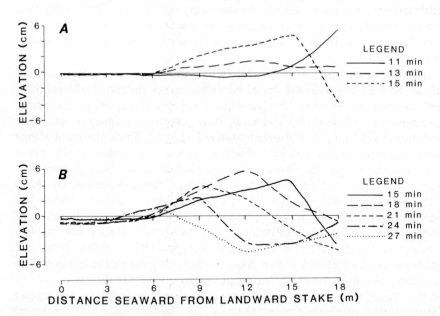

Fig.6. A. Shore-normal profiles of bed elevations showing how the crest of a sediment-level oscillation develops. B. Shore-normal profiles of bed elevations showing how a crest of a sediment-level oscillation migrates.

DISCUSSION

The type of sediment-level oscillation described above appears to be a feature common to swash zones. We have repeated our experiment at the same beach during flooding and ebbing tides and when the foreshore was undergoing net erosion and net accretion. In all of these experiments the results were similar, the features were always present and were always migrating landward. We have also conducted two smaller scale experiments on a different Monterey Bay beach. The beach was composed of much finer sand (mean ~ 0.3 mm) and was of gentler slope ($\sim 3.8°$) than the beach discussed above. The same type of feature appeared to be present in one of the experiments. In the other experiment, conducted on a day of low incident energy, the sediment level did not change appreciably. The reason that the sediment-level oscillations are not better known is probably because of their very low steepness making the oscillations difficult to detect visually. The cross-shore length for an oscillation is order 10 m (Figs.5 and 6) and height is about 6 cm (Fig.3) giving a steepness of 0.006.

The sediment-level oscillations described here are similar to those described by Waddell (1973, 1976) although there are important differences. Phase relationships between records at Waddell's two measurement locations indicated that his oscillations were migrating offshore rather than the onshore migrations observed in this study. Oscillations measured by Waddell had periods on the order of a minute whereas the oscillations observed in this study had much longer periods, in the range 6—15 min. With only two measurement locations, Waddell was unable to show, as we did in the present study, whether the migration of his oscillations involved a net sediment transport in the direction of migration.

The sediment-level oscillations were not sand waves, as suggested by Waddell, or any of the various types of swash zone morphologic features described previously. Sand waves generally have net transport in the direction of migration whereas the features observed here migrated landward while sediment was being transported seaward (Fig.6). Since the oscillations were landward progressive and had shore-normal length scales on the order of the swash-zone width, they resembled small-scale swash bars. Swash bars (called by some ridge and runnel topography) have been described by Davis et al. (1972), Owens and Frobel (1977), and others. However, swash bar migration involves a net landward sediment transport which is different from the mode of migration of oscillations described here. Backwash ripples, whose origin is discussed by Broome and Komar (1979), differ from the sediment-level oscillations described here in that they do not migrate extensively and have much shorter shore-normal length scales (50—70 cm).

When we first observed the net accretion that occurred in the upper swash zone during our experiment (Fig.3), we thought that the landward progressive oscillations may have contributed to the accretion. However, as discussed above, the migration of the sediment-level oscillations does not involve a net landward sediment transport. The observed accretion may have been related to tidal cycle sedimentation; with a rising tide the upper swash zone should accrete (see Duncan, 1964).

The origin of the sediment-level oscillations is not at all clear. As discussed in the Introduction, Waddell (1976) hypothesized that the oscillations were driven by low-frequency oscillations in ground water. Ground-water oscillations were supposedly caused by waves of the same frequency in the surf zone. However, this hypothesis cannot readily explain the observed change in frequency of sediment-level oscillations across the swash zone. Low-frequency waves may, however, be important. Using time-lapse photography, we measured time series of runup. In low-pass filtered runup records, there was some evidence of waves with periods of order 10 min. Such very low-frequency waves may be shelf waves or, since we were working in Monterey Bay, bay seiches. In a low-passed record, runup oscillations were apparent of the same frequency and 180° out of phase with sediment-level oscillations at stake 12. However, the relationship was not conclusive. Low-frequency waves may be important, but the exact processes are not clear.

CONCLUSIONS

Sediment-level oscillations with heights of 6 cm and shore-normal lengths roughly the swash zone width were measured in the swash zone of a high-energy, coarse-sand beach. The period of oscillation decreased across the swash zone ranging from 6 min below the mean swash level to 10—15 min above the mean swash level. The sediment-level oscillations were progressive landward; the oscillations in the middle to upper swash zone migrated at an average rate of roughly 0.8 m min^{-1}. The sediment-level oscillations differed from sand waves and the other swash-zone bedforms previously described in that only the form of an oscillation migrated landward. These results show that in a high-energy environment, the swash-zone profile changed its configuration in a surprisingly rapid and well-organized manner.

ACKNOWLEDGEMENTS

We thank Beth Laband, Jeff List, Bruce Jaffe, and Geof Caras for help in the field and Beth Laband and Jeff List for help in data reduction.

REFERENCES

Broome, R. and Komar, P.D., 1979. Undular hydraulic jumps and the formation of backwash ripples on beaches. Sedimentology, 26: 543—559.
Davis, R.A., Fox, W.T., Hayes, M.O. and Boothroyd, J.C., 1972. Comparison of ridge and runnel systems in tidal and non-tidal environments, J. Sediment. Petrol., 42: 413—421.
Duncan, J.R., 1964. The effects of water table and tide cycle on swash-backwash, sediment distribution, and beach profile development. Mar. Geol., 2: 186—197.
Owens, E.H. and Frobel, D.H., 1977. Ridge and runnel systems in the Magdalen Islands, Quebec. J. Sediment. Petrol., 47: 191—198.

Waddell, E., 1973. Dynamics of swash and implication to beach response. Coastal Studies Institute, Louisiana State University, Baton Rouge, La., Tech. Rep. 139, 49 pp.

Waddell, E., 1976. Swash—groundwater—beach profile interactions. In: R.L. Davis and R.L. Ethington (Editors), Beach and Nearshore Sedimentation. Soc. Econ. Paleontol. Mineral., Spec. Publ., 24: 115—125.

WAVE-FORMED STRUCTURES AND PALEOENVIRONMENTAL RECONSTRUCTION

H. EDWARD CLIFTON and JOHN R. DINGLER

U.S. Geological Survey, 345 Middlefield Road, Menlo Park, CA 94025 (U.S.A.)

(Received September 15, 1983; revised and accepted January 14, 1984)

ABSTRACT

Clifton, H.E. and Dingler, J.R., 1984. Wave-formed structures and paleoenvironmental reconstruction. In: B. Greenwood and R.A. Davis, Jr. (Editors), Hydrodynamics and Sedimentation in Wave-Dominated Coastal Environments. Mar. Geol., 60: 165—198.

Wave-formed sedimentary structures can be powerful interpretive tools because they reflect not only the velocity and direction of the oscillatory currents, but also the length of the horizontal component of orbital motion and the presence of velocity asymmetry within the flow. Several of these aspects can be related through standard wave theories to combinations of wave dimensions and water depth that have definable natural limits. For a particular grain size, threshold of particle movement and that of conversion from a rippled to flat bed indicate flow-velocity limits. The ratio of ripple spacing to grain size provides an estimate of the length of the near-bottom orbital motion. The degree of velocity asymmetry is related to the asymmetry of the bedforms, though it presently cannot be estimated with confidence. A plot of water depth versus wave height (h—H diagram) provides a convenient approach for showing the combination of wave parameters and water depths capable of generating any particular structure in sand of a given grain size. Natural limits on wave height and inferences or assumptions regarding either water depth or wave period based on geologic evidence allow refinement of the paleoenvironmental reconstruction. The assumptions and the degree of approximation involved in the different techniques impose significant constraints. Inferences based on wave-formed structures are most reliable when they are drawn in the context of other evidence such as the association of sedimentary features or progradational sequences.

INTRODUCTION

Sedimentary geologists have long sought to use depositional structures for interpreting ancient depositional environments. Quantitative analysis of paleo-processes based on such structures has proved at best only partly successful. Even where the structures can be related with reasonable precision to ancient processes, those processes commonly cannot be meaningfully incorporated into a broader environmental interpretation. For example, the flow-regime concept provides a comprehensive model for interpreting structures produced by unidirectional flow (Harms et al., 1982). Nonetheless, even where application of the concept generates specific data on such parameters as flow velocity and water depth, it commonly is unclear how these parameters contribute significantly to a paleoenvironmental reconstruction.

0025-3227/84/$03.00   © 1984 Elsevier Science Publishers B.V.

In contrast, parameters interpreted from wave-generated structures commonly can be linked clearly to important aspects of the depositional setting. Water depth, for example, which is of somewhat uncertain influence on the development of structures produced by unidirectional flow, bears in a direct, calculable way on the origin and nature of wave-formed structures.

Until recently, wave-generated sedimentary structures were poorly understood, and, consequently, they could not be used to successfully interpret depositional environments. Considerable data based on field and laboratory experiments and observations have accumulated in recent years (for example, Lofquist, 1978; Miller and Komar, 1980a, b; Dingler and Clifton, 1984, this volume), and several interpretative models have been proposed based on wave-generated structures (e.g., Allen, 1970; Tanner, 1971; Komar, 1974; Clifton, 1976; Allen, 1979, 1980, 1982; Harms et al., 1982). Although the relations between wave-generated structures and the associated fluid dynamics are still not fully understood, useful interpretations are possible. This paper summarizes the published research on wave-formed sedimentary structures and outlines a procedure whereby wave-formed structures can be used to interpret ancient depositional environments. The procedure is presented step by step, noting the physical basis for the parameters employed, and assessing the validity of the various approaches and techniques. In conclusion the procedure is applied to specific geological problems.

THE INTERPRETIVE PROCEDURE

The procedure of interpreting paleoenvironments from wave-generated structures requires three discrete steps. The first involves inferring flow parameters from specific aspects of the wave-generated structures using results from empirical investigations or experimental studies. The second step employs wave theory to determine the combinations of water depth and wave size and shape that could produce the inferred flow parameters. The third step utilizes the natural limits that exist for waves, geologic reasoning, or wave-hindcasting techniques to constrain the range of possible combinations of water depth and wave size and to relate those that are feasible to the paleoenvironment.

STEP 1. INFERRING FLOW PARAMETERS FROM WAVE-FORMED FEATURES

Wave-formed structures

Under wave action, the character of the flow and the composition of the bed (texture and mineralogy) combine to determine the general configuration of the bed (flat, hummocky, or rippled) and the size and shape of the bedforms themselves. Accordingly, aspects of size and shape of the bedforms can be used to infer previously existing flow parameters, which can in turn be applied to the interpretation of depositional environments.

Oscillation ripples are the predominant wave-generated bedforms. In profile, spacing λ, height η and symmetry β/λ characterize these ripples (Fig. 1). In

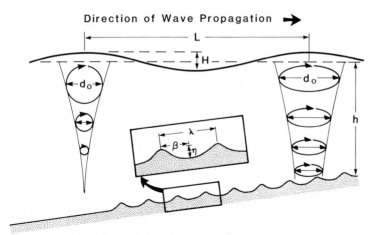

Fig.1. Geologically important parameters of waves, water motion and wave-formed ripples. Wave length (L) is the horizontal distance between successive wave crests; wave height (H) is the vertical distance between wave crest and trough; water depth (h) is the vertical distance from still water level to the seafloor; orbital diameter (d_0) is the maximum horizontal distance of excursion of water particles as a wave passes (a circular motion in deep water, an elliptical motion in shallow water); ripple spacing (λ) is the average horizontal distance between ripple crests; ripple height (η) is the average vertical distance between ripple crests and troughs; ripple asymmetry (β/λ) is the ratio between the average distance from ripple crest to leading trough (β) and the average ripple spacing (λ). Not shown: the wave period (T), the time required for successive wave crests to pass a given point; and the maximum orbital velocity (u_m), the maximum horizontal velocity in the direction of wave passage.

plan, crest length relative to spacing and crest sinuosity are primary characteristics; Inman (1957) called ripples short-crested, intermediate-crested, or long-crested if their crest-length to spacing ratio was less than 3, 3—8, or greater than 8, respectively. Crest pattern ranges from straight to sinuous; in the extreme they can take on a crescentic shape, such as the lunate megaripples of Clifton et al. (1971). Most oscillation ripples are transverse to the forming current, but a few types such as the cross ripples described by Clifton et al. (1971) are oblique to the flow.

The ratio of ripple height to wavelength η/λ is the ripple steepness; it and its inverse, the ripple index (Reineck and Singh, 1973) or vertical form index (Bucher, 1919), have been used to describe ripples (e.g., Dingler, 1974; Allen, 1980). Allen (1980) indicates that a wide range of ripple indices (steepnesses) is possible. Dingler and Inman (1977) showed that for fine sand near La Jolla, California, ripple steepness remained at a value of about 0.15 with increasing wave energy until, as sheet flow conditions were approached, the steepness decreased systematically to zero.

Symmetric ripples have a symmetry factor (β/λ) that approaches 0.5, or a ripple symmetry index [$(\lambda - \beta)/\beta$; Reineck and Singh, 1973] that approaches 1.0. The steeper side of most asymmetric ripples faces in the direction of ripple migration, making the symmetry factor less than 0.5 and the ripple symmetry index greater than 1.0. The maximum value of the ripple symmetry index acquired by asymmetric wave ripples is reportedly 3.8 (Reineck and Singh, 1973, p.27), equivalent to a symmetry factor of about 0.25.

The nature of the sediment that composes the bed is an important and potentially troublesome factor. Several different aspects of texture or composition such as mean grain size (D), sorting, shape, and particle density can bear on bedform development. The influence of mean grain size is fairly well known (Clifton, 1976), but very little is known about the effects of the other three factors. It would seem likely, however, that a bed of coquina would respond to a given type of flow differently than would a bed of quartz sand of similar mean grain size.

The recognition of structures as formed by waves is obviously critical to their use as interpretive tools. Symmetric ripples are generally accepted a priori as produced by wave activity, although the common presence of symmetrical ripples in deep-sea photographs (Heezen and Hollister, 1971, p.348) suggests other possible mechanisms. Asymmetric bedforms generated by waves may be difficult to distinguish from those formed by unidirectional currents. Tanner (1967), Reineck and Wunderlich (1968), Boersma (1970), and Reineck and Singh (1973) present criteria for recognizing wave-produced bedforms.

The problem of identifying effects of waves is further complicated in exposures where the bedforms themselves are poorly expressed. In such a case, the influence of waves must be inferred, often with difficulty, from the internal structure produced by migrating bedforms. Boersma (1970) and Allen (1982) offer a number of criteria for recognizing wave-formed ripples on the basis of internal structure. The orientation of the ripples can in some cases suggest their origin. Because of the general absence of shoreward-flowing currents, Clifton (1981) inferred that ripples that faced or migrated in a shoreward direction were solely the product of waves.

Most of the expressed relationships between wave-formed structures, flow parameters, and waves assume an absence of superimposed unidirectional current (Clifton, 1976; Allen, 1981a). Yet in natural environments, combined oscillatory and unidirectional flow, in the form of tidal, rip or longshore currents is fairly common. A few studies have described combined flow ripples (Reineck and Wunderlich, 1968; Harms, 1969; Bliven et al., 1977), but presently they cannot be used with confidence in environmental interpretation (Harms et al., 1982, pp.2—42). Distinguishing between purely oscillatory and combined-flow ripples may be very difficult.

The identification of flat bedding produced by oscillatory sheet flow can be particularly difficult. First, it may be impossible to recognize the contribution of superimposed unidirectional flow to the development of sheet flow conditions. Second, ripples that migrate across the seafloor may produce a very similar, nearly flat stratification that is analogous to the climbing translatent strata observed in eolian deposits (Hunter, 1977). Clifton (1976) suggests several criteria (mostly based on lithologic association) that may prove useful for distinguishing between sheet-flow laminae and subaqueous climbing translatent strata.

Flow parameters

The flow parameters that can be inferred from wave-generated structures are relatively straightforward. As a wave travels along the surface of the water, it sets the water particles in motion (Fig.1). If the water depth (h) is large relative to the length (L) of wave (that is, $h > L/4$), the wave form is sinusoidal and the induced water motion is essentially circular (Fig.1). The diameter of the circle (d_0) diminishes exponentially with depth, reaching zero above the bottom. In shallower water, where the wave interacts with the bottom, the wave may retain its nearly sinusoidal shape, but the water particles move in ellipses that become progressively flatter and smaller with depth (Fig.1). Just above the sea floor the elliptical motion becomes a horizontal oscillation, the length of which still is referred to as "orbital diameter". In very shallow water, just before breaking, the wave may lose its sinusoidal shape, and the water motion is nearly horizontal throughout the water column (Komar, 1976).

The velocity of the water particles, which is a critical parameter in the threshold of grain movement and in the shaping of bedforms, depends both on the magnitude of the orbital diameter and on the wave period (T). For deep-water waves (where water motion is circular), the maximum orbital velocity (u_m) equals the average orbital velocity, the circumference of the orbital motion (πd_0) divided by the time required to complete an orbit (T). In shallower water, the maximum orbital velocity above the bottom differs from the average velocity, but the relation, $u_m = \pi d_0/T$, remains valid.

As a wave approaches the shore, its form changes (Fig.2) such that the crest becomes increasingly narrow and peaked and the trough broad and flat. As the wave begins to break, it also becomes asymmetric about a vertical plane through and parallel to the crest, because its landward face steepens relative to its seaward face. These changes impart an asymmetry to the orbital motion.

Part of the physical basis for this asymmetry can be seen in Fig.2. If mass transport is assumed to be nil, the volume of water that moves forward under the crest of a wave must equal that which moves in the opposite direction under the trough. Because the crest of the wave is foreshortened relative to

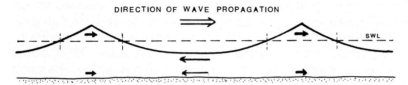

Fig.2. Typical form of a wave as it passes into shallow water. Note that the time available for movement of water in the direction of wave propagation under the crest of the wave is substantially less than that available for movement in the opposite direction under the wave trough. The result is a velocity—magnitude and velocity—time asymmetry whereby the forward motion of the water under the crest is strong but of short duration relative to the reverse motion under the trough.

the trough, water under the crest must move more rapidly to maintain mass balance. This condition causes the bottom orbital flow under the crest to be abrupt and strong relative to that under the trough. This onshore flow would be further reinforced by any shoreward mass transport.

The resulting orbital velocity asymmetry can be viewed as consisting of both a velocity-magnitude and a velocity—time component (Kemp, 1975). Velocity—magnitude asymmetry, as used here, refers to any difference between the peak or maximum velocity under the crest and trough of the wave. Velocity—time asymmetry refers to any difference between the duration of flow in the direction of wave propagation and that in the opposite direction. Figure 3 illustrates some of the conceivable velocity profiles that could be generated by shoaling waves. It should be noted that, in nature, asymmetry of flow is almost always due to a combination of velocity—magnitude and velocity—time asymmetry and is therefore complex.

The amount of water mass transport generated by asymmetric orbital motions seems variable and, under certain conditions, may be of minimal importance. The character of flow was qualitatively examined over a field of shoreward-facing lunate megaripples located seaward of the surf zone on the southern Oregon coast, using neutrally buoyant drifters, vertical streaks of dye, and clouds of sand thrown into suspension on the leeward side of the lunate megaripples (Clifton et al., 1971). In no case evidence was seen for shoreward water mass transport, even as the lunate megaripples migrated towards the shore.

In summary, orbital velocity asymmetry derives from differences in magnitude and duration of the back-and-forth components of oscillatory flow. Both aspects are important to the movement of sediment. Velocity-magnitude asymmetry is particularly important where only the stronger component exceeds the threshold velocity for movement of a given grain size (Kemp, 1975). Moreover, since bedload transport is thought to vary approximately with the third or fourth power of velocity (Inman, 1963; Wells, 1967), velocity—magnitude asymmetry may significantly influence onshore/offshore sediment transport. Net water transport is an additional factor that may be most important for the movement of suspended fine sand (Kemp, 1975).

Because of the complexities involved, an acceptable measure of velocity asymmetry is yet to be defined. Clifton's (1976) parameter Δu_m is the absolute difference in the peak orbital velocity under the crest and the trough of a wave; Kemp's parameter ν_m is the ratio between the two. Neither measure takes into account the duration of the opposing flows which must be accounted as important. Other authors (Dingler, 1974; Allen, 1979, 1980) measure orbital asymmetry in terms of the associated net drift of the water or the ratio of this drift to maximum orbital velocity. This approach does not accommodate the important influence of the velocity—magnitude asymmetry. Kemp (1975) suggests using the time—velocity curves to estimate the potential transport of a grain of a particular size. Such a process is laborious but should give the most reliable measure of the effects of orbital velocity asymmetry.

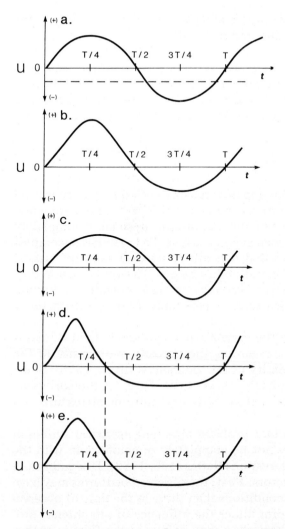

Fig.3. Possible profiles of velocity (u) over time (t) imparted by a wave of period T. Vertical axis = velocity (positive in the direction of wave approach); horizontal axis = time. (a) Neither velocity—magnitude nor velocity—time asymmetry; no net transport (typical symmetrical velocity profile under a sinusoidal wave; common in nature). (b) Velocity—magnitude asymmetry without velocity—time asymmetry; probable net transport in direction of stronger flow (not likely to occur in nature). (c) Velocity—time asymmetry without velocity—magnitude asymmetry; probable net flow in the direction of flow with longer duration (not likely to occur in nature). (d) Both velocity—magnitude and velocity—time asymmetry, balanced such that net transport is negligible (observed qualitatively over active highly asymmetric bedforms just seaward from the southern Oregon surf zone). (e) Both velocity—magnitude and velocity—time asymmetry, some net transport in direction of wave propagation. Note subtle difference from profile d (may be common under natural shoaling waves). Note that a superimposed unidirectional flow can impart both velocity—magnitude and velocity—time asymmetry to any of the profiles including profile a, where the effect can be visualized by adding a constant velocity to the curve shown (i.e., moving the curve up or down relative to the zero-velocity axis.

The velocity asymmetry induced by shoaling waves is extremely important in the sorting and transport of sand and in the development of sedimentary structures. It is, however, complex, involving both differences in the velocity components of oscillatory currents and a net transport of water. The issue can be further complicated by the presence of unidirectional flows such as rip currents or wind-driven flow, that are unrelated to the currents induced by the passing wave, but can further modify their character (e.g., Inman and Bowen, 1963).

Basis for inference of flow parameters

The basis for inferring the foregoing flow parameters from wave-generated structures lies largely in either empirical analysis of field data or experimental studies in the laboratory. Neither of these approaches produces completely satisfactory results relative to geological applications. Field studies encompass such a large number of variables that it is difficult to ascertain the critical relationships, and the spectre of metastability haunts the results. Laboratory experiments can reduce the number of variables and can generate equilibrium conditions; commonly, however, such experiments cannot satisfactorily duplicate natural conditions.

One approach to dealing with the variables encountered in field studies is to make simplifying assumptions regarding the viscosity and density of the water and the size, shape, and density of the sediment (Clifton, 1976). Some of these assumptions, unfortunately, have little basis. Although wave-winnowed sand typically is well-sorted, it is not uniform in texture and composition.

Metastability can be an important problem attending empirical studies in that the observed bedforms may not be completely in equilibrium with the processes active at the time of observation (Harms et al., 1982). Disequilibrium may result from two different factors. First, the observed bedforms may have developed under more energetic conditions than those at the time of observation and retained their initial form under the influence of less intense processes. Second, the bedforms themselves may influence the flow in such a way as to retain their original character. Commonly it is impossible to establish in the field if the observed structures are responding completely to on-going processes without prior influence.

Laboratory studies circumvent many of these problems because sand of uniform size can be used as bed material and the waves, or flow, carefully controlled. Although these experimental studies can do much to define the influence of specific flow parameters, they can duplicate only a small range of natural environmental conditions. Specifically, laboratory studies have yet to replicate conditions imposed by large, long-period oceanic waves. Moreover, certain types of experimental techniques (specifically the use of an oscillating bed) may produce misleading results (Miller and Komar, 1980a; Harms et al., 1982).

Empirical and experimental studies of wave-generated bedforms provide the basis for interpreting maximum orbital velocity, orbital diameter, wave period, and questionably, orbital velocity asymmetry. Estimates of orbital velocity are based on threshold criteria for grain movement or for sheet flow. Estimates of orbital diameter are predicated on the relation of ripple spacing or steepness to grain size. Estimates of orbital velocity asymmetry derive from the degree of asymmetry of the depositional structures.

Threshold velocities

Two threshold velocities can be defined for oscillatory flow: that required to initiate grain movement and that required to produce sheet flow. Under the oscillatory currents produced by surface gravity waves, ripples form quickly upon the initiation of grain movement (Dingler, 1974); the lower flat-bed regime that occurs in unidirectional flow appears to be largely suppressed. Bagnold (1946), Komar and Miller (1973, 1975), and Dingler (1979) are among several investigators who have studied the threshold of grain motion in oscillatory flow. The relation for the onset of grain motion under oscillatory flow resembles the Shields (1936)—Bagnold (1966) relationship for onset under unidirectional flow (Madsen and Grant, 1976; Dingler, 1979).

Threshold criteria are most accurately presented in terms of shear stress, τ, which is related to the mean velocity by the equation $\tau = f\rho u_m^2/2$ (Jonsson, 1967) where ρ is the fluid density and f is an empirically obtained friction factor. Because the friction factor is hard to determine, most investigators present threshold curves using the calculated near-bottom maximum orbital velocity. Komar and Miller (1973) defined the threshold for movement of grains smaller than 0.5 mm with the dimensionless equation:

$$\frac{\rho u_m^2}{(\rho_s - \rho)gD} = 0.21 \left(\frac{d_0}{D}\right)^{1/2} \tag{1}$$

where ρ_s is sediment density and g is the gravitational constant. For quartz sand in water, the relationship $u_m = \frac{\pi d_0}{T}$ gives:

$$u_m = 0.337(g^2 TD)^{1/3} \tag{2}$$

which, in units of centimeters and seconds is equivalent to $33.3\,(TD)^{1/3}$ cm s^{-1} (Clifton, 1976). For movement of grains coarser than 0.5 mm, Komar and Miller define threshold conditions by the dimensionless equation:

$$\frac{\rho u_m^2}{(\rho_s - \rho)gD} = 0.46\,\pi \left(\frac{d_0}{D}\right)^{1/4} \tag{3}$$

which for quartz sand in water reduces to:

$$u_m = 1.395(g^4 TD^3)^{1/7} \tag{4}$$

In units of seconds and centimeters, this is equivalent to $71.4(TD^3)^{1/7}$ cm s^{-1} (Clifton, 1976). The threshold curves of Komar and Miller (1975) for motion of sediment equivalent in density to quartz are based on eqs. 1 and 3 and shown in Fig. 4.

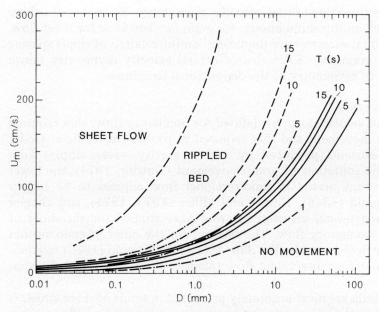

Fig.4. Velocity thresholds for grain movement and sheet flow of quartz sand in water. Solid lines are threshold curves of Komar and Miller (1975) for grain movement. Dashed lines connected by dots are threshold curves of Dingler (1979) in the range of experimental evidence; dots are absent where curves are extrapolated. Threshold curve for sheetflow, from Dingler and Inman (1977), is solid in the size range of experimental evidence and dashed where extrapolated. Note effect of differing wave period on threshold of motion.

Dingler, in a separate experimental study (1979), found that motion for grain sizes between 0.18 and 1.454 mm commenced when the dimensionless criterion:

$$\frac{(\rho_s - \rho)gT^2}{\rho D} = 240 \left(\frac{d_0}{D}\right)^{4/3} \left[\frac{\rho(\rho_s - \rho)gD^3}{\mu^2}\right]^{-1/9} \tag{5}$$

is satisfied, where μ is the fluid viscosity and the other terms as defined above. In terms of threshold velocity, eq. 5 reduces to:

$$u_m = 0.052 \left[\frac{g^5(\rho_s - \rho)^5}{\rho^4 \mu}\right]^{1/6} (TD)^{1/2} \tag{6}$$

which for quartz sand in water, in units of grams, centimeters, and seconds, is equivalent to $52.4\ (TD)^{1/2}$ cm s^{-1} (Clifton, 1976). Threshold curves based on this equation are also shown in Fig.4.

It should be noted that all of the above threshold equations show a dependence on wave period. For any particular grain size a longer period wave requires a higher velocity to initiate grain movement than does a shorter period wave. The basis for this relationship is unresolved. Possibly it derives from the more abrupt onset of flow that occurs under a shorter period wave and the gradient of stress that is associated with greater acceleration.

The threshold curves for onset of grain motion are plotted in Fig.4. In the range of fine sand, the sets of curves are fairly consistent, but they deviate markedly when extrapolated to coarser bed material. Unfortunately, it is the larger size ranges where threshold values are most useful for defining minimum possible wave size and water depth combinations.

Dingler and Inman (1977) determined that sheet flow occurs in fine sand under a relationship whereby $\rho u_m^2/(\rho_s - \rho)gD = 240$. For quartz sand in water this equation reduces to:

$$u_m = 19.9(gD)^{1/2} \tag{7}$$

which, in units of centimeters and seconds is equivalent to $623 D^{1/2}$ cm s^{-1}.

The threshold curve for sheetflow derived therefrom is shown on Fig.4. It should be noted that this curve was observed only in a narrow range of grain size (0.0128—0.0158 cm) and extrapolation beyond this range must be done with caution.

All the above threshold equations assume spherical grains of uniform size on a flat bed. Bagnold (1963) suggested that sand grains on a previously rippled bed would move at somewhat lower flow velocities, and Southard and Dingler (1971) showed that ripples under unidirectional flow could propagate downstream of a disturbance under subthreshold conditions. Hallermeier (1981) notes that the velocity required to initiate movement over a rippled bed may be half of that required for the same material on a flat bed. In the absence of a definitive study on this problem, reliance must be placed on the flat-bed thresholds noted in the foregoing.

When a range of grain sizes occurs, as is the case outside the laboratory, most people use the median or arithmetic mean diameters, which are easily calculated. Some evidence exists that the effective size for consideration of movement of poorly sorted sediment is less than the median diameter (Hallermeier, 1981). Bagnold (1966) recommended that the geometric mean diameter be used because it more realistically weights the size distribution. Inasmuch as wave-worked sands are typically well-sorted, the median diameter probably suffices.

Threshold values commonly have their greatest use in calculating the forces required to move the coarsest clasts available. In many cases, the size of these largest clasts substantially exceeds that of the bulk of the bed material. The assumption of uniformly sized particles in the foregoing equations casts doubt on their applicability to isolated large clasts on a smaller bed. Fahnestock and Haushild (1962) suggest that isolated cobbles would move under unidirectional flow as easily on a plane sand bed as on a bed of gravel. But would the threshold velocity thereby be significantly reduced? Preliminary experiments in a water-tunnel indicate that the threshold for movement of subspherical quartz grains about 1 cm in diameter on a bed of about 0.05 cm sand is not greatly less than that predicted by Komar and Miller (1975). Under the highest oscillatory velocity possible with the apparatus (85 cm s^{-1} at $T = 16$ s), the clasts remained immobile (R. August, pers. commun., 1983). According to the curves of Komar and Miller (1975), threshold velocity of such clasts under 16 s waves is slightly more than 100 cm s^{-1} (Fig.4).

The effect of grain shape on the threshold curve has not been determined. However, using the equivalent sphere diameter is satisfactory in most situations involving terrigenous detrital material. Shelly or other non-spherical biogenic material would almost certainly require different threshold equations.

Relations between ripple spacing and orbital diameter

Two disparate views exist regarding the relation of ripple spacing and orbital diameter. Inman (1957) suggested that for a given grain size, ripple spacing is directly proportional to orbital diameter until some critical maximum orbital diameter is reached, whereupon spacing becomes inversely proportional to orbital diameter, diminishes and ultimately reaches a constant intermediate value. Dingler (1974), plotting both Inman's (1957) field data and original field and laboratory measurements found a similar relationship. In contrast, Allen (1979), after plotting a large amount of existing data (mostly laboratory), found no well-defined relation between orbital diameter and ripple spacing for a given size of sand. He therefore concluded that Inman's bell-shaped curve was spurious. Miller and Komar (1980a), after analysing much of the same data, concluded that there were differences in the data sets that could be attributed to the type of laboratory device used to generate the oscillatory motions. In particular, the results of oscillating bed experiments, which dominate Allen's data, are different from water-tunnel, wave-channel, and presumably, field results. Oscillating bed experiments indicate that, for a given grain size, ripple spacing increases with orbital diameter until it reaches a maximum and then remains constant. Plots of water tunnel and wave channel experiments, in contrast, show a tendency toward the bell-shaped curve (Miller and Komar, 1980a).

Figure 5a is a dimensionless plot of λ/D against d_0/D for a number of field and laboratory studies. Oscillating bed experiments are specifically omitted. The field data tend to dominate the right side of the diagram (high d_0/D values), whereas the laboratory experiments dominate the plot at low d_0/D values. As might be expected the field data are more broadly scattered, but both sets of data show the bell-shaped relationship.

Using a similar plot that incorporated Inman's (1957) and Dingler's (1974) data, Clifton (1976) subdivided symmetric ripples into three types based on the relationship between ripple spacing and orbital diameter. This subdivision (Fig.5b) appears to be valid for the larger data set presented here in Fig.5a.

Orbital ripples are those on the left side of Fig.5b where ripple spacing is proportional to orbital diameter in the approximate relationship (Miller and Komar, 1980a):

$$\lambda = 0.65 \, d_0 \qquad (8)$$

Such ripples can form under conditions where the d_0/D ratio lies in the range of 100—3000 or more (Fig.5b). Their spacing-to-grain-size ratio (λ/D)

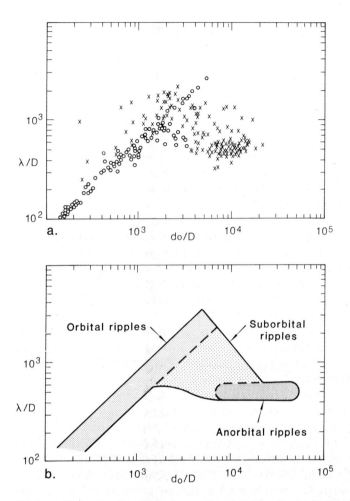

Fig.5. a. Plot of ratio of ripple spacing to grain size against ratio of orbital diameter to grain size. Data include field observations (crosses) and experimental (wave channel, water tunnel) data (circles). Field data from Inman (1957), Dingler (1974), Miller and Komar (1980b) and Dingler and Clifton (this volume). Experimental data from Carstens et al. (1969), Mogridge and Kamphuis (1972), and Miller and Komar (1980a). b. Classification of ripples based on the distribution shown in a.

ranges from less than 100 to more than 2000. Because of the requirement for short oscillatory motion, orbital ripples occur most commonly in very shallow water under short-period waves. Long-period waves can generate similarly short orbital flow at the bottom in deeper water, but because of the relationship $u_m = \pi d_0/T$, the velocity will be reduced and threshold conditions less likely to be reached. The spacing of orbital ripples tends to increase in a shoreward direction, paralleling the increase in d_0 as a wave shoals (Komar, 1974). The spacing appears to be independent of grain size.

Ripple spacing remains proportional to orbital diameter until the d_0/D ratio reaches the range of 1000–3000 (Clifton, 1976, and calculations from

Miller and Komar, 1980a). Under such conditions (Fig.5b), the ripple spacing decreases as orbital diameter increases. Ripples formed under these conditions were accordingly termed "suborbital" (Clifton, 1976). The ripple spacing appears to depend both on orbital diameter and grain size in some undefined relationship.

At d_0/D values in excess of 5000, ripple spacing stabilizes at a value that is independent of orbital diameter (Fig.5b). Termed "anorbital ripples" by Clifton (1976), such ripples are most commonly observed in fine sand where they have a spacing of 5—10 cm. Typically their λ/D ratio lies in the range of 400—600 under conditions of a single train of waves. Recent field studies indicate that, under a polymodal wave spectrum (more than one wave train present), the spacing-to-grain size ratio of anorbital ripples may be on the order of 1200 (Miller and Komar, 1980b). Anorbital ripples are probably the only type to form in fine sand under very long period (>12 s) waves (assuming a threshold velocity of 15 cm s^{-1} for sand 0.125 mm in median diameter, a 12 s wave will induce a threshold orbital diameter of nearly 60 cm, and a d_0/D ratio of 4800). Anorbital ripples include the "reversing" ripples described by Inman (1957), which alternate their direction of asymmetry with each reversal of the oscillatory flow.

An intriguing relation exists between anorbital ripples and the "maximum ripples" produced by an oscillating bed. Both have been described using the dimensional parameter $\lambda/D^{1/2} \simeq 60$ cm$^{1/2}$ (Clifton, 1976, for anorbital ripples; Bagnold, 1946, for "maximum" or "natural pitch" ripples). If this relationship is not entirely coincidental, it may provide insight into a fundamental difference between ripples formed on an oscillating bed and those formed under oscillating fluid. Both ultimately generate ripples for which the spacing is independent of orbital diameter and can be defined as $\lambda \simeq 60 D^{1/2}$ if both λ and D are in cm. Ripples on oscillating beds reach this spacing by continuously increasing their size; ripples formed by fluid motion such as those occurring in nature seem to have the potential to grow as orbital ripples beyond the size of the maximum ripple of the oscillating bed. Miller and Komar (1980a) suggest that this growth ceases at the point whereby $\lambda = 14.7 D^{1.68} \times 10^3$ (both λ and D measured in cm). Further increase in orbital diameter causes the spacing to shrink (suborbital ripples) until the "natural pitch" is achieved (anorbital ripples).

Except for reversing ripples, the relationships between λ and d_0 described in the foregoing paragraphs appear to be valid only for symmetrical ripples. The spacing of asymmetric ripples appears to follow a different pattern, which remains to be resolved (Clifton, 1976). The continuum that appears to exist between small and large symmetric bedforms is lacking for asymmetric bedforms. The marked difference in size between wave-formed lunate megaripples and associated long-crested asymmetric ripples near the high-energy surf zone (Clifton et al., 1971) suggests a discontinuity in the scale of asymmetric wave-formed bedforms similar to that within the lower regime of unidirectional flow.

Nielsen (1981) relates ripple spacing to a "mobility number" ψ (Brebner, 1981) that is equivalent to the relative stress of Komar and Miller (1972)

$(\pi d_0/T)^2 \rho/(\rho_s - \rho)gD$. Nielsen proposes that the spacing of naturally occurring ripples is best described by the equation:

$$\lambda = \frac{d_0}{2} \exp\left(\frac{693 - 0.37 \ln^8 \psi}{1000 + 0.75 \ln^7 \psi}\right) \quad (9)$$

This expression is complicated by the presence of three variables (d_0, D, T), but can be solved as a set of curves relating λ and d_0 for a specific grain size (Fig.6) or a specific wave period. As shown in Fig.6, Nielsen's equation predicts a λ/d_0 relation similar to that described by Inman (1957) and Dingler (1974) in which orbital, suborbital and anorbital ripples can be readily identified. It should be noted that on this diagram, the ripple type is no longer simply a function of d_0/D. Nielsen (1981) suggested that the spacing of ripples formed in the laboratory follows a somewhat different pattern (Fig.7) indicated by the equation:

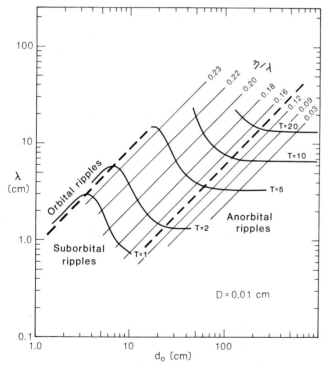

Fig.6. Plot of spacing of naturally occurring quartz sand (D = 0.01 cm) ripples against orbital diameter for waves of different period based on eq. 9 (Nielsen, 1981). The curve representing the indicated wave period is terminated on the left side of the diagram at the threshold orbital diameter (using eq. 19) and on the right side of the diagram at the maximum possible orbital diameter (using a combination of Fig.9 and eq. 16). Fields for orbital, suborbital and anorbital ripples are indicated as a function of the nature of the relation between spacing and orbital diameter. Lines of equal ripple steepness (λ/η) are drawn based on eq. 11 (Nielsen, 1981). Data set includes no waves with periods less than 6 s.

$$\lambda = \frac{d_0}{2}(2.2 - 0.345\,\psi^{0.34}) \tag{10}$$

This equation is of limited validity at larger values of ψ, where the corresponding values of λ become negative (see Fig.7). The difference between the spacing relations of naturally occurring and laboratory ripples is attributed without elaboration to the irregularity of natural waves (Nielsen, 1981).

The question of the influence of wave period on ripple spacing remains unresolved. Bagnold's (1949) statement that, in all of his experimental studies, the pitch (spacing) was independent of the speed of oscillation implies that period was not a factor. The spacing of anorbital ripples described by Miller and Komar (1980b) does not change significantly under unimodal waves of periods that ranged from 8 to 16 s. Nonetheless, as noted before, they found that the spacing of anorbital ripples under a polymodal wave spectrum was more than twice that of ripples formed in sand of the same size under unimodal waves of similar periods. On the basis of experimental evidence, Nielsen (1981) suggests that spacing depends on wave period, particularly at the shorter periods (in the range of 1—2 s). Nielsen's expression for the spacing of naturally occurring ripples (eq. 9) contains wave period as a variable, the effect of which can be seen in Fig. 6. It should be noted, however, that the data base from which Nielsen derives his expression contains no

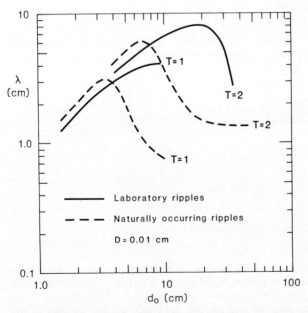

Fig.7. Comparison of spacing-orbital diameter relations for laboratory and naturally occurring ripples in quartz sand ($D = 0.01$ cm) under waves with periods of 1 and 2 s using eqs. 9 and 10 (Nielsen, 1981). Curves drawn for naturally occurring waves are suspect because of the absence of data for waves of 1 and 2 s. Curve for laboratory ripples formed under 2-s waves is suspect at the higher values of d_0 where it begins a precipitous decline.

waves with periods of less than 6 s. For waves of shorter period, the relation based on laboratory ripples (Fig. 7) may be of greater validity, particularly at the less extreme values of d_0 for waves of a given period.

In summary, field observations and laboratory experiments using wave channels and water tunnels suggest that orbital diameter can be estimated from the ratio of ripple spacing to grain size, both readily measured parameters. At λ/D values less than 400, ripple spacing seems to depend on orbital diameter in the approximate relationship $\lambda = 0.65\, d_0$ (orbital ripples). A ratio in the range of $\lambda/D = 400$—600 indicates either orbital ripples or anorbital ripples ($d_0 > 5000\, D$). Spacing-to-grain size ratios in excess of 600 suggest either orbital ripples or suborbital ripples ($d_0 = 1000$—$5000\, D$). Where the type of ripple is ambiguous (orbital or anorbital, orbital or suborbital), the complete process of interpretation may indicate which is more likely. Orbital diameter cannot presently be estimated from the spacing-to-grain size relationship of asymmetric wave-formed ripples. Variation in wave period may further complicate the interpretive process in ways that are not yet fully understood.

Relations between ripple steepness and orbital diameter

Another approach to determining orbital diameter is based on the ratio of ripple height to wavelength (η/λ) or ripple steepness. Several investigators have described three types of asymmetrical ripples based on steepness: rolling-grain ripples, vortex ripples, and post-vortex ripples. Bagnold (1946) gave the name rolling-grain ripples to low-amplitude ripples that form on flat beds under oscillatory flows just above the threshold for grain motion. Sleath (1976) and Allen (1979) apply this name to all low-amplitude, wave-generated ripples. The steepness (η/λ) of rolling-grain ripples ranges from zero to about 0.12 (a VFI or λ/η of about 8; Allen, 1979); they are too low in amplitude for a vortex to form in the ripple troughs. At least in the lower part of the ripple regime these ripples are metastable; they readily convert to vortex ripples if there are many large disturbances on the bed or if the flow velocity increases. The discussion by Miller and Komar (1980a) suggests that rolling-grain ripples are stable bedforms only on oscillating beds.

Vortex ripples occupy much of the ripple regime under oscillating fluid, extending from near the threshold of grain motion to near the onset of sheet flow (Miller and Komar, 1980). The large, sediment-laden vortex, which forms in the lee of each crest, gives these ripples their name (Bagnold, 1946). Ripple steepness is essentially constant throughout the ripple regime, having a typical value of about 0.15 (Dingler and Inman, 1977) and a range of about 0.12—0.22 (or VFI between 4.5 and 8; Bagnold, 1946).

As flow velocity increases over the vortex ripples, a point is reached where sand is stripped from the ripple crests. Orbital diameters are very large when this velocity is attained, and the ripple wavelength is unchanged by increasing flow. The net result is post-vortex ripples (Dingler and Inman, 1977) or rolling-grain ripples (Allen, 1979) that show a systematic decrease in ripple steepness from 0.15 to 0 as sheet flow conditions are approached (Dingler and Inman, 1977).

The relationship between ripple steepness and ripple spacing for given values of orbital diameter and sediment grain size is not completely clear. Several workers (Allen, 1979; Allen, 1981a, b) equate vortex ripples with orbital ripples. By this interpretation, ripples that have steepness values in the range of 0.12—0.20 can be used to calculate orbital diameter from eq. 8. This approach may be overly simplistic, however. A plot of field and wave tank data (Fig.8) indicates that vortex ripples ($\eta/\lambda > 0.1$) exist for d_0/D values of less than 5000. This plot confirms that orbital ripples ($d_0/D < 1000$) are vortex ripples, but demonstrates that the converse may not be true. Vortex ripples also form at d_0/D values of 1000—5000, where ripple spacing may be inversely related to orbital diameter. Post-vortex ripples appear to be stable only under conditions where anorbital ripples form, and therefore indicative of d_0/D values >5000.

Nielsen (1981) proposes that ripple steepness, like spacing, is a function of the mobility number ψ. Using the same data set as incorporated into Fig.7, he suggests that for naturally occurring ripples,

$$\eta/\lambda = 0.342 - 0.34(1/2\, f_w \psi)^{1/4} \tag{11}$$

where f_w is a friction factor equivalent to $\exp[5.213\,(5D/d_0)^{0.194} - 5.977]$. Curves of equal steepness for ripples in quartz sand 0.01 cm in diameter on Fig.6 also indicate that the transition from vortex to post-vortex ripples (in the range of $\eta/\lambda = 0.12$) accompanies the transition to anorbital ripples.

It should be noted that the steepness of vortex ripples may be reduced by faunal activity, compaction, or other post-depositional processes (Reineck and Wunderlich, 1968; Boersma, 1970; Allen, 1981a). Therefore, low values of ripple steepness may not necessarily reflect anorbital conditions. For

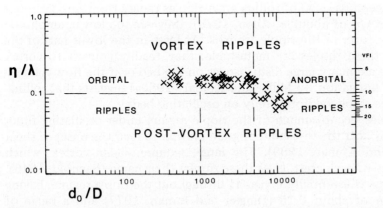

Fig.8. Plot of ripple steepness (η/λ) against the orbital diameter to grain size ratio (d_0/D), showing the relationship between vortex and post-vortex ripples (defined by ripple steepness) and orbital and anorbital ripples (defined by the d_0/D ratio). Vertical form index (VFI) scale on right side of plot. Ripples with $d_0/D < 1000$ are assumed to be orbital; those with d_0/D ratios >5000 are assumed to be anorbital. Ripples with d_0/D ratios between 1000 and 5000 are considered to be transitional between orbital and anorbital ripples (suborbital ripples). Data from Inman (1957) and Dingler (1974).

these reasons, we feel that ripple steepness by itself presently is not a reliable indicator of orbital diameter in ancient deposits.

Asymmetry of bedforms

Little is known about the degree of orbital velocity asymmetry that is required to generate asymmetric bedforms. Clifton (1976) used several lines of reasoning to suggest that a velocity—magnitude difference Δu_m of a few centimeters per second could produce asymmetry in small ripples, and Allen (1979) concluded that the degree of ripple asymmetry is proportional to the ratio of the wave-drift velocity to the near-bottom velocity maximum. This conclusion is highly tentative and virtually no data exist regarding the nature of flow that causes asymmetry of larger ripples or wave-formed lunate megaripples (which, as noted in the foregoing, have been observed to migrate in the absence of an observable wave-drift current). Tietze (1978) produced asymmetric ripples in a small experimental wave tank under measured Δu_m values between 1 and 11 cm s^{-1}. He demonstrates a relationship between the degree of ripple asymmetry and the ratio between drift velocity and maximum orbital velocity, but does not indicate where in the water column the net drift was measured. His observation that ripple asymmetry is increased by adding coarse sand to the bed strongly suggests an influence by velocity—magnitude asymmetry rather than by net drift (Kemp, 1975). Field observations show that shoreward-facing, wave-induced asymmetry of bedforms is most common in shallow water near the breaker zone (Davidson-Arnott and Greenwood, 1974; Clifton, 1976). Without a documented relationship, however, between ripple shape and some measure of flow asymmetry, quantitative estimates of the paleo-asymmetry of oscillatory flow are presently questionable.

STEP 2. ESTIMATING POSSIBLE COMBINATIONS OF WAVE PARAMETERS AND WATER DEPTHS FROM FLOW PARAMETERS

Once flow parameters, such as orbital diameter or maximum velocity, have been derived from the sedimentary structures, a range of wave conditions and water depths can be estimated. The normal complexity of the sea surface forces an investigator to undertake a great deal of simplification; commonly a "typical wave form" is identified that can be represented by mathematical equations from an appropriate wave theory. Such simple waves can be approximated in experimental studies. Spectral analysis provides a more accurate approach to describing natural waves (Dingler, 1974; Miller and Komar, 1980b; Dingler and Clifton, 1984, this volume). It is worth noting that a complicated wave field, composed of several different trains of waves can profoundly influence the development of bedforms (Clifton et al., 1971; Miller and Komar, 1980b).

Various wave theories can be used to relate flow parameters to basic wave parameters. The four most commonly cited are the: (1) Airy; (2) Stokes; (3)

cnoidal; and (4) solitary wave theories. Each is most applicable under a specific set of conditions of wave height, wave period and water depth (Fig. 9). Each has associated disadvantages and each should be considered only an approximation. A brief description of these theories is presented here; for further information, the reader is directed to the useful summary provided by Komar (1976).

Airy theory, which treats waves as sinusoidal forms, is the simplest in application. It is applicable to small amplitude waves in a wide range of water depths (relative to wave length) and provides for reasonable approximation of measured orbital diameter and near-bottom maximum velocities for real waves in shallow water (LeMehaute et al., 1969). It does not, however, provide for asymmetric flows.

The other theories noted above apply to waves with peaked crests and flattened troughs, a shoaling transformation of the sinusoidal wave. All predict asymmetric oscillatory motion. The Stokes wave theory is relatively simple, but, according to Komar (1976), becomes inaccurate for large waves when extended into shallow water. Figure 10 indicates the combinations of water depth and wave height under which the Airy and Stokes theories apply for waves of different periods using the criterion employed by Komar (1976) whereby the expression $HL^2/h^3 = 32 \pi^{2/3}$ defines the boundary between cnoidal and Airy or Stokes waves, and $H/L = 0.0625 \tanh(2\pi h/L)$ defines the boundary between Airy and Stokes waves. This figure shows, for example, that the Stokes wave theory can describe wave form and water motions for a 10 s wave in 3 m of water, provided the wave height does not exceed 1 m. It should be noted that in water depths greater than about 7 m, cnoidal wave theory does not apply regardless of wave height. At greater depths and/or

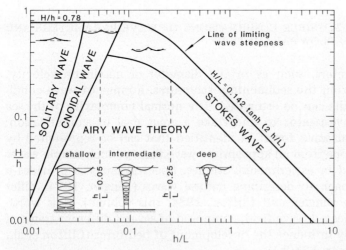

Fig.9. Conditions of wave length (L), wave height (H) and water depth (h) for which different wave theories are most applicable (from Komar, 1976). Approximate theoretical waveform shown within each field.

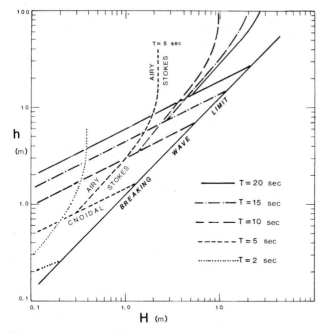

Fig.10. Areas of applicability (in terms of water depth and wave height) for Airy, Stokes, and cnoidal wave theories for waves of different period. Limiting relations shown for 5-s wave are similar for the waves of other periods. Limits between Stokes and Airy theory defined by $H/L = 0.0625 \tanh(2\pi h/L)$ and between Stokes-Airy and cnoidal theory by $HL^2/h^3 = 32\pi^{2/3}$ (Komar, 1976).

smaller waves, the less cumbersome Airy theory provides an equally valid approximation.

Cnoidal wave theory may be more accurate than Stokes theory for large waves in shallow waters (Wiegel, 1960), but its complexity severely limits its use. In many cases the Stokes or Airy theory may sufficiently approximate the water motion under conditions where cnoidal wave theory is otherwise indicated (Komar, 1976, p.62).

Solitary waves are individual progressive waves composed of a single crest. Waves very close to shore commonly resemble the solitary wave shape although such waves can be described by their wavelength and period, terms that are not appropriate for true solitary waves. Although solitary wave theory is relatively accessible, it does not truly describe periodic oscillatory motion of real wind-generated waves. This and the deviation of predicted results from measured parameters casts doubt on the use of solitary wave theory for nearshore studies (Komar, 1976, p.59).

Wave length and period

The flow parameters described in Step One (u_m, d_0, etc.) result from different combinations of wave size, shape, and water depth. The size of a wave is most readily perceived in terms of its height H and length L (Fig.1). The significant wave height (average of the highest 1/3 of the waves) is often used to represent a wave field (Dingler, 1974), even though the root-mean-square wave height is naturally associated with spectral analysis. The length of a wave is a variable parameter inasmuch as it changes significantly as a wave shoals. The wave period T, the time required for one oscillation, is related to the wave length by $T = L/C$, where C is the phase velocity (velocity of propagation) of the wave in question. Both phase velocity and wave length decrease progressively at the same rate as a wave shoals. The wave period remains unchanged and is therefore a more useful description of a wave than either length or velocity of propagation.

Airy wave theory relates the length of a given wave to its period and to the water depth h by the equation:

$$L = \frac{gT^2}{2\pi} \tanh\left(\frac{2\pi h}{L}\right) \tag{12}$$

Equation 12 can be simplified in deep water ($h/L_0 > 1/4$) where the hyperbolic term approaches unity:

$$L_0 = \frac{gT^2}{2\pi} \tag{13}$$

or, in mks units, $L_0 = 1.56\, T^2$ m. This relationship implies that, for waves of any period, there is an associated, easily calculated deep-water wave length. The deep-water wave length can be introduced into eq. 12 whereby:

$$L = L_0 \tanh(kh) \tag{14}$$

where k, the wave number, equals $2\pi/L$. Expressed this way, the hyperbolic term can be viewed as a shoaling factor that is applied to the deep-water wave length to give the wave length at any water depth. Equation 14, however, remains complicated by the presence of the unknown (L) on both sides of the equation.

This problem can be resolved by dividing both sides of eq. 14 by the water depth h and rearranging such that:

$$\frac{h}{L_0} = \frac{h}{L} \tanh(kh) \tag{15}$$

This expression has been solved by Wiegel (1954) and presented in tabular form in the Shore Protection Manual (U.S. Army Coastal Engineering Research Center, 1973). Using these tables the solution of many of the wave equations is greatly simplified. For any ratio of water depth to deep-water wave length (or, by inference, any combination of water depth and wave period), the tables indicate the corresponding values of h/L, kh, the different hyperbolic and trigonometric functions of (kh), and other parameters. Such relations allow the wave size at any water depth to be readily expressed in terms of wave height and wave period.

Wave height

Some natural limits to wave height provide useful constraints to the combination of solutions possible from the wave equations. One such limit occurs in shallow water whereby waves become unstable and break at some critical water depth. The ratio of wave height to water depth at which breaking occurs depends on the beach slope and the initial wave steepness (Iverson, 1952). A value of 0.78 has been most widely accepted for this ratio (Komar, 1976). In deeper water, waves will break if their height exceeds the value $L/7$ (Miche, 1944). Although the largest possible wind wave could theoretically exceed 65 m in height (Bascom, 1980, p.58), the largest recorded remains the 34-m wave observed from the U.S.S. "Ramapo" in 1933. Accordingly, 40 m seems a reasonable maximum height for a set of wind-generated waves. Figure 11 illustrates the maximum stable height for waves of different periods.

Orbital diameter and maximum velocity

In Airy wave theory the orbital diameter at the sea floor d_0 is:

$$d_0 = \frac{H}{\sinh(kh)} \tag{16}$$

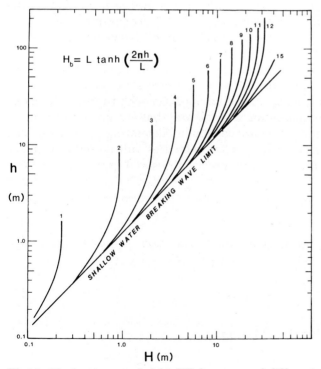

Fig.11. Maximum wave height (H) for waves of different period as a function of water depth. Curves for waves of different period terminate at approximately $h = L_0$ (= $1.56\ T^2$ m).

and the maximum orbital velocity at the seafloor u_m under the crest of the wave is:

$$u_m = \frac{\pi d_0}{T} = \frac{\pi H}{T \sinh (kh)} \tag{17}$$

The velocity under the trough of an Airy wave is of similar magnitude but opposite in direction. Although Airy theory is specifically applicable to conditions of relatively small waves in deep water, LeMehaute et al. (1969) showed that it provides a reasonable approximation to measured orbital diameters and near-bottom maximum velocities for finite amplitude waves in shallow water. Equations 16 and 17 are valuable for calculating the combinations of wave height, period, and water depth that will generate a particular wave-formed bedform or internal structure. One can recast these equations in terms of wave height

$$H = d_0 \sinh (kh) = \frac{u_m T \sinh (kh)}{\pi} \tag{18}$$

and then, using the structurally indicated value of d_0 or u_m, solve the equation for a series of selected water depths under waves of several different periods. In these calculations (shown in more detail in the first example, following), the chosen combination of wave period and water depth determines the value of h/L_0 (equal to $h/1.56\ T^2$ in meters), which then can be used to enter the wave tables to compute the appropriate value of h/L. The results can then be plotted as curves of equal wave period on an "H–h" (wave height vs. water depth) diagram. Figure 12 is an example of an H–h diagram that shows the combinations of wave height and water depth required to produce sheet flow in medium-grained sand under waves of several different periods.

One can also combine estimates of orbital diameter with those of threshold velocity to determine the maximum wave period that can produce the given combination. Longer period waves are capable of generating the same orbital diameter in deeper water, but the orbital velocities will be reduced below threshold level. Dingler (1979) combines his threshold equation for grain movement of quartz sand with the relation $u_m = \pi d_0/T$ (all measurements in cm and s) to derive the expression for maximum or threshold wave period:

$$T = 0.17 (d_0^2/D)^{1/3} s \tag{19}$$

The threshold equations of Komar and Miller (1975) provide for similar expressions of threshold wave period (in units of cm and s):

$$T = 0.17 \left(\frac{d_0^3}{D}\right)^{1/4} s \qquad \text{(for } D < 0.05 \text{ cm)} \tag{20}$$

$$T = 0.065 \left(\frac{d_0^7}{D^3}\right)^{1/8} s \qquad \text{(for } D > 0.05 \text{ cm)} \tag{21}$$

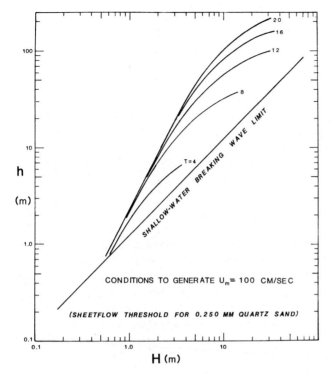

Fig.12. Combinations of wave height and water depth that will generate sheetflow of 0.250 mm quartz sand (u_m = 100 cm s^{-1}) under waves of different period. Curves for waves terminated at maximum stable wave height.

Velocity asymmetry

As noted in the foregoing section, the Airy theory cannot provide estimates of velocity asymmetry, and a higher order theory must be invoked. Stokes second-order theory provides the easiest calculable estimate of velocity asymmetry in shallow-water. Asymmetries can be calculated from the other shallow-water wave equations, but their inherent complexity generally limits their application in paleoenvironmental interpretations.

Stokes' (1847) solution for waves of finite height results in water-particle motion that is asymmetric with respect to both maximum velocity and time. The velocity of the water moving forward under the wave crests exceeds the reverse velocity under the wave trough. The duration of forward flow is less than that of the reverse flow, but the net result is an onshore migration of the water particles.

Stokes second-order wave theory provided equations both for velocity-magnitude asymmetry and for the net drift velocity of the water particles. The complete equation for near-bottom orbital velocity under the crest of a Stokes wave is:

$$u_m^c = \frac{\pi H}{T \sinh (kh)} + 3/4 \left[\frac{(\pi H)^2}{LT \sinh^4 (kh)} \right] \tag{22}$$

whereas that under the trough of the wave is:

$$u_m^t = -\left[\frac{\pi H}{T \sinh (kh)} \right] + 3/4 \left[\frac{(\pi H)^2}{LT \sinh^4 (kh)} \right] \tag{23}$$

The first term in eqs. 22 and 23 is the expression for maximum orbital velocity under an Airy wave; the second term can be viewed as a correction factor imposed by the Stokes wave. Clifton's (1976) expression for velocity magnitudes asymmetry is the sum of eqs. 22 and 23 (in effect, twice the correction factor):

$$\Delta u_m = \frac{3(\pi H)^2}{2 LT \sinh^4 (kh)} = \frac{14.8 H^2}{LT \sinh^4 (kh)} \tag{24}$$

The net drift velocity $\overline{U}$ generated by a Stokes wave is derived by integrating (over a wave period) the Stokes second-order equation for water particle displacement. The result involves assumptions (infinite channel length, constant depth, absence of viscosity) that are inappropriate for most natural situations. Accordingly, Longuet-Higgins (1953) developed a wave drift relationship for the case of a Stokes wave in a channel of finite length with a real viscous fluid. The resulting equation:

$$\overline{U} = \frac{5(\pi H)^2}{4 LT \sinh^2 (kh)} \tag{25}$$

describes a slow onshore mass transport of water due to wave passage.

Unfortunately, as noted previously, neither Δu_m nor $\overline{U}$ as defined in eqs. 24 and 25 is completely satisfactory for expressing the velocity asymmetry imparted by shoaling waves. Compounding this problem is the fact that the Stokes second-order equation can be of uncertain validity at the shallow, near-breaker-zone depths where asymmetry of flow is most important. Unfortunately, cnoidal wave theory is generally too complicated for geologic application and solitary wave theory has been shown to be unreliable under real conditions (Komar, 1976, p.59). It thus appears that, despite its obvious importance in determining bedform shape and sediment transport, there presently is no satisfactory means of quantitatively estimating velocity asymmetry.

STEP 3. DERIVING PALEOENVIRONMENT INTERPRETATIONS FROM INFERRED COMBINATIONS OF WAVE CHARACTER AND WATER DEPTH

Typically, the results of an analysis of ancient, wave-generated structures take the form of an $H-h$ diagram on which are plotted the various combinations of wave height, period and water depth that will generate a particular flow parameter. Commonly, these combinations span a broad range of water depth and wave size. To apply these data to a paleoenvironmental problem requires that the combinations be restricted to a smaller range. The natural limits to wave height described in the foregoing section provide one such restriction. Further constraints can be placed by an application of wave hindcasting techniques, geological reasoning, or a combination of the two. Wave hindcasting is a technique for estimating the height and period of waves on the basis of past speeds and duration of winds, fetch length and water depth. (Shore Protection Manual, U.S. Army Corps of Engineers Coastal Research Center, 1973). The inferred period and height of ancient waves can be thus related to basin size and wind speed and direction. Paleogeographic reconstruction may thereby impose limits on the waves, or conversely, the reconstructed waves may indicate size and shape of the depositional basin.

Geological evidence can also place constraints on water depth. Evidence for water depth may be drawn from paleoecologic analyses, the nature of associated facies, or the vertical distance to the inferred base-of-beach in a prograding shoreline deposit (see Dupré, 1984, this volume). Directional features may indicate the direction of wave approach relative to the shoreline and the presence of wave-driven currents near the shoreline. The use of geological reasoning is limited only by the availability of critical data and the resourcefulness of the investigator.

EXAMPLES

The following examples illustrate different ways in which wave-formed structures can be used to interpret aspects of the paleoenvironment. In each case the procedure is outlined in detail, noting particularly each of the three steps involved.

Example 1. Estimating wave size from threshold velocity

This example has been published (Hunter and Clifton, 1982) but is included here for analysis and to illustrate the procedure. The problem is determining whether or not storm waves were involved in the formation of hummocky cross-stratification in sandstone of Late Cretaceous age exposed at Cape Sebastian on the coast of southern Oregon. The hummocky cross-stratification occurs in the lower part of sediment cycles typically tens of centimeters thick. The hummocky cross-stratified sandstone overlies an erosional surface and grades upward into horizontally stratified sandstone that in turn grades upward into thoroughly bioturbated sandstone. The cycles are composed of slightly graded fine sand. Small pebbles are scattered in the hummocky cross-stratified sandstone and a few lie within the overlying horizontally stratified sand. Exposures of bedding surfaces of the horizontally stratified sandstone commonly show straight-crested symmetrical ripple marks.

The largest pebble, about 5 cm in diameter, found in the horizontally stratified sandstone forms the basis for the calculations presented here. The common presence of symmetrical ripple marks in this facies implies the presence of waves, which suggests the possibility of using the threshold for movement to calculate maximum orbital velocity.

Step 1. The threshold velocity for moving a 5 cm pebble can be estimated from Fig.4. The curves of Komar and Miller (1975) provide a more conservative value at wave periods larger than 5 s and are therefore employed. These curves indicate an orbital velocity on the order of 200 cm s^{-1}. It should be noted that the fact that the pebble moved on a bed of fine sand rather than on a bed of similarly sized pebbles (assumed for the threshold curves) introduces a measure of uncertainty.

Step 2. To determine the combination of wave size and water depth that would produce the orbital velocity derived in Step 1 requires selection of the appropriate wave theory. Since asymmetry of flow is not involved, Airy theory should provide for reasonable calculations based on the indicated maximum orbital velocity using eq. 18. A table can subsequently be constructed to determine the wave height that at given water depths will generate an orbital velocity of 200 cm s^{-1}. Table I, for example, indicates the wave heights that will generate this orbital velocity at a variety of water depths for a 12-s wave. Similar tables can be constructed for waves of other periods and the results plotted on an $H-h$ diagram (Fig.13) that indicates the combination of water depth and wave height at which waves of several specific periods will generate a near-bottom maximum orbital velocity of 200 cm s^{-1}. Note that the curves for the waves of smaller period are terminated at their maximum stable wave height.

Step 3. Figure 13 indicates that the specified orbital velocity of 200 cm s^{-1} can be generated by waves of widely varying size. Constraints, however, can be imposed by geologic reasoning. The absence of unidirectional crossbedding and shoreline progradational sequences from the cyclic part of the Upper Cretaceous sandstone at Cape Sebastian led Hunter and Clifton (1982) to conclude that deposition did not occur in very shallow water close to a

TABLE I

Computation of wave heights (H) required to generate u_m = 2.0 m s^{-1} at various water depths (h) under a 12 s wave, using Airy wave theory (eq. 18). L_0 = 1.56 T^2m = 225 m; Hm = 0.142 L_0 = 32 m

h[1]	h/L_0[2]	$\sinh\left(\dfrac{2\pi h}{L}\right)$[3]	H[4]
5	0.022	0.390	3.0
10	0.044	0.579	4.4
20	0.089	0.922	7.0
30	0.133	1.27	9.7
50	0.222	2.21	17.0
70	0.311	3.72	28.0
100	0.444	8.27	63.0

[1]Selected arbitrarily; [2]calculated from indicated values; [3]computed from Wave Data tables (U.S. Army Coastal Engineering Research Center, 1973); [4]calculated from eq. 18.

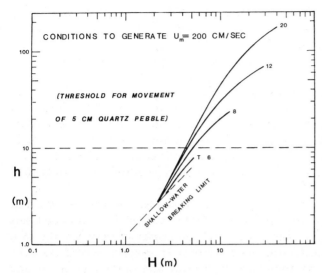

Fig.13. Combinations of wave height and water depth at which a 5-cm quartz pebble will move (estimated $u_m = 200$ cm s^{-1}) under waves of different period. Curves of waves terminated at maximum stable wave height.

shoreline. Accordingly, they felt that a depth of 10 m was a conservative minimum for deposition of this part of the unit. Applying this limit to Fig.13 eliminates the shorter period waves and implies wave heights of at least 4 m. Since these conditions presumably prevailed during the waning phase of the event that created the sedimentary cycle, even larger waves attended the development of the hummocky cross-stratified sandstone.

Example 2. Determining depositional environment on the basis of grain size and ripple spacing

The Berea Sandstone is a paralic deposit of Mississippian Age that crops out in a north-south belt across the state of Ohio (Pepper et al., 1954; Coogan et al., 1982). In exposures southeast of the city of Cleveland the unit consists of fine sandstone in which sets of trough crossbedding 1—3 m thick alternate with thin (typically less than 1 m thick) sequences of flat-bedded or wavy-bedded sandstone. Bedding surfaces of the flat- or wavy-bedded sandstone commonly exhibit small symmetrical ripple marks. The crossbeds dip to the west and northwest, presumably an onshore direction, and channelling is conspicuously uncommon (Coogan et al., 1982). The deposit has been variously attributed to accumulation in a subaerial delta, river system, barrier, bar, lagoon, wind-tidal flat or tidal channel system (Coogan et al., 1982).

Step 1. The symmetrical ripples prove particularly useful in resolving the depositional environment of this unit. The smallest ripples, with a spacing of 1.5 cm, were composed of sand of about 0.01 cm diameter. The λ/D ratio of 150 indicates that the ripples are orbital ripples ($\lambda = 0.65\, d_0$) formed under an orbital diameter of 2.3 cm. This interpretation is consistent with $\lambda - d_0$ relations predicted by Nielsen's equation for laboratory ripples (Fig.7), which may be more applicable than that for naturally occurring ripples at low values of λ. Equations 19 and 20 indicate maximum wave periods of 1.3 and 1.0 s, respectively, for forming these ripples.

Step 2. The combinations of water depth and wave height whereby a wave with a period of 1.3 s (the more conservative at the indicated values) will generate a near-bottom orbital diameter of 2.3 cm can be estimated using Airy wave theory (eq. 18, this paper). Following the procedure outlined in the previous example, an $H-h$ diagram can be constructed (Fig.14), which indicates that the observed ripples formed in very shallow water (less than 2 m deep).

Step 3. The inference that the ripple-marked sandstone was deposited at water depths smaller than the thickness of the intercalated cross-bed sets imposes special constraints. The nature of the contacts between the two lithologies becomes critical to the interpretation. In every observed case, the planar- or wavy-bedded sandstone sharply overlies the sandstone below and extends gradationally into overlying cross-bedding foresets above. This relation implies that the bedforms that produced the crossbedding migrated over the planar- or wavy-bedded sandstone with the small ripples. The truncated top to the cross-beds indicates that the bedform relief exceeded the thickness of the cross-bedding unit, which in turn generally is greater than the water depth implied by the small symmetric ripples. The implication that the height of the bedforms exceeded (perhaps substantially) the depth of water into which they migrated and the scarcity of current ripples relative to symmetric ripples effectively eliminate a subaqueous origin for the crossbedding. The bedforms are best interpreted as aeolian dunes that migrated over a surface that at least part of the time was the site of interdune ponds.

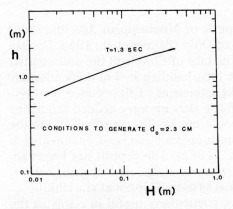

Fig.14. Combinations of water depth and wave height whereby a wave with a period of 1.3 s will generate a near-bottom orbital diameter of 2.3 cm (threshold for forming orbital ripples with a spacing of 1.5 cm in 0.100 mm quartz sand). Wave curve terminated at maximum stable wave height.

Further examination of the deposit supports this interpretation. The wide variability of the trend of the ripple marks (sets on the same bedding surface may diverge by nearly 90°) suggests local winds blowing over very shallow water. Mud cracks on a few of the ripple surfaces are consistent with a pond environment. The crossbedded sand locally shows subtle bedding features (climbing translatent strata, sandflow tongues) attributed to aeolian deposition (Hunter, 1977, 1981).

Other examples. Several recent papers use wave-generated structures to reconstruct paleoenvironments. Allen (1981) uses an approach similar to that analysed in Example 2 to estimate the size and depth of a Devonian lake on southeast Shetland. Clifton (1981) draws heavily on the orientation of wave-formed structures to develop a detailed reconstruction of a Miocene shoreline in the southern Coast Range of California. Papers by Allen and Dupré in this volume provide additional examples.

CONCLUSIONS

Wave-formed sedimentary structures can be a valuable tool for interpreting paleo-environments. A number of caveats should be noted relative to the implementation of these tools:

(1) The results are approximations. The use of wave equations typically gives precise solutions, but the equations themselves are approximations in a real environment. The indicated values should be considered as reasonable estimates only.

(2) Much uncertainty exists regarding the nature of the structures and the hydrodynamic processes involved in their formation. Threshold equations should be extrapolated with care, particularly to the coarser grain sizes. The threshold of movement of large isolated clasts on a bed of smaller material requires further study. The relations between ripple spacing, grain size and orbital diameter need to be defined more closely particularly in terms of the influence of wave period. The basis for asymmetry of wave-formed structures is controversial and awaits further study before being applicable in any hydrodynamic sense.

(3) The interpretation of wave-formed structures is best done in conjunction with other geologic evidence. Inferences of water depth or wave size based on wave-formed features are most credible when supported by other observations.

Much further field observation and experimental study must be done before wave-formed structures reach their full potential for paleoenvironmental interpretation. Presently, they can provide powerful, but often tantalizing, clues regarding ancient environmental conditions.

REFERENCES

Allen, J.R.L., 1970. Physical Processes of Sedimentation. Allen and Unwin, London, 248 pp.

Allen, J.R.L., 1979. A model for the interpretation of wave ripple-marks using their wavelength, textural composition, and shape. J. Geol. Soc. London, 136: 673—682.
Allen, J.R.L., 1980. Sand waves: a model of origin and internal structure. Sediment. Geol., 26: 281—328.
Allen, J.R.L., 1982. Sedimentary Structures — Their Character and Physical Basis, Vol. 1. (Developments in Sedimentology, 30A). Elsevier, Amsterdam, 593 pp.
Allen, P.A., 1981a. Some guidelines in reconstructing ancient sea conditions from wave ripples. Mar. Geol., 43: M59—M67.
Allen, P.A., 1981b. Wave-generated structures in the Devonian lacustrine sediments of south-east Shetland and ancient wave conditions. Sedimentology, 28: 369—379.
Allen, P.A., 1984. Miocene waves and tides from the Swiss Molasse. In: B. Greenwood and R.A. Davis, Jr. (Editors), Hydrodynamics and Sedimentation in Wave-Dominated Coastal Environments. Mar. Geol., 60: 455—473 (this volume).
Bagnold, R.A., 1946. Motion of waves in shallow water. Interaction of waves and sand bottoms. Proc. R. Soc. London, Ser. A, 187: 1—16.
Bagnold, R.A., 1963. Mechanics of marine sedimentation. In: M.N. Hill (Editor), The Sea, Vol. 3. Wiley-Interscience, New York, N.Y., pp. 507—528.
Bagnold, R.A., 1966. An approach to the sediment transport problem from general physics. U.S. Geol. Surv., Prof. Pap., 422-I.
Bascom, W.A., 1980. Waves and Beaches — Revised and updated. Anchor Press/Doubleday, Garden City, N.Y., 366 pp.
Bliven, L., Huang, N.E. and Janowitz, G.S., 1977. An experimental investigation of some combined flow sediment transport phenomena. North Carolina State Univ., Center for Marine and Coastal Studies, Rept. No. 77-3.
Boersma, J.R., 1970. Distinguishing features of wave-ripple cross-stratification and morphology. Thesis, University of Utrecht, Utrecht, 65 pp.
Brebner, A., 1981. Sand bedform lengths under oscillatory motion. Proc. 7th Conf. Coastal Engineering, Sydney, N.S.W., pp.1340—1341.
Bucher, W.H., 1919. On ripples and related sediment surface forms and their paleogeographical interpretations. Am. J. Sci., 47: 149—210, 241—269.
Clifton, H.E., 1976. Wave-generated structures — a conceptual model. In: R.A. Davis and R.L. Ethington (Editors), Beach and Nearshore Processes. Soc. Econ. Paleontol. Mineral., Spec. Publ., 24: 126—148.
Clifton, H.E., 1981. Progradational sequences in Miocene shoreline deposits, southeastern Caliente Range, California. J. Sediment. Petrol., 51: 165—184.
Clifton, H.E., Hunter, R.E. and Phillips, R.L., 1971. Depositional structures and processes in the high-energy nonbarred nearshore. J. Sediment. Petrol., 41: 651—670.
Coogan, A.H., Heimlich, R.A., Malcuit, R.J., Bork, K.B. and Lewis, T.L., 1982. Early Mississippian deltaic sedimentation in central and northeastern Ohio. In: T.G. Roberts (Editor), G.S.A. Cincinnati '81 Field Trip Guidebooks, Vol. 1: Stratigraphy, Sedimentology. Am. Geol. Inst., Falls Church, Va., pp.113—152.
Davidson-Arnott, R.G.D. and Greenwood, B., 1974. Bedforms and structures associated with bar topography in the shallow-water wave environment. J. Sediment. Petrol., 44: 698—704.
Dingler, J.R., 1974. Wave-formed ripples in nearshore sands. Thesis, Univ. Calif. at San Diego, San Diego, Calif., 136 pp.
Dingler, J.R., 1979. The threshold of grain motion under oscillatory flow in a laboratory wave channel. J. Sediment. Petrol., 49: 287—294.
Dingler, J.R. and Clifton, H.E., 1984. Tidal-cycle changes in oscillation ripples on the inner part of an estuarine sand flat. In: B. Greenwood and R.A. Davis, Jr. (Editors), Hydrodynamics and Sedimentation in Wave-Dominated Coastal Environments. Mar. Geol., 60: 219—233 (this volume).
Dingler, J.R. and Inman, D.L., 1977. Wave-formed ripples in nearshore sands. Proc. 15th Conf. on Coastal Engineering, Honolulu, Hawaii, pp.2109—2126.

Dupré, W.R., 1984. Reconstruction of paleo-wave conditions from Pleistocene marine terrace deposits, Monterey Bay, California. In: B. Greenwood and R.A. Davis, Jr. (Editors), Hydrodynamics and Sedimentation in Wave-Dominated Coastal Environments. Mar. Geol., 60: 435—454 (this volume).

Fahnestock, R.K. and Haushild, W.L., 1962. Flume studies of the transport of pebbles and cobbles on a sand bed. Geol. Soc. Am., 73: 1431—1436.

Hallermeier, R.J., 1981. Critical wave conditions for sand motion initiation. Coastal Eng. Tech. Aid No. 81-10, U.S. Army Corps Engineers Coastal Eng. Res. Center, 16 pp.

Harms, J.C., 1969. Hydraulic significance of some sand ripples. Geol. Soc. Am. Bull., 80: 363—396.

Harms, J.C., Southard, J.B. and Walker, R.G., 1982. Structures and sequences in clastic rocks. S.E.P.M. Short Course No. 9, Soc. Econ. Paleontol. Mineral., Tulsa, Okla., 249 pp.

Heezen, B.C. and Hollister, C.D., 1971. The Face of the Deep. Oxford Univ. Press, New York, N.Y., 659 pp.

Hunter, R.E., 1977. Basic types of stratification in small aeolian dunes. Sedimentology, 24: 361—387.

Hunter, R.E., 1981. Stratification styles in eolian sandstones: some Pennsylvanian to Jurassic examples from the western interior, U.S.A. In: F.G. Ethridge and R.M. Flores (Editors), Recent and Ancient Nonmarine Depositional Environments: Models for Exploration. Soc. Econ. Paleontol. Mineral., Spec. Publ., 31: 315—329.

Hunter, R.E. and Clifton, H.E., 1982. Cyclic deposits and hummocky cross-stratification of probable storm origin in Upper Cretaceous area, southwestern Oregon. J. Sediment. Petrol., 52: 127—146.

Inman, D.L., 1957. Wave generated ripples in nearshore sands. Tech. Memo. Beach Erosion. Bd. U.S. Army Corps of Engineers, 100.

Inman, D.L., 1963. Ocean waves and associated currents. In: F.P. Shepard (Editor), Submarine Geology (2nd ed.). Harper and Row, New York, N.Y., pp.49—81.

Inman, D.L. and Bowen, A.J., 1963. Flume experiments on sand transport by waves and currents. Proc. 8th Conf. Coastal Engineering, Berkeley, Calif., pp.137—150.

Iverson, H.W., 1952. Studies of wave transformation in shoaling water, including breaking. Natl. Bur. Stand., Circ., 521: 9—32.

Jonsson, I.G., 1967. Wave boundary layers and friction factors. Proc. 10th Conf. Coastal Engineering, Tokyo, pp.127—148.

Kemp, P.H., 1975. Wave asymmetry in the nearshore zone and breaker area. In: J. Hails and A. Carr (Editors), Nearshore Sediment Dynamics and Sedimentation. Wiley, New York, N.Y., pp.47—65.

Komar, P.D., 1974. Oscillatory ripple marks and the evaluation of ancient wave conditions and environments. J. Sediment. Petrol., 44: 169—180.

Komar, P.D., 1976. Beach Processes and Sedimentation. Prentice-Hall, Englewood Cliffs, N.J., 417 pp.

Komar, P.D. and Miller, M.C., 1973. The threshold of movement under oscillatory water waves. J. Sediment. Petrol., 43: 1101—1110.

Komar, P.D. and Miller, M.C., 1975. The initiation of oscillatory ripple marks and the development of plane-bed at high shear stresses under waves. J. Sediment. Petrol., 45: 697—703.

LeMehaute, B., Divoky, D. and Lin, A., 1969. Shallow water waves: a comparison of theories and experiments. Proc. 11th Conf. Coastal Engineering, London, pp.86—96.

Lofquist, K.E.G., 1978. Sand ripple growth in an oscillatory-flow water tunnel. Tech. Pap. Coastal Eng. Res. Center, U.S. Army Corps Eng., 75-8.

Longuet-Higgins, M.S., 1953. Mass transport in water waves. Philos. Trans. R. Soc. London, Ser. A, 245: 535—581.

Madsen, O.S. and Grant, W.D., 1976. Sediment transport in the coastal environment. Rept. no. 209, Ralph M. Parsons Lab., Mass. Inst. Tech., 105 pp.

Miche, R., 1944. Undulatory movements of the sea in constant and decreasing depth. Ann de Ponts et Chaussées, May—June, July—August, pp.25—78, 131—164, 369—406.

Miller, M.C. and Komar, P.D., 1980a. Oscillation sand ripples generated by laboratory apparatus. J. Sediment. Petrol., 50: 173—182.

Miller, M.C. and Komar, P.D., 1980b. A field investigation of the relationship between oscillation ripples spacing and the near-bottom water orbital motions. J. Sediment. Petrol., 50: 183—191.

Nielsen, P., 1981. Dynamics and geometry of wave-generated ripples. J. Geophys. Res., 86: 6467—6472.

Pepper, J.F., DeWitt, W. and Desmarest, D.F., 1954. Geology of the Bedford shale and Berea sandstone in the Appalachian Basin. U.S. Geol. Surv., Prof. Pap., 259, 106 pp.

Reineck, H.E. and Singh, I.B., 1973. Depositional Sedimentary Environments. Springer, New York, N.Y., 439 pp.

Reineck, H.E. and Wunderlich, F., 1968. Zur Unterscheidung von asymmetrischen Oszillationrippeln und Strömungsrippeln. Senkenbergiana Lethaea, 49: 321—345.

Shields, A., 1936. Anwendung der Ähnlichkeitsmechanik und der Turbulenzforschung auf die Geschiebebewegung. Mitt. Preuss. Versuchsanst. Wasserbau Schiffbau, 26, 20 pp.

Sleath, J.F.A., 1975. A contribution to the study of vortex ripples. J. Hydraul. Res., 13: 315—328.

Sleath, J.F.A., 1976. On rolling-grain ripples. J. Hydraul. Res., 14: 69—81.

Southard, J.B. and Dingler, J.R., 1971. Flume study of ripple propagation behind mounds on flat sand beds. Sedimentology, 16: 251—263.

Stokes, G.G., 1847. On the theory of oscillatory waves. Trans. Cambridge Philos. Soc., 8: 441.

Tanner, W.F., 1967. Ripple mark indices and their uses. Sedimentology, 9: 89—104.

Tanner, W.F., 1971. Numerical estimates of ancient waves, water depth and fetch. Sedimentology, 16: 71—88.

Tietze, K.W., 1978. Zur Geometrie von Wellenrippeln in Sanden unterschiedlicher Korngrösse. Geol. Rundsch., 67: 1016—1033.

U.S. Army Coastal Engineering Research Center, 1973. Shore Protection Manual, 3 vols.

Wells, D.R., 1967. Beach equilibrium and second-order wave theory. J. Geophys. Res., 72: 497—504.

Wiegel, R.L., 1954. Gravity Wave, Tables of Functions. Council on Wave Research, Eng. Found. Univ. California, Berkeley, Calif., 30 pp.

Wiegel, R.L., 1960. A presentation of cnoidal wave theory for practical application. J. Fluid Mech., 7: 273—286.

BOUNDARY ROUGHNESS AND BEDFORMS IN THE SURF ZONE

DOUGLAS J. SHERMAN* and BRIAN GREENWOOD

Department of Geography, Scarborough Campus, University of Toronto, Scarborough, Ont. M1C 1A4 (Canada)
Departments of Geography and Geology, Scarborough Campus, University of Toronto, Scarborough, Ont. M1C 1A4 (Canada)

(Received April 16, 1983; revised and accepted August 31, 1983)

ABSTRACT

Sherman, D.J. and Greenwood, B., 1984. Boundary roughness and bedforms in the surf zone. In: B. Greenwood and R.A. Davis, Jr. (Editors), Hydrodynamics and Sedimentation in Wave-Dominated Coastal Environments. Mar. Geol., 60: 199—218.

Hydrodynamical models of the nearshore system frequently assume that a single friction coefficient is sufficient to represent flow conditions at a point in the surf zone. Furthermore, models attempting to relate bed configuration to surf zone flows have relied primarily upon the wave orbital velocity as an indicator of potential bedforms, and thus as the control on boundary roughness. The data presented here point out potential errors arising from either of these approaches. The results of a field experiment conducted at Wendake Beach, Ontario, show that at a single location in an active surf zone, the Darcy-Weisbach friction coefficient, f, varied by approximately 250% (in this case between 0.016 and 0.041).

It is also shown that existing bedform models, based upon primary wave motions alone, do not accurately predict conditions at this study site. For a relatively constant wave orbital velocity and velocity asymmetry, it is found that changes in bed roughness, as a result of bedform development, are reflected mainly in the vertical profile of the longshore current velocity. A sequence of bedforms, from oscillatory ripples through flat bed, is inferred from the data, and found to be supported by diver observations and preserved primary sedimentary structures.

INTRODUCTION

Considerable time and effort has been expended on research into the nature of the complex interactions at the fluid—sediment interface, especially concerning the development of bedforms. These interactions are important to both the accurate modeling of modern prototype fluid systems, and the analysis of bedding genesis and interpretation of sedimentary environments. However, in nearshore systems dominated by wave-generated oscillatory flows and wave-induced, quasi-steady currents, our knowledge of fluid—

---

*Present address: Department of Ocean Engineering, Woods Hole Oceanographic Institution, Woods Hole, MA 02543, U.S.A.

0025-3227/84/$03.00 © 1984 Elsevier Science Publishers B.V.

sediment interaction remains rudimentary. According to Allen (1982, Vol. I, p.444), "Little attempt has been made to define the existance fields of bedforms in oscillatory flows", and about wave-current ripples he goes on to say (1982, Vol. I, p.448), "These forms are less well known and understood than their more symmetrical relatives".

To a large extent, this lack of understanding can be attributed to the paucity of prototype data describing these bedforms. Although considerable advances have been made in the last decade or so in formulating conceptual models of bedform sequences in oscillatory flows (e.g., Clifton, 1976; Davidson-Arnott and Greenwood, 1974, 1976; Allen, 1982; Clifton and Dingler, 1984) there is still relatively little empirical evidence to support these models (see, for example, Clifton et al., 1971; Davidson-Arnott and Greenwood, 1976; or Dabrio and Polo, 1981, for some prototype results). Thus, a primary motivation for this research was to improve the data base, attempt to determine the sequential development of bedforms through a storm, and recognize the key fluid—sediment relationships. Toward this end, a field experiment was conducted at Wendake Beach, Ontario, in May and June, 1980.

STUDY SITE AND EXPERIMENTAL DESIGN

The Wendake Beach study site was located on the southeastern shore of Nottawasaga Bay, in Lake Huron (Fig.1). The nearshore slope is gentle ($\simeq 0.015$), with three low-amplitude, nearshore bars. The mean sediment diameter by weight frequency distribution, D, is 0.21 mm. Additional details of this site are presented in Randall (1982), Sherman (1982) and Greenwood and Sherman (1983, and 1984, this volume). For this field experiment, six bi-directional, fast-response electromagnetic water current meters (Marsh-McBirney models 511 and 512) were deployed in two vertical arrays (of three current meters each) along the edges of the outer nearshore trough. In vertical array 1 (VA1), about 50 m from the still water line, the current meter elevations were 0.10, 0.60, and 1.00 m above the bed, and in vertical array 2 (VA2), about 105 m offshore, the installation elevations were 0.10, 1.00, and 1.45 m (Fig.2). Water depths at the two arrays were 1.60 and 1.70 m, respectively. The purpose of the vertical arrays was to measure an assumed logarithmic velocity profile in the longshore current. It has long been recognized that the structure of these vertical profiles should reflect the magnitude of boundary roughness (e.g., Nikuradse, 1933; Wooding et al., 1973), although in the nearshore there are complications arising primarily from the presence of the wave boundary layer (Grant and Madsen, 1979) and near-bed sediment transport (Smith and McLean, 1977; Grant and Madsen, 1982). Stratification due to temperature or suspended sediments was not considered to be a factor at this site. There was occasional direct observation of bed configuration during the storm by divers, and indirect monitoring through post-storm box core samples taken at the vertical arrays. The evidence from these measurements and observations allow at least a qualitative description of the nearshore flow regime.

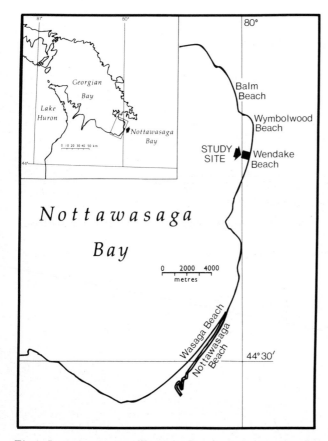

Fig.1. Location map of Wendake Beach study site.

Nine sets of measurements were obtained from the current meters during a storm that occurred on June 8, 1980, with specific record lengths of either nine or eighteen minutes, and with sampling at approximately 0.5 s intervals, or less. Thus each file for each current meter comprises about either 1000 or 3000 samples.

FLOW STRUCTURE AND BOUNDARY ROUGHNESS

Mean flow conditions through the storm were determined in the first stage of data analysis. From each of the vertical arrays, all of the longshore current velocity measurements from the set of three current meters were averaged to obtain a depth-integrated, mean velocity at that location. The velocities at each elevation were then averaged and compared to the depth-integrated mean. These results are illustrated in Fig.3. The forms of these profiles indicate the presence of the type of near bed deformation expected, although the anomalous surface decrease in velocity is not readily explainable. Also note that the relatively small amount of vertical variability (plus or minus

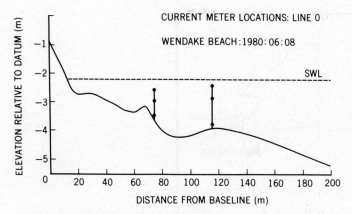

Fig.2. Current meter locations in the vertical arrays, 1980:06:08. *SWL* is still water level.

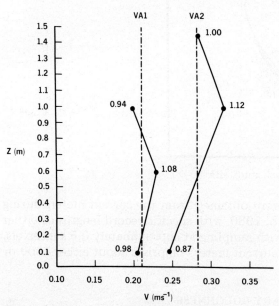

Fig.3. Relative longshore current profiles through a storm, 1980:06:08. Vertical dashed lines at *VA1* and *VA2* represent depth integrated mean current velocities. Points are mean velocities relative to the array mean at different elevations above the bed and the associated numbers are the ratios of point velocities to the depth-integrated average.

about 10% of the overall average) supports the general longshore current modeling assumption of velocity homogeneity through the water column (see discussion in Basco, 1982). Nevertheless, the two bottom current meters in VA2 consistently produced results that were consistent with the presence of a logarithmic boundary layer. The data from VA1 do not fit the logarithmic model because the velocity gradient between the two elevations is too steep to yield physically plausible boundary roughness lengths. This is believed to reflect, in part, the relative insensitivities of the two lower current meters in VA1 (model 511). These instruments have a calibration precision of only 3 cm s^{-1}, whereas the other four current meters (model 512) calibrate at ±0.3 cm s^{-1}. These terms do not include the ±10% measurement errors of the instruments. Thus these data are suspect and not considered further.

Using the 0.10 and 1.00 m current meters (CM6 and CM4, respectively) in VA2, the apparent roughness length z_o, as felt by the longshore current, may be derived using analytic geometry:

$$\ln z_o = \frac{(V_b \cdot \ln z_a) - (V_a \cdot \ln z_b)}{V_b - V_a} \quad (1)$$

where V and z refer to the longshore current velocity and elevation above the bed, respectively, and the subscripts a and b refer to values for the lower and upper current meters, respectively.

The first analytic complication arises in the determination of the elevation for the current meter sensors. Although they were originally installed at 0.10 and 1.00 m above the bed, it is known, through data obtained from locally emplaced, depth-of-disturbance rods (Greenwood et al., 1980; Greenwood and Hale, 1980) and box core data, that the bed elevation decreased a maximum of 0.16 m through the storm. Thus, at some stage of the storm, presumably at its peak, the elevations of the lower current meters, CM6 and CM4, were 0.26 and 1.26 m above the bed, respectively. These changes in the values of z have important effects in the derivation of z_o. For example, the natural log of 10 cm is 2.30, whereas ln 26 cm is 3.26. This change in ln z can result in apparent changes in the roughness length of several orders of magnitude. It was therefore decided to attempt to model bed elevation change through the storm based upon several assumptions.

First it must be assumed that the maximum amount of bed depression is coincident with the largest longshore current velocities. For the June 8 storm, this velocity was 0.55 m s^{-1} at 1130 h (Eastern Daylight Saving Time). Thus, at this time it is assumed that the elevation of CM4 is 1.26 m above the bed. It is also, somewhat arbitrarily, assumed that bed depression begins when the longshore current velocity exceeds 0.10 m s^{-1}. Thus, when the current velocity is less than 0.10 m s^{-1} the bed elevation is a constant. Above 0.10 m s^{-1}, the bed elevation is assumed to vary as a function of the square of the velocity. This assumption is based upon bed shear stress increasing solely with V^2 and ignoring potential changes associated with wave effects.

Given the above qualifications, changes in the value for z for each instrument were calculated through the storm. These changes are shown in Fig.4. Of the ten values shown, three are measured; the two end points and the maximum depression. The result of the predictions for the other values is a reasonable sequence of elevations through the storm. Thus values for both the mean longshore current velocities (averaged over a record length) and the

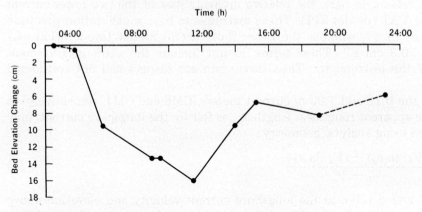

Fig.4. Predicted sequence of bed elevation changes through the storm.

projected height above the bed are available for each of the current meters. These values are substituted, in turn, into eq.1 to calculate the apparent roughness lengths at VA2 through the storm.

An independent estimate of the minimum physical roughness of the bed may be obtained by using grain size information to calculate the Nikuradse (1933) equivalent sand grain roughness, k_s, and thus derive a value for z_o for plane bed conditions. It is assumed here that $k_s = 2D_{50}$, where D_{50} is the mean grain size by weight (0.21 mm at Wendake Beach), and the constant of proportionality represents the effect of an uneven surface packing of the sediment on an otherwise plane bed (Yalin, 1972). Further assuming that:

$$k_s = 30 z_o \qquad (2)$$

(Schlichting, 1968), the minimum value of z_o for the Wendake Beach surf zone should be about 0.014 mm (minimum k_s = 0.42 mm).

Table I presents a summary of the VA2 data used in these calculations and the results. Note that the difference between the k_s estimates of 6.73 × 10^{-2} m at 0910 and 5.26 × 10^{-2} m at 0935 may not be significant because the measurements used for these calculations are at or near the limits of the current meter accuracy. For example, the only difference in velocities for the file pair WAV20 and VERT5 is the 0.005 m s^{-1} velocity at CM6. Because of the instrument calibration, however, the smallest increment of velocity that can be measured is 0.003 m s^{-1}. Thus these values may be virtually identical. Note that even this small difference in velocity results in a change of about 25% in the estimate of k_s. Therefore all values of k_s presented here must be considered as approximations only. Figure 5 illustrates the velocity profiles and the resulting estimates of z_o, including the minimum estimate based upon the grain size procedure. Note also the dashed lines labeled 10 and 100 cm. These indicate the unadjusted instrument elevations. It can be seen that the failure to account for bed elevation changes could greatly reduce the derived estimates of z_o. Indeed, most estimates would then fall below the minimum physical limit set by $2D_{50}/30$.

TABLE I

Summary of data used for the derivation of bed roughness estimates, Wendake Beach surf zone, 1980:06:08. The velocities from CM6 and CM4 are denoted by subscripts a and b, respectively. The predicted changes in bed elevation are Δz. VERT files are of 18 min duration, WAV files are 9 min long

File name	Time (h)	v_a (m s^{-1})	v_b (m s^{-1})	Δz (m)	z_0 (m × 10^{-3})	k_s (m × 10^{-2})
VERT2	0420	0.137	0.180	−0.01	0.09	0.28
WAV19	0530	0.313	0.405	−0.07	0.33	0.98
VERT3	0600	0.359	0.450	−0.10	0.24	0.72
WAV20	0910	0.381	0.512	−0.13	2.24	6.73
VERT5	0935	0.386	0.512	−0.13	1.75	5.26
WAV21	1130	0.460	0.551	−0.16	0.14	0.42
WAV22	1400	0.267	0.346	−0.09	0.52	1.56
VERT8	1515	0.136	0.181	−0.07	0.65	1.96
WAV24	1900	0.199	0.288	−0.08	3.28	9.83

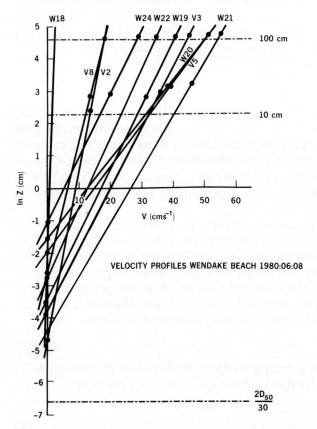

Fig.5. Graphical estimation of boundary roughness length from measurements of average longshore current velocity and estimated elevations above the bed. V and W are file designators here, representing record lengths of 18 and 9 min, respectively. $2D_{50}/30$ is a sand grain roughness estimate of the minimum physical roughness length.

The values of z_o shown here range from 0.09 to 3.28 mm. Using the already mentioned relationship between z_o and k_s, the range of equivalent sand grain roughness is thus from 2.8 to 98 mm. These values of k_s can be used to obtain graphic estimates of the Darcy-Weisbach f from the familiar Moody diagram, where specific values of f are found as a function of the Reynolds number and relative roughness. Plotting the Wendake Beach data against this relationship, we obtain the results shown in Fig.6. The arrows and numbers are to indicate the sequence of findings. The data show a minimum f of 0.016 and a maximum estimate of 0.041, representing a variability of about 250%. Most previous attempts at quantifying bed friction in the surf zone have assumed either a constant value for the entire nearshore or constant friction coefficient at a given location with spatial variability according to changes in local slope and relative roughnesses (e.g., Wright et al., 1982). This is clearly not the case here, where a large variability is found at a point. These changes may arise from several sources: a change in the turbulent structure of the flow; a change in the wave boundary layer thickness; a change in the thickness of the near bed sediment transport layer, changes in bed configuration; or combinations of the above. It is believed that changes in bed configuration are primarily responsible for the range of f found in these data, for the reasons presented below.

RESULTS AND DISCUSSION

For the Wendake Beach data, most values of the Reynolds number are shown to be near or beyond the limit for fully developed turbulent flow (this limit is indicated by the dashed line curving from upper left to lower right in Fig.6). For fully rough flow, the Darcy-Weisbach friction factor depends solely upon the relative roughness (e.g., Vennard, 1961). Therefore it is not likely that the variability in f is attributable to changes in Reynolds number.

Grant and Madsen (1979) have theoretically proposed that the presence of the wave boundary layer within the current boundary layer will result in an increase in apparent boundary roughness (that felt by the current) from the physical boundary roughness and enhanced stress. Thus changes in the thickness of the wave boundary layer would be reflected in changes in the apparent bottom roughness length. According to Grant and Madsen (1979), the thickness of the wave boundary layer (δ) may be approximated by:

$$\delta = 2\kappa |u_{*cw}|/\omega \qquad (3)$$

where κ is the Von Karman constant (0.4), ω is the radian frequency, $2\pi/T$, and $|u_{*cw}|$ is the shear velocity due to wave and current interaction:

$$|u_{*cw}| = \left(\frac{1}{2}f_{cw}\alpha\right)^{1/2} u_m \qquad (4)$$

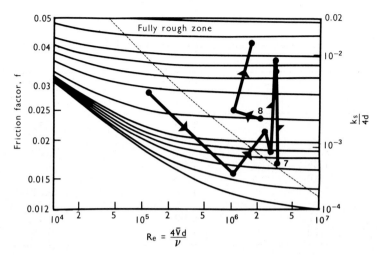

Fig.6. Wendake Beach storm data plotted on a Moody Diagram. Arrows and numbers indicate the sequence of f. Values of f range from about 0.016 to about 0.041.

where f_{cw} is the combined wave and current friction coefficient, and for trans-directional flow:

$$\alpha = 1 + (v/u_m)^2 \tag{5}$$

where v is the steady current velocity and u_m is the wave orbital velocity:

$$u_m = \gamma(gh)^{1/2} \tag{6}$$

where g is the gravitational constant, h is water depth, and γ is the wave (amplitude) breaking criterion (0.4). Grant and Madsen (1979) also propose that f_{cw} is a function of v/u_m and k_s/A_b, where A_b is the horizontal orbital amplitude, $A_b = u_m/\omega$. From these relationships it can be seen that, for a constant water depth and bottom roughness, δ will depend only upon T and v.

Changes in v affect the ratio of the steady current to the oscillating current, v/u_m. This ratio is one control on f_{cw} (Grant and Madsen, 1979). However, for trans-directional flow, relatively large changes in v/u_m change f_{cw} only slightly. For the range of v/u_m measured at Wendake Beach (Table II), estimates of f_{cw} change by only about 10% and the ratio is almost constant through the peak of the storm. It is therefore presumed that for these results, the wave boundary layer thickness is dependent primarily upon wave period. As wave period increases, so should the boundary layer thickness. For the Wendake Beach data, through the storm, the peak wave period varied from a minimum of 2.3 s to a maximum of 6.4 s, a variability of about 69%. However, through the middle of the storm, when there were large changes in apparent roughness, the mean period varied only about 17%,

TABLE II

Predicted and measured orbital velocities and relative velocities at VA2, 1980:06:08, Wendake Beach

File	U_{max}* (m s^{-1})	U_{rms} (m s^{-1})	V_a/U_{rms}
VERT2	1.24	1.09	0.13
WAV19	1.03	1.08	0.29
VERT3	1.53	1.12	0.32
WAV20	1.26	1.07	0.34
VERT5	1.40	1.12	0.35
WAV21	1.37	1.18	0.39
WAV22	1.37	1.12	0.24
VERT8	1.15	0.96	0.15
WAV24	1.18	1.05	0.19

*For d = 2.00 m, U_m (predicted) = 1.73.

between 5.3 s and 6.4 s. Further, T increased from 5.3 s at 0910 to 6.4 s at 1130 while k_s dropped from 6.73 to 0.42 cm. Thus, while there is no doubt that the presence of the wave boundary layer is reflected in the measurements leading to the derivation of z_o (or k_s), it alone cannot account for either the direction or magnitude of variability in the boundary roughness.

Smith and McLean (1977), and Grant and Madsen (1982), have shown how apparent roughness is also affected by the thickness of the near-bed sediment transport layer. According to Grant and Madsen (1982) the thickness of this layer, as reflected in its contribution to the total roughness, is approximated by:

$$k_{bs} = 160 (s + C_m) D \psi_c [(\psi'/\psi_c)^{1/2} - 0.7]^2 \tag{7}$$

where k_{bs} is the sediment transport induced roughness, $s = \rho_s/\rho$, where ρ_s is the density of the sediment and ρ is the fluid density, C_m is an added mass coefficient (taken as 1/2 for spheres), ψ_c is the critical value of the Shields parameter for the initiation of motion and ψ' is the maximum value. Grant and Madsen (1982, p.471) define:

$$\psi' = \tau'_{bm}/\rho(s-1)gD \tag{8}$$

where τ'_{bm} is the maximum value of the skin friction under the wave:

$$\tau'_{bm} = \frac{1}{2} \rho f'_w u_m^2 \tag{9}$$

where f'_w is based upon k_s and T. Therefore, for constant sediment characteristics, and with an otherwise unchanging bed roughness, the thickness of the near bed sediment transport layer depends upon u_m and T only, and as noted above (eq.4), u_m is a function of water depth in the surf zone, and T is relatively stable through the storm. Thus, in an otherwise constant environment, changes in k_s or z_o cannot be solely accounted for by changes in k_{bs}.

It should be noted that even under optimum conditions, i.e., with several high-resolution current meters in an array and the bottom precisely defined, large errors in z_o (hence k_s) are to be expected. Grant et al. (1983) working on the continental shelf, used four acoustic current meters to measure water velocities and an echo sounder to sense the bottom. Regression lines fit to the data with $R^2 > 0.99$ showed estimates of z_o to vary on the order of ±100% at the 95% confidence level. Uncertainty as to the bed elevation at Wendake Beach, and the overall precision of measurement required to accurately locate z_o, imply that values of bed roughness reported here be used only as indicators, not absolutes. However, the order of magnitude variability in roughness estimates (Table I) is much greater than the potential error terms alone can account for. Therefore, although specific values of k_s (or z_o) may be inaccurate, the temporal variability is real.

As expected, attention may therefore be focused upon changes in the surf zone bed configuration as being the primary agent affecting measurements of z_o. It is assumed that the major influence on bed deformation is the wave orbital velocity, with the current adding a secondary, but important contribution to the total stress. Indeed most attempts to formulate models of bed form development due to periodic waves (Clifton, 1976; Davidson-Arnott and Greenwood, 1976; Allen, 1982; Clifton and Dingler, 1983) rely, at least in part, on the use of orbital velocity and mean grain diameter as the main variables. Using Clifton's (1976) model, the bed form thresholds are found to be functions of u_m, D, and, in some cases, T. For the threshold of sheet flow and flat bed, Clifton has used Dingler's (1974) data to set this limit based upon grain size. The critical orbital velocity for transition to sheet flow is determined from the relationship:

$$u_{mc} = (3.88D \times 10^5)^{1/2} \tag{10}$$

where u_{mc} is the threshold velocity. Using this relationship and the Wendake Beach mean grain diameter of 0.021 mm, the critical orbital velocity for plane bed configuration is about 0.90 m s^{-1}.

Estimates of the Wendake Beach orbital velocities are obtained from linear wave theory, using eq.4, and from the current meter data that provide a maximum measured velocity (u_{max}) and a predicted velocity, $u_{rms} = 2.8$ s, where s is the standard deviation of the current meter record and 2.8 is an H_{rms} analogy (after C.E.R.C., 1977, pp.3—12). This yields a conservative estimate for the orbital velocity. From eq.4, the predicted value of u_m is 1.73 m s^{-1}. The results obtained from the current meter record through the storm are presented in Table II. For all data records summarized in Table II, the values of u_m exceed the threshold velocity given by eq.8. Thus, according to Clifton's (1976) model, the bed configuration throughout the storm is predicted as flat, and therefore could not account for the measured variability in z_o.

Clifton's (1976) model also attempts to define the effects of wave orbital asymmetry, Δu_m, on bedform development. His work suggests that for a given wave period and sediment grain diameter, the nature of bed deformation is a function of the relationship between u_m and Δu_m. Again, however, the current meter data indicate that there was little variability in Δu_m through the storm. Indeed this factor was almost constant when changes in z_0 were greatest, near the peak of the storm.

Of the surf zone velocities measured, only the longshore current showed a substantial variation through the monitoring period. Figure 7 illustrates the measurements of V, u_{rms}, and Δu_m (Δu_m is plotted as the mean flow asymmetry, after Dingler, 1974) through the storm. Given the relative uniformity of the latter two variables, they are considered constants against which changes associated with current can be evaluated.

Diver observations and box core data, taken across the surf zone, clearly indicate that plane bed conditions were not present at all times throughout the storm. There was instead a sequential development of bedforms, and this sequence seems to qualitatively correspond with changes in longshore current velocities (implicitly superimposed on a constant u_m). Figure 8

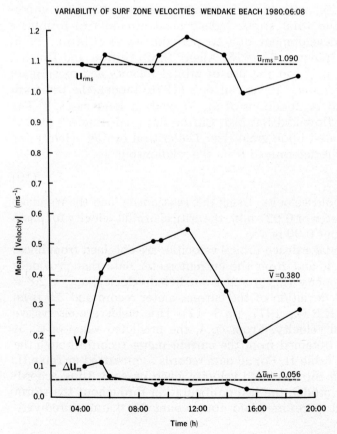

Fig.7. Sequence of surf zone velocity measurements through the storm, 1980:06:08.

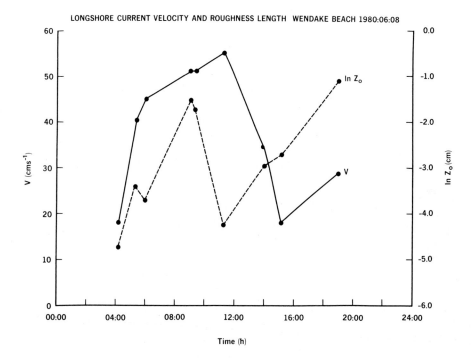

Fig.8. Concurrent changes in longshore current velocity and apparent roughness length, 1980:06:08.

shows the concurrent changes in the current velocity, V, and $\ln z_0$, through the storm. There is a general, coincident increase in V and $\ln z_0$ up to about 50 cm s^{-1}. Then the incremental increase in velocity (at 1130 h) is accompanied by a large decrease in the apparent roughness length. Thereafter, the lower velocities are associated with increasing roughness. These changes in bed roughness are primarily attributed to changes in bed configuration, a finding in contradiction with nearshore bedform development models employing U_m only.

A tentative means of identifying the bedforms is offered in Fig.9. Here, the natural log of the equivalent sand grain roughness is plotted through the storm. The upper pair of lines represent predicted roughness values obtained using relationships between ripple height and k_s. From a limited data set obtained by direct measurement and from box core data, a typical ripple height, η, for Wendake Beach varied between 1.5 and 2.5 cm. According to Stefanick (as cited in Jonsson, 1980) a reasonable approximation of the bedform induced boundary roughness is $k_s = 2.5\ \eta$. This relationship is plotted for the pair of estimated η. Working backwards, these lines, therefore, serve as indicators of the expected bed configurations, given a measured estimate of k_s. There are pairs of points at the beginning and end of the series to represent the variability in k_s that arises from a 2 cm error in the estimate of z used to obtain z_0. These represent maximum errors for the

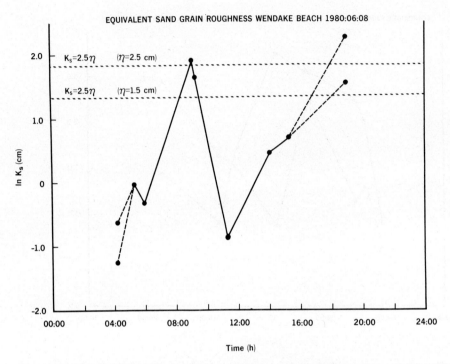

Fig.9. Equivalent sand grain roughness estimates and roughness values attributed to specific bedforms. The upper dashed lines are ripple values.

series because the relative change in ln z is greatest for the smaller values of z. The value of 2 cm was selected as being approximately the potential ripple height migration effect on bed elevation.

The information presented in Fig.9 suggests that for most observations ripples were present. This is not surprising, as the ripples occur at the beginning and end of the storm sequence. The lower and upper end points of the first and last values of k_s, respectively, are considered errors based upon physical considerations of what was occurring at those times. A larger than rippled bed is present at 0910 h, with a large decrease in roughness at 1130 h. This drop in roughness, associated with increased velocities, is assumed to indicate a change to a flat bed configuration. The difference between this roughness derived from the velocity profiles and the grain roughness (4.2 and 0.42 mm, respectively) is attributed to nearbed sediment transport. These results ignore any possible effects of bedform orientation (relative to the flow) on boundary roughness.

Based upon this analysis, and field observations, inferred bed form types are plotted with shear stress, τ_o ($\tau_o = 1/2\,\rho\,fV^2$; with f found from Fig.6) and V (Fig.10). The numbers at each point are to identify the sequence of the data. This figure shows two clusters of supposed ripples, two megaripple predictions and one value at plane bed. The change of position between the

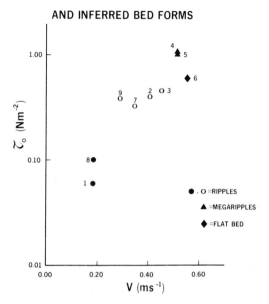

Fig.10. Boundary shear stress, longshore current velocity and inferred bedforms. Numbers indicate the sequence of data.

ripple clusters is attributed to changes in the thickness of the near-bed sediment transport layer due to changing bed roughness and velocity. Note also that among the six ripple observations, there is a distinct segregation between points from the waxing and waning limbs of the storm (as indicated by the point numbers). Points 1, 2 and 3 lie almost in a straight line, and points 7, 8 and 9 are all above that line. This corresponds well with what is found in studies of river flow regime, where the rating loop effect is well documented (Allen, 1982). This effect is a function of the response time of the bed to changes in flow. With decreasing current velocities, the bedforms change more slowly, response time is increased, and thus relict roughnesses may be present during velocity measurements. The rating loop effect also seems to be indicated in the Fig.8 data.

Figure 11 is Southard's (from Middleton and Southard, 1977) conceptual model of the relationships between bed form, velocity (U), and boundary shear stress (τ_0). In this model it is shown that a given current velocity is associated with a unique bedform. For increases in velocity within a specific bedform class, there is a relatively uniform change in bed shear stress [e.g., increasing in the ranges of ripples (R) and sand dunes (SD)]. There are also relatively abrupt changes in τ_0 associated with small changes in U in the transition between bedform regimes. Of particular interest is the rapid decrease in τ_0, despite an increase in U, with the change from dunes to flat bed (F). Note that in general, the data shown in Fig.10 follow a similar form of curve. Figure 12 is the Wendake Beach data plotted in a manner

RELATIONSHIP BETWEEN MEAN FLOW VELOCITY U AND BOUNDARY SHEAR STRESS τ_o

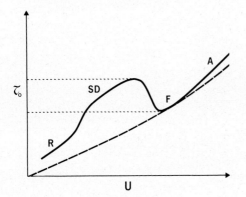

Fig. 11. Southard's (Middleton and Southard, 1977) conceptual model of the relationship between mean flow velocity, boundary shear stress and bedforms. R is ripple, SD is sand dunes, F is flat bed, and A is antidunes.

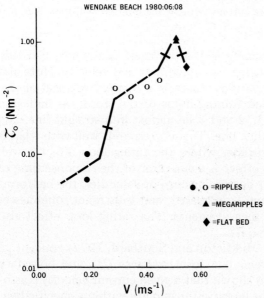

Fig.12. Predicted bedform transitions through a storm sequence, as interpreted from the Wendake Beach data and Southard's conceptual model (Fig.11). All gradients and cut-off points are estimates.

corresponding with that used by Southard. These data show changes in τ_o with changes in V that are gradual within bedform groups and with steep gradients between the inferred groups. The cross-bars indicate the estimated breaks between groups. This analysis is speculative, of course, because of the small size of the data set and the uncertainties with some of the measurements. Nevertheless, the concurrence of this progression with that suggested by Southard indicates that these findings are not unreasonable.

More concrete evidence of the propriety of the relationship suggested in Fig.12 is obtained from analysis of an epoxy relief peel of a box core taken at VA2 after the storm. This peel (Fig.13) clearly shows a quasi-planar structure at −0.16 m. This is at the maximum depth of disturbance, as indicated above. The presence of the plane lamination at −0.16 m (equivalent to a flat bed configuration) reinforces the validity of the method used for measuring bed roughness, as this is the configuration predicted from the data. Further, if bedforms produced during the falling limb of the storm at least partially reflect the rising limit sequence, the supposition of the sequential development of ripples through flat bed is also supported. Although the presence of megaripples is not clearly indicated in the peel (see Davidson-Arnott and Greenwood, 1976, figs.7e and 8b for some examples), some form of large-scale, rhythmic roughness is apparent. Allen (1981) describes similar structures, undulatory laminations, that he attributes to the existence of low-amplitude, rolling grain ripples, occurring subsequent to flat bedding and before vortex ripples appear (in a waning sequence). This was perhaps the case at Wendake Beach. Small scale ripple cross-lamination is present in the upper 3 or 4 cm of the peel. These last forms are the product of oscillatory ripple formation toward the end of the storm.

CONCLUSIONS

Although wave orbital velocity is the primary agent responsible for generating the shear stress that causes bed deformation in most wave-dominated nearshore zones, the magnitude and variability in wave parameters are not sufficient to correlate directly with changing bedforms. Indeed, as Grant and Madsen (1979) suggest, wave-current interaction can substantially enhance the stress over that predicted for waves alone. However, present models relating nearshore bedforms to flow conditions consider only the direct wave effects. For example, Clifton and Dingler (1984, this volume) note that their model is designed to operate only in the absence of secondary, unidirectional flows, and is thus inappropriate for the description of conditions in nearshore zones where such currents are present. The formulation of a comprehensive wave and current model may be possible using a format similar to that shown in Figs. 11 and 12 where a total contribution to τ_0 and V by the waves and current can be determined. This is, however, not attempted here, because it is beyond the intent of the present study. Where wave and current characteristics are unknown, approximations using existing theories will have to be employed, but they must also predict the magnitude of the longshore current. Further, a great deal of additional field data is required to accurately fix the relationships and limits shown in Fig.12 (indeed, to see if they are real). Finally, it has been shown that apparent bed roughness, as reflected by the Darcy-Weisbach friction coefficient, can vary at least 250% at a given

Fig.13. Epoxy peel from box core taken at VA2 after the 1980:06:08 storm. Near-plane bed configuration is present at maximum depth of disturbance as predicted from data. Large-scale undulatory laminations may be associated with post-vortex, rolling grain ripples (Allen, 1981). Small-scale ripple cross-stratification is present near the top of the peel. Slight landward slope at −0.16 m is due to core position on landward slope of the outer bar. Some distortion due to sampling appears along the edges of the peel. Area shown is approximately 35 × 25 cm.

location in the surf zone (at Wendake Beach, f values ranged from 0.016 to 0.041 in 2 m of water). This result implies that care must be taken in applying models that assume a uniform friction coefficient to the solution of surf zone flow problems. Ideally, future models will not consider this parameter to be a constant.

ACKNOWLEDGEMENTS

This study forms a part of a continuing research programme at Scarborough College in Nearshore Hydrodynamics and Sedimentation supported by both capital (E39218, E6614) and operating grants (A7956) from the Natural Sciences and Engineering Research Council of Canada awarded to B.G. University of Toronto Open Fellowships and an Ontario Graduate Scholarship are acknowledged by D.J.S. Assistance with computer costs was given by both Scarborough College and the Department of Geography, University of Toronto. The Academic Workshops at Scarborough College assisted with instrument design and construction, while the Graphics and Photography Division of the College produced the illustrations. We gratefully acknowledge field and laboratory assistance from M. Rollingson, P. Christilaw, J. McDonnell and R. Sutherland, Scarborough College and R.G.D. Davidson-Arnott and D. Randall, Department of Geography, University of Guelph. Gretchen McManamin, Woods Hole Oceanographic Institution, typed the manuscript. J.R. Dingler (USGS) and W.D. Grant (WHOI) are thanked for their constructive criticism. The results and interpretations presented in this paper are, however, the sole responsibility of the authors.

REFERENCES

Allen, J.R.L., 1982. Sedimentary Structures: Their Character and Physical Basis, Vols. I and II. Elsevier, Amsterdam, Vol. I, 593 pp., Vol. II, 663 pp.
Allen, P.A., 1981. Wave-generated structures in the Devonian lacustrine sediments of south-east Shetland and ancient wave conditions. Sedimentology, 28: 369—379.
Basco, D.R., 1982. Surf Zone Currents, Vol. I: State of Knowledge. U.S. Army Corps of Engineers, Coastal Engineering Research Center, Miscell. Rep. 82-7 (I), 243 pp.
C.E.R.C., 1977. Shore Protection Manual. U.S. Army Corps of Engineers, Coastal Engineering Research Center, Ft. Bevoir, Va.
Clifton, H.E., 1976. Wave formed sedimentary structures — a conceptual model. In: R.A. Davis and R.L. Ethington (Editors), Beach and Nearshore Sedimentation. Soc. Econ. Paleontol. Mineral., Spec. Publ., 24: 126—148.
Clifton, H.E. and Dingler, J.R., 1984. Wave-formed structures as interpretive tools. In: B. Greenwood and R.A. Davis, Jr. (Editors), Hydrodynamics and Sedimentation in Wave-Dominated Coastal Environments. Mar. Geol., 60: 165—198 (this volume).
Clifton, H.E., Hunter, R.E. and Phillips, R.L., 1971. Depositional structures and processes in the non-barred high energy nearshore. J. Sediment. Petrol., 41: 651—670.
Dabrio, C.J. and Polo, M.D., 1981. Flow regime and bedforms in a ridge and runnel system, S.E. Spain. Sediment. Geol., 28: 97—110.
Davidson-Arnott, R.G.D. and Greenwood, B., 1974. Bedforms and structures associated with bar topography in the shallow-water wave environment, Kouchibouguac Bay, New Brunswick, Canada. J. Sediment. Petrol., 44: 698—704.

Davidson-Arnott, R.G.D. and Greenwood, B., 1976. Facies relationships on a barred coast, Kouchibouguac Bay, New Brunswick, Canada. In: R.A. Davis and R.L. Ethington (Editors), Beach and Nearshore Sedimentation. Soc. Econ. Paleontol. Mineral., Spec. Publ., 24: 149—168.

Dingler, J.R., 1974. Wave-formed ripples in nearshore sands. Ph.D. Dissertation, University of California, San Diego, Calif., 136 pp.

Grant, W.D. and Madsen, O.S., 1979. Combined wave and current interaction with a rough bottom. J. Geophys. Res., 84 (C4): 1797—1808.

Grant, W.D. and Madsen, O.S., 1982. Moveable bed roughness in unsteady oscillatory flow. J. Geophys. Res., 87 (C1): 469—481.

Grant, W.D., Williams III, A.J., Glenn, S.M., Cacchione, D.A. and Drake, D.E., 1983. High frequency bottom stress variability and its prediction in the CODE Region. Woods Hole Oceanographic Institution, Tech. Rep. WHOI-83-19, 72 pp.

Greenwood, B. and Hale, P.B., 1980. Depth of activity, sediment flux and morphological change in a barred nearshore environment. In: S.B. McCann (Editor), The Coastline of Canada: Littoral Processes and Shore Morphology. Geol. Surv. Can. Pap., 80-10: 89—109.

Greenwood, B. and Sherman, D.J., 1983. Shore-parallel flows in a barred nearshore. Proc. 18th Coastal Engineering Conf., Cape Town, A.S.C.E., pp.1677—1696.

Greenwood, B. and Sherman, D.J., 1984. Waves, currents, sediment flux and morphological response in a barred nearshore system. In: B. Greenwood and R.A. Davis, Jr. (Editors), Hydrodynamics and Sedimentation in Wave-Dominated Coastal Environments. Mar. Geol., 60: 31—61 (this volume).

Greenwood, B., Hale, P.B. and Mittler, P.R., 1980. Sediment flux determination in the nearshore zone: prototype measurements. In: Workshop on Instrumentation for Currents and Sediments in the Nearshore Zone, National Research Council of Canada, Associate Committee for Research on Shoreline Erosion and Sedimentation, Ottawa, Ont., pp. 99—120.

Jonsson, I.G., 1980. A new approach to oscillatory rough turbulent boundary layers. Ocean Eng., 7: 109—152.

Middleton, G.V. and Southard, J.B., 1977. Mechanics of Sediment Transport. Soc. Econ. Paleontol. Mineral., Short Course Notes, 3.

Nikuradse, J., 1933. Laws of Flow in Rough Pipes. National Advisory Committee for Aeronautics Tech. Memo. 1292 (1950 translation).

Randall, D.C., 1982. Changes in the form and energy spectra of storm waves across a nearshore bar system, Wendake Beach, Nottawasaga Bay, Ontario. M.Sc. Diss., University of Guelph, Guelph, Ont., 133 pp.

Schlichting, H., 1968. Boundary Layer Theory. McGraw-Hill, New York, N.Y., 748 pp.

Sherman, D.J., 1982. Longshore currents: A stress balance approach. Ph.D. Diss., University of Toronto, Scarborough, Ont., 208 pp.

Smith, J.D. and McLean, S.R., 1977. Spatially averaged flow over a wavy surface. J. Geophys. Res., 82: 1735—1746.

Vennard, J.K., 1961. One-dimensional flow. In: V.L. Streeter (Editor), Handbook of Fluid Dynamics. McGraw-Hill, New York, N.Y., pp.3-1—3-30.

Wooding, R.A., Bradley, E.F. and Marshall, J.K., 1973. Drag due to regular arrays of roughness elements of varying geometry. Boundary Layer Meteorol., 5: 285—308.

Wright, L.D., Nielsen, P., Short, A.D. and Green, M.O., 1982. Morphodynamics of a macrotidal beach. Mar. Geol., 50: 97—128.

Yalin, M.S., 1972. Mechanics of Sediment Transport. Pergamon, New York, N.Y., 290 pp.

TIDAL-CYCLE CHANGES IN OSCILLATION RIPPLES ON THE INNER PART OF AN ESTUARINE SAND FLAT

J.R. DINGLER and H.E. CLIFTON

U.S. Geological Survey, 345 Middlefield Road, Menlo Park, CA 94025 (U.S.A.)

(Received March 14, 1983; revised and accepted December 17, 1983)

ABSTRACT

Dingler, J.R. and Clifton, H.E., 1984. Tidal-cycle changes in oscillation ripples on the inner part of an estuarine sand flat. In: B. Greenwood and R.A. Davis, Jr. (Editors), Hydrodynamics and Sedimentation in Wave-Dominated Coastal Environments. Mar. Geol., 60: 219—233.

Oscillation ripples form on subaqueous sand beds when wave-generated, near-bottom water motions are strong enough to move sand grains. The threshold of grain motion is the lower bound of the regime of oscillation ripples and the onset of sheet flow is the upper bound. Based on the relation between ripple spacing and orbital diameter, three types of symmetrical ripples occur within the ripple regime. In the lower part of the ripple regime (orbital ripples), spacing is proportional to orbital diameter; in the upper part (anorbital ripples) spacing is independent of orbital diameter. Between these regions occurs a transitional region (suborbital ripples).

Oscillation ripples develop on a sandy tidal flat in Willapa Bay, Washington, as a result of waves traversing the area when it is submerged. Because wave energy is usually low within the bay, the ripples are primarily orbital in type. This means that their spacing should respond in a systematic way to changes in wave conditions. During the high-water parts of some tidal cycles, ripples near the beach decrease in spacing during the latter stage of the ebb tide while ripples farther offshore do not change. Observations made over several tidal cycles show that the zone of active ripples shifts on- or offshore in response to different wave conditions.

Detailed bed profiles and current measurements taken during the high-water part of spring tides show the manner in which the oscillation ripples change with changes in orbital diameter. Changes in ripple spacing at the study site could be correlated with changes in orbital diameter in the manner suggested by the criterion for orbital ripples. However, there appeared to be a lag time between a decrease in orbital diameter and the corresponding decrease in ripple spacing. Absence of change during a tidal cycle could be attributed to orbital velocities below the threshold for grain motion that negated the effects of changes in orbital diameter.

Because changes in sand-flat ripples depend both upon changes in orbital diameter and upon the magnitude of the orbital velocity, exposed ripples were not necessarily produced during the preceding high tide. In fact, some ripples may have been just produced, while others, farther offshore, may have been produced an unknown number of tides earlier. Therefore, when interpreting past wave conditions over tidal flats from low-tide ripples, one must remember that wave periods have to be short enough to produce velocities greater than the threshold velocity for the orbital diameters calculated from the observed ripple spacings.

0025-3227/84/$03.00 © 1984 Elsevier Science Publishers B.V.

INTRODUCTION

Wave size, wave shape and water depth, in concert with grain size, determine the size and shape of wave-formed ripples (Inman, 1957; Dingler, 1974; Clifton, 1976; Miller and Komar, 1980). Little is known, however, about the way ripple-form responds in the field to changing wave conditions. For example, do ripples continuously adjust and thereby remain in constant equilibrium with changing wave conditions, or do they retain a metastable form until some threshold for change is exceeded? In the process of changing to meet new conditions, does the spacing continuously expand or contract or do the old ripples influence the size of the new ones? Such questions are highly relevant to the interpretation of wave-formed ripples in paleoenvironmental analysis (Clifton and Dingler, 1984, this volume). The specific relation between flow parameters and ripple size and shape is also important to studies of the movement of nearshore sand.

This paper chronicles the change in oscillation-ripple size that attends the rise and fall of the tide on a tidal flat of an estuary in the United States Pacific Northwest. We document the character of the change and relate it to the variation of wave character during the tidal excursion.

BACKGROUND

Active oscillation ripples exist for a definable range of wave conditions, called the ripple regime (Dingler and Inman, 1977). With respect to sediment transport, the threshold of grain motion is the lower boundary of the ripple regime, and the onset of sheet flow the upper boundary. Bagnold (1946) first quantified the threshold of grain motion for an oscillating flow, and several investigators (e.g., Komar and Miller, 1973; Dingler, 1974) have carried that work to the point where the threshold can be accurately calculated for a wide range of grain sizes and wave conditions. Dingler and Inman (1977) presented a dimensionless criterion for the onset of sheet flow based on field work in fine sand.

Three parameters — spacing λ, height η, and asymmetry β/λ — define the size and shape of an oscillation ripple for sand of median grain size D_m and density ρ_s. Wave period T and either orbital diameter d_0 or maximum orbital velocity u_m define the near-bottom flow conditions. Based on the work of Dingler (1979) the equation

$$d_0 = 14.3 \ (T^3 D_m)^{0.5} \tag{1}$$

defines the minimum orbital diameter required to move quartz sand in sea water.

Inman (1957) empirically defined the ripple regime in a plot of ripple spacing versus orbital diameter for various grain sizes. Clifton (1976) sectioned a form of that plot into orbital, suborbital, and anorbital ripples. Orbital ripples occur in the lower part of the ripple regime, and their spacing is directly proportional to orbital diameter. Anorbital ripples occur in the upper part of the ripple regime, and their spacing is independent of orbital diameter. Suborbital ripples are a transitional form between orbital and anorbital ripples; some investigators, though, question their existence (e.g., Allen, 1979).

The spacing of orbital ripples is given by

$$\lambda = Cd_0 \qquad (2)$$

where C is a constant between 0.65 (after Miller and Komar, 1980) and 0.8 (Komar, 1974). The maximum ripple spacing, which defines the upper limit for orbital ripples, can be calculated from the equation

$$\lambda = 14.7 \times 10^3 D_m^{1.68} \qquad (3)$$

where λ and D_m are both measured in centimeters (Miller and Komar, 1980). Dingler and Inman (1977) stated that the ratio of ripple height to spacing for orbital ripples was a constant such that

$$\eta/\lambda = 0.16 \qquad (4)$$

LOCATION

The study area is part of a sandy tidal flat on the west side of the Bay Center Peninsula in Willapa Bay, Washington (Fig.1). The bay, which is about 40 km long, is about 6.5 km wide at the study area. Because sediment moves northward along this section of coast, the entrance to Willapa Bay lies near the north end of the bay, west northwest of the study area.

A bluff composed of Pleistocene terrace deposits runs along the west side of the Bay Center Peninsula, and, in many places, a narrow beach sits at the base of the bluff. A well-defined break in slope and a concentration of coarse debris marks the toe of the beach. From this toe a sandy tidal flat extends bayward for about 1 km with a gradient of 10—20 cm per 100 m. The median diameter of the sand on the inner part of the flat is 0.0165 cm. Figure 2 shows a profile across the beach from the bluff to a distance of 270 m offshore from the toe of the beach.

Our experiments took place during spring tides in the month of August 1976 and 1977, a time when the local winds were variable. Tides were semi-diurnal with a maximum range of 2.7 m; at high water on observational days, one to one-and-a-half meters of water covered the study site. Under fair-weather conditions, when our experiments took place, the wind was from the north and northwest with wind speeds of 5 m s^{-1} or less recorded about 50 km south at Astoria, Oregon.

Most of the waves traversing the sand flat are generated within Willapa Bay because the energy of oceanic waves dissipates over the shoals at the mouth of the bay. Under particularly calm conditions, waves with periods of 8—10 s, breaking on the flats or beach, reflect the remnants of oceanic swell. These long-period waves, however, are typically of very small amplitude and generate smaller near-bottom water velocities than those produced within the bay by wind blowing over fetches of 5—10 km. The waves observed at the study site were mostly of a period of a few seconds and no more than a few tens of centimeters high.

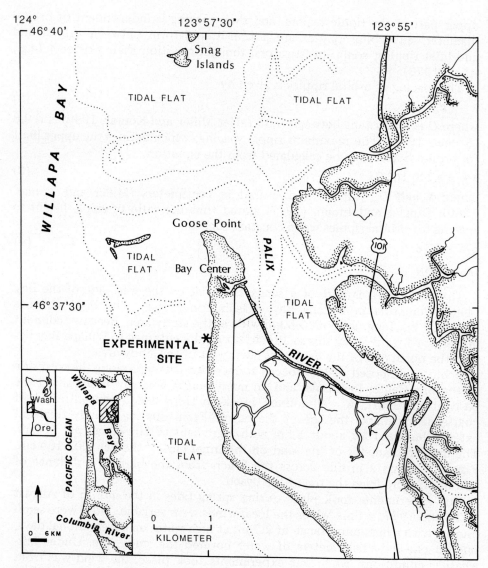

Fig.1. Location of study site in Willapa Bay, Washington.

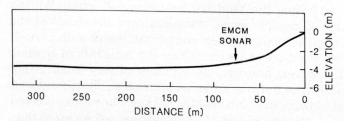

Fig.2. Profile of beach and tidal flat at site of ripple study. *EMCM* means electromagnetic current meter. Elevations are relative to a marker at the base of the cliff.

PRELIMINARY OBSERVATIONS

From preliminary observations of sedimentary processes on the sand flat, we discovered that the ripples present when the innermost flats were submerged often differed in size from those present when the flat was next exposed. We confirmed this observation by comparing ripples encased at high water with those exposed in the same area during the ensuing low water (Fig.3A, B). Also, at high water the positions of successive ripple crests were marked with short segments of copper wire; when the sand flat was next exposed the spacings between wires were greater than those of the exposed ripples.

These observations verified that ripple spacing commonly decreased significantly on the tidal flats near the toe of the beach during the fall of the tide (Fig.3A). Rather surprisingly, the spacing of ripples a few tens of meters farther onto the flat did not diminish (Fig.3B). The ripples exposed there at low tide were flatter and more rounded, but typically neither migrated nor decreased in spacing during the fall of the tide.

FIELD METHODS

To establish the basis for and the nature of the change in ripple spacing, and to understand why spacing decreased selectively near the toe of the beach, we carefully monitored the ripples and the wave-generated, near-bottom flow parameters during several tidal cycles. The measuring equipment consisted of a high-resolution sonar to make detailed profiles of the bottom and an electromagnetic current meter to measure near-bottom flow velocities. The interpretations hereinafter rely on the sonar and current-meter data in conjunction with visual observations.

The current meter, which is a commercially available, two-axis unit with a 2.5 cm diameter sensor mounted at the end of a rod, was mounted to the sonar frame. The center of the sphere was 20 cm above the bed, which is above the vortex layer for ripples of the size found on the sand flat (Dingler, 1974).

Dingler et al. (1977) detailed the high-resolution sonar used in this study. The unit, which spanned two meters of the sand bed, consists of a transducer mounted in a frame and a small electronics package. The skeletal nature of the frame (Fig.4) permits easy underwater transport and minimizes flow disturbance. The sonar head, which contains the transducer mounted in a plastic housing, moves in a channel by means of a manually operated chain drive. During this study, the transducer sat 28 cm above the bed, and the frame was oriented with its long dimension in a shore-normal direction. Resolution is of the order of a millimeter, which is necessary because ripples in the study area typically had spacings of less than 10 cm and heights of less than 2 cm.

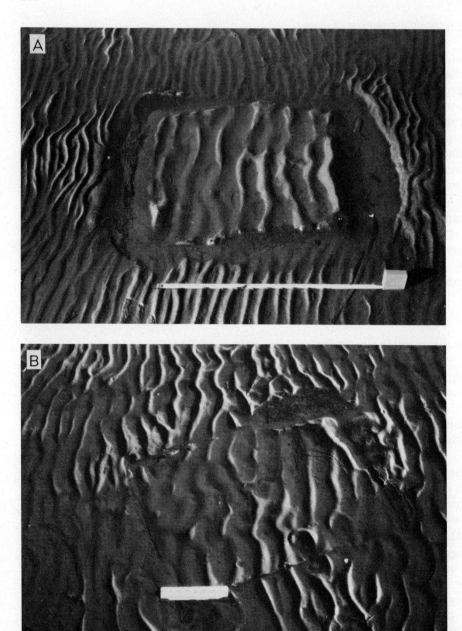

Fig.3. "High-tide" ripples compared with ripples normally exposed at low tide. A box placed upside-down on the bottom (outline of box is in center of photographs) at high water preserved the ripples through the falling tide. Beach is to the left of both photographs. A. Ripples close to the toe of the beach. Note dramatic change in spacing. Scale is 60 cm long, including the case. B. Ripples a few tens of meters bayward from A. Note that neither ripple spacing nor position changed during the fall of the tide. Scale is 16 cm long.

Fig.4. High-resolution sonic profiler that was used to measure ripples during the study. Beach is to the right. The frame is two-meters long; it has mounted upon it an electromagnetic current meter (A). The transducer, which rides in the aluminum channel, is driven by the hand crank (B).

Electrical cables connected the sensors to a pulse-code-modulated encoder onshore. Wave data were digitized at either 16 (1976) or 8 samples per s (1977) and sonar profiles at 64 samples per s. Data collection took place aperiodically over an approximately four-hour span when the sensors were underwater; each set of measurements consisted of a minimum of 10 sonar scans and/or a 5-min (minimum) time series of the wave field. All of the encoded data were recorded on quarter-inch magnetic tape for subsequent analysis.

DATA ANALYSIS

Spectral analysis of the current-meter time series produced estimates of the dominant wave period and near-bottom velocity. Reformatting the data occurred, when necessary, before spectral analysis; each 1976 current-meter record was subsampled so that all time series were 2048 points long at 8 samples per s.

Each of the U (shore-normal) and V (shore-parallel) spectra were derived by ensemble averaging four complete spectral functions, each computed from a different quarter of the appropriate time series. This procedure produced spectra with 256 spectral coefficients; these were then smoothed with a five-point, running binomial filter (Otnes and Enochson, 1978) and plotted as in Fig.5. The dominant wave period occurs at the highest spectral

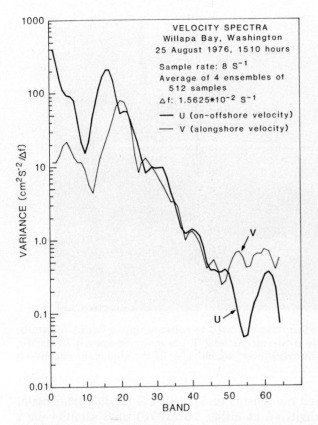

Fig.5. Wave-velocity spectra (shore-normal and shore-parallel components of velocity) taken during the period of high tide at observation site on 25 August, 1976 beginning at 1510 h.

peak, and orbital velocity is proportional to the area under the spectral curve in the region of the dominant period. Because the current meter may have been oriented at an angle to the wave orthogonals, the real velocity was taken to be the sum of the shore-normal and alongshore velocity components, especially because spectral analysis removes the mean current.

Because the ripple data had been recorded on two encoder channels — elevation on one and distance along the bed on the other — the profiles were plotted onto graph paper for analysis (Fig.6). Counting the peaks on a ripple profile and dividing that number into the distance between the first and last peak gave an average ripple spacing. Then, the ripple traces were redigitized to produce 512 points spaced 0.371 cm apart, and wave-number spectra were calculated using a similar procedure to the one used with the current-meter data. This produced a plot (Fig.7) on which the peaks represented the ripple spacings that dominated the profile.

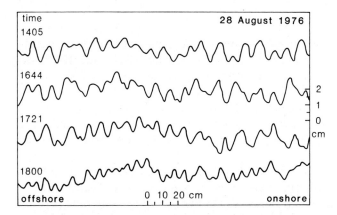

Fig.6. Ripple profiles from the study site at four times on 28 August 1976. High water occurred at 1621 h.

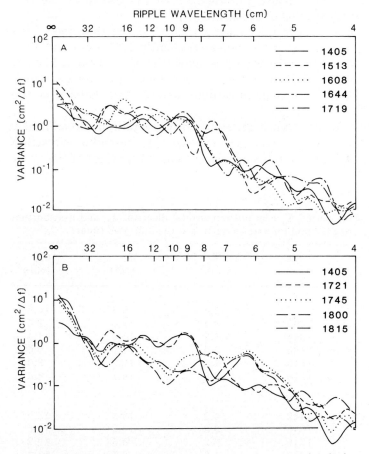

Fig.7. Spectral analysis of ripple spacing at the same site during a four-hour period on 28 August 1976. First five spectra appear in A; the last four spectra along with the first one appear in B. High water occurred at 1621 h.

RIPPLE CHANGES

We observed that sand-flat ripples only change under certain wave conditions; twice, during 10 spring tidal cycles in August 1976 and 1977, decreases in ripple spacing were documented. The measurements of 28 August 1976 (Tables Ib and IIb) show the response of the sand bed to changes in the wave field; those of 25 August 1976 (Tables Ia and IIa) show no significant changes. Figure 6 shows selected bed profiles and Fig.7A, B compare ripple spectra from various times during the tidal cycle on 28 August.

The bed was profiled 9 times on 28 August, thus obtaining a picture of the bedforms from just after the water rose above the sonar head (1405 h) to just before the water dropped below the sonar (1815 h). The nine profiles were irregularly spaced relative to the 1621 h high water of about 1.5 m — three occurred before high water and six afterwards. Before the tide reached the area, a measurement of 26 ripples under the sonar frame yielded an average spacing of 7.3 cm. Scaled photographs of the bed after the tide receded show an average spacing of about 6 cm.

On 28 August the average ripple spacing was 9.3 cm at the time of the first profile (1405 h). The spacing increased slightly to 10 cm and then dropped to 8.8 cm at 1644 h. The average spacing was still 8.8 cm at 1721 h, but then it decreased to 6.6 cm where it remained. This change, which Fig.8A shows, paralleled the change in orbital diameter calculated from the velocity spectra.

The ripple spectra for 28 August generally showed multiple, broad peaks. When the profiles were only frequency averaged (no ensemble averaging) as in Fig.7, each spectral peak encompassed a range of spacings of the order

TABLE I

Spectrally dominant wave period T, near-bottom orbital diameter d_0, and near-bottom maximum velocity u_m for 8.53 min time series starting at the indicated times

	Time (h)	T (s)	d_0 (cm)	u_m (cm s^{-1})
a. 25 August 1976				
	1210	2.4	8.3	10.8
	1300	3.4	13.4	12.4
	1346	3.2	11.6	11.4
	1425	3.6	11.0	9.6
	1510	4.0	10.2	8.0
	1546	3.8	10.1	8.3
b. 28 August 1976				
	1442	4.0	16.8	13.2
	1615	3.2	14.9	14.6
	1715	2.2	8.5	12.2
	1733	1.6	5.7	11.3
	1804	1.6	4.1	8.0

TABLE II

Average λ and spectral λ_s ripple spacings for profiles taken at the indicated times. Spectral spacings λ_{s_1} and λ_{s_2} are ranked relative to the height of the spectral peak with the higher one first (see text for the discussion of this ranking)

	Time (h)	λ (cm)	λ_{s_1} (cm)	λ_{s_2} (cm)
a. 25 August 1976				
	1320	6.9	9.1	6.4
	1400	6.5	9.6	7.4
	1445	6.8	6.6	9.6
	1501	6.6	6.6	9.6
	1527	6.8	6.6	8.0
	1545	6.3	6.6	8.0
	1601	5.7	10.1	6.6
b. 28 August 1976				
	1405	9.3	9.1	13.7
	1513	10.0	12.8	7.7
	1608	10.1	12.0	9.6
	1644	8.8	12.0	8.7
	1709	8.5	9.1	7.4
	1721	8.8	9.1	7.4
	1745	6.6	6.2	8.0
	1800	6.5	6.2	8.4
	1815	6.4	6.2	9.1

of 1—2 cm. Over time the trend, as with the average spacing, was for the peaks representing longer spacings to predominate early in the tide and for the peaks at shorter spacings to emerge during the falling tide. None of the spectra were unipeaked, as would be expected if the ripples were in complete equilibrium with the waves at the measurement time.

The ripple data for 25 August were much more static (Fig.8A); the average spacing remained constant during the experiment except for a slight decrease at the end of the day. Early in the tidal cycle, the ripple spectra developed a dominant peak at 6.6 cm that existed throughout the experiment.

DISCUSSION

Because the ripple height for orbital ripples is related to its spacing (eq.4), only one equilibrium ripple height exists for a given spacing. Therefore, a switch from one ripple spacing to a shorter one produces a decrease in the spectral variance, which is proportional to ripple height. The dependence of the maximum height of a spectral peak on spacing makes the interpretation of ripple spectra difficult. Thus, the presence of peaks of equal size, but at two different spacings, means that the shorter ripples are closer to their equilibrium size than are the longer ones. Such appears to be the case in the ripple spectra from 28 August; the late emergence of the spectral peak at 6.2 cm suggests that the ripples are slowly changing to that shorter spacing.

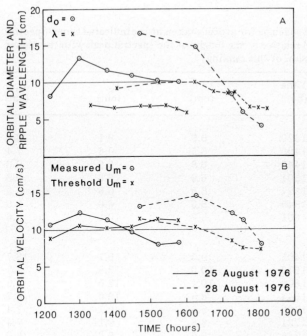

Fig.8. Comparison of observed and calculated wave and ripple parameters from 25 August 1976 and 28 August 1976. A. Observed ripple spacing (λ) and orbital diameter (d_0). B. Observed maximum velocity and calculated threshold velocity calculated using eq.1.

The existence of multiple peaks in the ripple spectra indicates that the transition from a longer spacing to a smaller one is a gradual process, and, under a gentle wave climate, that remnants of the larger ripples often exist up to the time the flat is exposed.

The average spacing, which is calculated directly from the profile, does not differentiate the various spacings making up a bed. However, as the ripples change, the average spacing also changes. Figure 8A is a plot of the average spacing and orbital diameter against time. Not only do the changes in ripple size appear to be discontinuous, Fig.8A suggests that there is a time lag between the time when the orbital diameter decreases to when the spacing reaches a smaller value. This time lag is indicative of the difficulty that a less energetic oscillatory flow has in destroying the older bedforms.

Both the magnitude of the orbital diameter—grain-size ratio (Clifton and Dingler, 1984, this volume) and the parallelism between the spacing and orbital diameter curves, demonstrate that these are orbital ripples. Furthermore, the spacing to orbital diameter ratio falls between 0.65 and 0.8 values of eq.2 during much of the tidal cycle with the points that fall above the 0.8 d_0 curve representing the time lag between the decrease in d_0 and that in λ. Rather than the exact value of that ratio, the parallelism of the curves is the important factor here.

Thus, as expected, ripple spacing follows orbital diameter for orbital ripples. On 28 August 1976 this meant a slight increase in spacing early in the incoming tide, and a greater decrease during the latter part of the falling tide. This does not occur at the same depth every tidal cycle, or even at different depths during the same tidal cycle because there is a second, often overlooked, control to ripple formation — the threshold of grain motion.

The threshold of grain motion depends on both wave period and orbital diameter for sand of a given grain size (eq.1). In Fig.9, arrows mark the minimum orbital diameters required to move 0.165 mm sand (the size found on the sand flat) for various wave periods; these arrows are superimposed on a schematic of the λ vs. d_0 relationship. Thus, longer periods require larger orbital diameters to initiate grain motion, and periods greater than 5 s do not even fall within the range of orbital ripples. Figure 8 compares observed orbital velocity with the calculated threshold velocity, which is based on the dominant wave period for each record. Figure 8A and B together show that ripple spacing follows the orbital diameter when the observed velocity exceeds the threshold velocity. When, as occurs near the end of the ebb tide or during days of longer dominant periods (e.g., 25 August 1976), the actual velocity approximates or is less than the threshold velocity, the ripples do not change even though the orbital diameter continues to fluctuate. Therefore, ripples on the sand flatt will respond to changes in orbital diameter only when the near-bottom oscillatory flow exceeds the threshold velocity.

This control of ripple change by threshold velocity may explain why some ripples remained unchanged through a falling tide while the spacing of ripples further inshore decreased noticeably (Fig.3A, B). Ripples at both

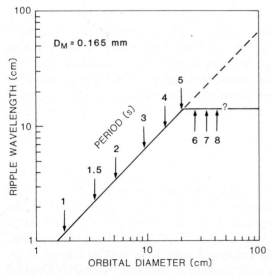

Fig.9. Threshold orbital diameters (arrows) superimposed on a plot of ripple spacing (λ) versus orbital diameter (d_0) for 0.165-mm-diameter sand.

locations experienced a similar decrease in water depth as the tide level fell. Apparently, however, when the water depth over the outer ripples reached the level at which the spacing of ripples at shallower depths diminished a short time earlier, the orbital velocity was subthreshold. This situation could be caused by an increase in wave period during the fall of the tide, a phenomenon that was observed, but which did not seem to occur consistently (Table I).

An alternate possibility relates to the tidal-flat profile and the changes in wave energy that attend the falling tide. Within about 60 m of the toe of the beach, the slope of the profile increases. The ripples that decreased in spacing occur on the upper part of this slope, and the unchanged ripples on the lower part. By the time these latter ripples were exposed to the water depths at which the inshore ripples changed, the general depth over the flats would be significantly reduced. The loss of wave energy due to frictional interaction with the bottom would be greater, resulting in reduced wave heights. Accordingly, the associated orbital velocities would have more quickly become subthreshold.

The ripple response process outlined here has important implications for estimating past wave conditions from low-tide ripples. In many cases interpretations will be inaccurate because ripples exposed at low tide could have been produced anytime during the previous tidal cycle or even during an earlier one.

CONCLUSIONS

In low-energy wave environments both wave period and orbital diameter control ripple size. Ripple spacing and, therefore, ripple height, are proportional to orbital diameter only when the near-bottom orbital velocity exceeds the threshold velocity; threshold velocity is proportional to the ratio of orbital diameter to wave period. Ripples exposed at low tide may have formed at different times during the previous high tide, or even during earlier tidal cycles. The decrease in ripple spacing sometimes observed during the falling tide comes about when decreasing orbital diameters occur with short wave periods above the threshold. The decrease in ripple spacing with the decrease in orbital diameter is discontinuous with a time lag caused by the existence of larger ripples.

ACKNOWLEDGMENTS

We thank the many field assistants who participated in this study. Special thanks go to Jeff Hedenquist and Curt Peterson, who spent hours in the water operating the sonar. The junior author thanks his family, who spent several nights with him on the exposed, wind-swept tide flats of Willapa Bay, searching with flashlights for ripples marked with little bits of copper wire. The senior author started this project while a National Research Council Postdoctoral Fellow with the U.S. Geological Survey, working with E.D. McKee.

REFERENCES

Allen, J.R.L., 1979. A model for the interpretation of wave ripple-marks using their wavelength, textural composition, and shape. J. Geol. Soc. London, 136: 673—682.
Bagnold, R.A., 1946. Motion of waves in shallow water. Interaction between waves and sand bottoms. Proc. R. Soc. London, Ser. A, 187: 1—18.
Clifton, H.E., 1976. Wave-formed sedimentary structures — a conceptual model. In: R.A. Davis, Jr. and R.L. Ethington (Editors), Beach and Nearshore Sedimentation. Soc. Econ. Paleontol. Mineral., Spec. Publ., 24: 126—148.
Clifton, H.E. and Dingler, J.R., 1984. Wave-generated structures and paleoenvironmental reconstruction. In: B. Greenwood and R.A. Davis, Jr. (Editors), Hydrodynamics and Sedimentation in Wave-Dominated Coastal Environments. Mar. Geol., 60: 165—198 (this volume).
Dingler, J.R., 1974. Wave-formed ripples in nearshore sands. Unpubl. Ph.D. thesis, University of California, San Diego, Calif., 136 pp.
Dingler, J.R., 1979. The threshold of grain motion under oscillatory flow in a laboratory wave channel. J. Sediment. Petrol., 49: 287—294.
Dingler, J.R. and Inman, D.L., 1977. Wave-formed ripples in nearshore sands. Am. Soc. Civ. Eng., Proc. 15th Coastal Engineering Conf., 2109—2126.
Dingler, J.R., Boylls, J.C. and Lowe, R.L., 1977. A high-frequency sonar for profiling small-scale subaqueous bedforms. Mar. Geol., 24: 279—288.
Inman, D.L., 1957. Wave-generated ripples in nearshore sands. Dept. of the Army, Corps of Engineers, Tech. Memo, 100, 65 pp.
Komar, P.D., 1974. Oscillatory ripple marks and the evaluation of ancient wave conditions and environments. J. Sediment. Petrol., 4: 169—180.
Komar, P.D., 1976. Beach Processes and Sedimentation. Prentice-Hall, Englewood Cliffs, N.J., 429 pp.
Komar, P.D. and Miller, M.C., 1973. The threshold of sediment movement under oscillatory water waves. J. Sediment Petrol., 43: 1101—1110.
Miller, M.C. and Komar, P.D., 1980. Oscillation sand ripples generated by laboratory apparatus. J. Sediment. Petrol., 50: 173—182.
Otnes, R.K. and Enochson, E., 1978. Applied time series analysis, Vol. 1. Wiley, New York, N.Y., 449 pp.

BEDFORMS AND DEPOSITIONAL SEDIMENTARY STRUCTURES OF A BARRED NEARSHORE SYSTEM, EASTERN LONG ISLAND, NEW YORK

R. CRAIG SHIPP*

Marine Systems Laboratory, Smithsonian Institution, Washington, DC 20560 (U.S.A.)

(Received July 15, 1983; revised and accepted October 18, 1983)

ABSTRACT

Shipp, R.C., 1984. Bedforms and depositional sedimentary structures of a barred nearshore system, eastern Long Island, New York. In: B. Greenwood and R.A. Davis, Jr. (Editors), Hydrodynamics and Sedimentation in Wave-Dominated Coastal Environments. Mar. Geol., 60: 235—259.

The depositional sedimentary structures and textures of a single-barred nearshore system on the Atlantic coast of eastern Long Island, New York, were studied along seven shore-normal transects. Data along these transects consisted of textural analysis of 160 sediment samples, temporal bedform observations, and 42 can cores for the analysis of sedimentary structures.

Six sedimentary subenvironments were observed, based on distinct combinations of sediment color and texture, bedforms, physical, and biogenic sedimentary structures, and benthic infaunal communities. The shoreface environment is divided into the upper shoreface, the longshore trough, and the longshore bar. The divisions of the inner shelf environment are the shoreface-inner shelf transition, the offshore, and the coarse-grained deposit. The first five subenvironments are arranged in bands parallel to the shoreline, whereas the coarse-grained deposit occurs in patches across the inner shelf.

The location of fair-weather wave base, coinciding with a reduction in slope (3.0—0.3°) from the shoreface to the inner shelf, is characterized by the cessation of debris surge in the troughs of ripples, the formation of a "rust layer" of microorganisms over the bedform surface, and a sediment color change caused by an increase in organic detritus. The sequence of bedforms and physical sedimentary structures observed in this system fits well with existing wave-generated (oscillatory) flow regime models. These models explain the observed sequences as a response to the degree of asymmetric flow created by shoaling waves. Distribution of biogenic structures and assemblages of infaunal organisms is influenced by the distance landward or seaward of fair-weather wave base.

The overall relationships of this nearshore system can then be summarized as a hypothetical prograding stratigraphic sequence. The entire sequence is underlain by organic-rich, bioturbated, offshore deposits. Overlying the offshore is the planar-laminated sediments of the transition. Grading upward from the transition are the cleaner, planar-laminated, seaward slope deposits of the longshore bar. Above this, is a distinct erosional surface indicating the base of the massive to cross-laminated coarse sediments of the

---

*Present address: Oceanography Program, 14 Coburn Hall, University of Maine, Orono, ME 04469, U.S.A.

0025-3227/84/$03.00 © 1984 Elsevier Science Publishers B.V.

longshore trough. Capping the sequence are the cross- to planar-laminated, clean sands of the upper shoreface and foreshore.

INTRODUCTION

The study of the nearshore zone has been long neglected due to poor accessibility. The historical concentration of research efforts on the beach and offshore has created a large gap in the understanding of nearshore processes and sedimentation. Concomitant with difficult access, investigation of the nearshore zone has presented a broad array of logistical problems including sampling strategy, equipment deployment, and weather dependence.

Only in the last two decades have a handful of detailed geologic nearshore studies emerged. Initially, these studies defined nearshore subenvironments by examining sediment texture, physical sedimentary structures, and biogenic features (Reineck, 1963; Reineck and Singh, 1971; Howard and Reineck, 1972a, b, 1981). Later, these initial ideas were developed further by relating nearshore subenvironments to "an oscillatory flow regime concept" (Clifton et al., 1971; Davidson-Arnott and Greenwood, 1974, 1976; Clifton, 1976). More recently, investigations have dealt either with the morphology or sedimentology of prominent nearshore features such as intertidal and submarine bars (Greenwood and Mittler, 1979; Hunter et al., 1979; Greenwood and Hale, 1980; Goldsmith et al., 1982), or with the sedimentary and biogenic trends from the supratidal zone across the intertidal and through the nearshore zone (Kent, 1976; Hill and Hunter, 1976). Few studies have explored the characteristics of the nearshore zone from mean low water (MLW) across the shoreface to the inner shelf.

The present study describes the morphologic variation and sedimentary subenvironments of a single-barred nearshore system off eastern Long Island, New York. Even though most data were collected during fair-weather conditions, the overall sedimentological setting (fair-weather and storm conditions) of this nearshore system are suggested. Finally, the depositional relationships are summarized as a hypothetical prograding stratigraphic sequence.

Of all the published studies to date, the conceptual model of Clifton (1976) seems the most useful in explaining nearshore dynamics and the sequence of bedforms generated by shoaling waves. This model integrates Airy (linear) and solitary wave theory and empirical measurements to explain sedimentologic and stratigraphic observations. An important element in Clifton's model is the change from symmetric to asymmetric oscillatory flow as the wave shoals. This change in flow symmetry of a shoaling wave, in turn, alters the symmetry of the bedforms. The depth at which the asymmetry develops is a function of wave height, wave period, grain size, slope, and bottom roughness (Newton, 1968). Clifton applies this principle by suggesting a flow sequence of wave-generated bedforms (Fig.1A). In addition, the work of Davidson-Arnott and Greenwood (1976) have modified the Clifton sequence for a barred nearshore system. They suggest that the Clifton

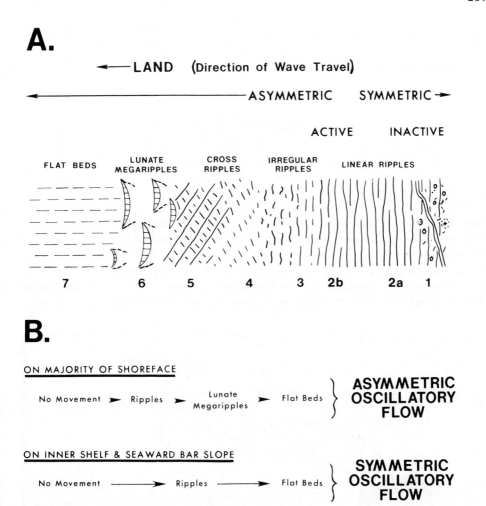

Fig.1. Bedform sequences in the nearshore zone. A. Hierarchical sequence of bedforms produced by shoaling waves (modified from Clifton, 1976). Numbers along the bottom correspond to the bedforms named above. B. Suggested flow regime sequences for barred nearshore systems (modified from Davidson-Arnott and Greenwood, 1976).

sequence is valid over the majority of the shoreface for strong asymmetric oscillatory flow during fair-weather swell conditions, while a distinct symmetric oscillatory regime would be produced during periods of local storms on the inner shelf and the seaward slope of the bar (Fig.1B).

Because of its peculiar geometry, a particularly interesting form in the sequence is the cross ripple (5 on Fig.1A). This bed configuration consists of two sets of ripples oriented obliquely to the oscillatory current (wave approach observed directly from the bed). One set is longcrested, while the second set is composed of short ripples occupying the troughs of the longer set. Cross ripples seem to represent a structural transition between asymmetric ripples and lunate megaripples (Clifton et al., 1971) and are shown in the present study to be an easily recognized field indicator of asymmetric oscillatory flow.

STUDY AREA

The study was conducted off southeastern Long Island in Suffolk County, New York (Fig.2A). The study area, adjacent to the villages of Beach Hampton, Amagansett, and East Hampton, extends for a total shore-parallel distance of 13 km. The limit of the investigation was 3.25 km seaward of the shoreline, approximately at the 22 m isobath (Fig.2B).

The beach and nearshore deposits of Long Island consist of reworked outwash deposits and sediments transported from the updrift Montauk Till Member (located 25 km to the east at Montauk Point) of the late Wisconsin Ronkonkoma Moraine (Taney, 1961). The entire eastern Long Island nearshore system is dominated by a linear to slightly irregular subtidal bar. The bar crest reaches within 3.5 m of the water surface and varies from 100 to 300 m seaward of MLW. The gross morphology of the bar varies from a

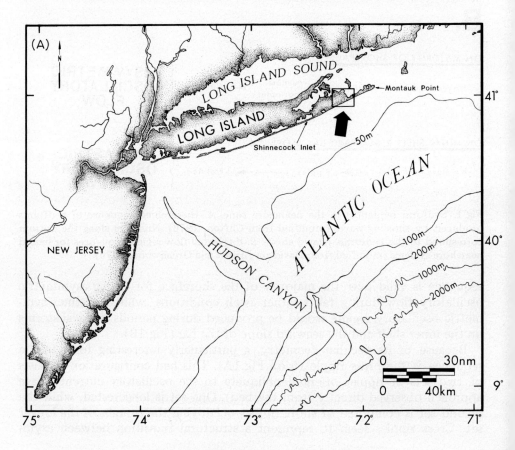

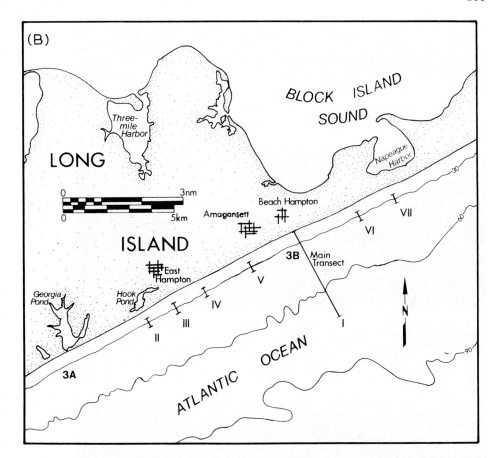

Fig.2. A. Location of the study area within the New York Bight. The box on the southern fork of eastern Long Island is the approximate geographic limits of B. B. Location of the seven transects off the south shore of eastern Long Island. The six short transects established along the shoreface—inner shelf contact vary from 100 to 300 m and are not drawn to scale. 3A and 3B refer to location of photographs in Fig.3. Isobaths in feet.

straight and narrow, landward asymmetric form (Fig.3A) to a slightly irregular and wider, symmetric form (Fig.3B).

Prevailing winds in the study area are southwest in the summer, with a 3.6—5.2 m s^{-1} average hourly velocity range, and northwest in the winter, with a 4.7—6.8 m s^{-1} average range (Lettau et al., 1976). The dominant wind generally blows from the northeasterly to easterly sectors during storms that are either tropical (predominantly hurricanes) or extratropical (northeastern). Tropical storms are most prevalent from August through October, whereas extratropical storms usually occur most frequently between October and April (U.S. Army Corps of Engineers District, 1977).

Visual and remotely measured data from eastern Long Island show a mean annual breaker height of 0.7 m and wave period of 8.0 s (DeWall, 1979). According to the wave hindcasting calculations of Saville (1954), 72% of all

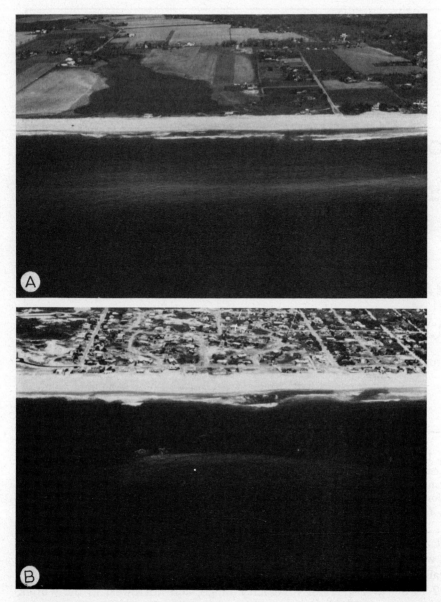

Fig.3. Oblique aerial photographs of the longshore bar in the study area. A. Straight, landward asymmetric form. B. Slightly irregular, symmetric form. Locations of photographs on Fig.2B.

deep-water waves in western Long Island approach from the east-northeast through south-southeast. The largest predicted waves are 7.6—9.1 m. The wave dominance from the easterly quadrants continues in the study area, which is located 170 km to the east. In addition, a secondary dominance from the southwest is caused by the increased fetch between the New Jersey shoreline and the eastern Long Island study site (Fig.2A).

The semidiurnal spring tide in eastern Long Island is 0.7 m at Montauk Point and increases to 1.1 m at Shinnecock Inlet (U.S. Department of Commerce, 1978). An estimated spring range for the study area is therefore 0.85 m. This range places the study area well within the microtidal classification of Davies (1964).

METHODS

Transect lines

In the study area, one long intermittent transect was established from MLW to a distance of 3.25 km offshore. Six shorter transects (100—300 m in length) were surveyed over the change in slope at the seaward base of the longshore bar (Fig.2B). With the aid of Scuba, polypropylene-line transects were placed shore-normally across the bottom, anchored every 3.0 m with a 0.5 m section of hooked reinforcing bar (rebar). Transect orientation was maintained by compass. With this method, 200 m of line could be deployed at a time. All transects were surveyed once and were usually removed the same day, except for the main transect at Beach Hampton (Fig.2B). This transect was buoyed and maintained for the duration of the study (two months).

Along the transects, sampling intervals were chosen wherever bedform or textural changes occurred. The interval of sampling varied from 3 to 6 m, on the landward slopes of the longshore bar, to over 1 km on the inner shelf. At each sample site, bedforms were examined for planimetric geometry, height, spacing, degree of asymmetry, presence of wave surge, and amount and type of biological activity.

Sediment textural analysis

Sediment samples were analyzed for size (mean) and sorting (standard deviation) utilizing the Hydraulic Equivalent Sediment Analyzer (HESA) described by Anan (1972) that was later modified by replacement of the original pressure transducer system with an electrobalance. The modified HESA is interfaced with a HP 9825A calculator and a HP 9872A graphic plotter (Hewlett-Packard Calculator Products Division). The system is programmed to tabulate grain-size distribution and graphically depict size-frequency and cumulative probability curves for individual samples. Determination of mud content ($>4\ \phi$) was made by wet sieving the sediment samples. A complete dispersion of silts and clays was assured by agitating the sample for 15 min in an ultrasonic disruptor after the methods of Kravitz (1966).

Sediment cores

A total of 42 can cores were taken at selected sites on four of the seven transects. The can coring technique is a modification of the procedure used by Howard and Frey (1975). The corer consists of a 7.6 l metal gas can (measuring 26.9 × 21.7 × 14.7 cm) with the bottom removed. A slurry of sand, water, and red lead (NL Industries) was mixed in a plastic bag and distributed over the area to be cored. The incorporation of the red lead into the sediment revealed structures that are caused by the coring process (Howard and Reineck, 1972a) and also aided in the preservation of the surface bedforms (Davidson-Arnott and Greenwood, 1974).

The cores were extruded in 2 cm thick slabs into plastic trays for transport to the laboratory. The leftover sand from the extruded cores was elutriated and the infaunal invertebrates collected and preserved. In the laboratory, the slabbed cores were first X-rayed using routine radiographic techniques for unconsolidated sediments (e.g., Howard, 1968; Bouma, 1969; Hamblin, 1971). Next, two different methods were used to make relief peels of the core slices. For coarse sand and gravel the epoxy-resin technique of Howard and Frey (1975) was employed. A second technique utilizing a lacquer cement (Bouma, 1969) was found to be more suitable for finer sediments.

Some structural information in the relief peels was lost due to post-core amphipod bioturbation. This generally occurred only in the upper 10 cm of the core. The problem was eliminated in later cores by saturating the sediment with a 10% solution of $MgCl_2$.

RESULTS

In the Long Island nearshore system, six sedimentary subenvironments can be defined. These areas are described by distinct combinations of sediment color and texture, physical sedimentary structures, and, occasionally, infaunal assemblages and/or biogenic structures. Beginning at MLW, the subenvironments defined are: (1) upper shoreface; (2) longshore trough; (3) longshore bar including the subdivisions: (a) landward bar slope, (b) bar crest, and (c) seaward bar slope; (4) shoreface-inner shelf transition (hereafter referred to as transition); (5) offshore; and (6) inner shelf coarse-grained deposit. The first five subenvironments occur in continuous bands parallel to the shoreline. The coarse-grained deposit is unique in that it occurs in small patches scattered across the inner shelf. Additionally, the sedimentary subenvironments of the Long Island nearshore zone are grouped into two major environments dependent on their location relative to fair-weather wave base (Fig.4). The first three subenvironments are landward of fair-weather wave base, thus comprising the shoreface. The last three subenvironments are seaward of wave base and, therefore, are on the inner shelf.

Fair-weather wave base

A change in the offshore slope from approximately 3.0—0.3° occurs at the seaward base of the longshore bar at a depth of approximately 10 m (Fig.4). This slope change marks the seaward extent of the shoreface and the land-

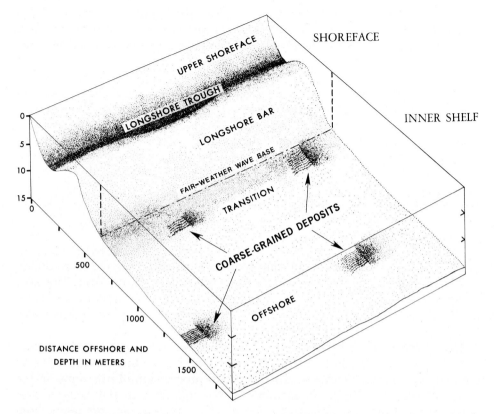

Fig.4. Location of the sedimentary environments in the Long Island nearshore system. Coarse-grained deposits are located in scattered patches on the inner shelf and are not necessarily drawn to scale.

ward boundary of the inner shelf. Observations of this area during fair weather indicate that this contact is the limit of wave influence, and, therefore, the greatest depth of active sand transport during these conditions. This transitional region has been defined as the effective or fair-weather wave base (Curray and Moore, 1964; Dietz and Fairbridge, 1968; Elliott, 1978). It is characterized in the study area by:

(1) The point at which debris in the ripple troughs ceases to surge and becomes motionless.

(2) A distinct color change from a clean, yellow-tan sand on the shoreface to a gray-brown sand on the inner shelf. This color change is caused by an increase in organic detritus seaward of the contact.

(3) A layer of rust colored microbenthos covering the bedforms seaward of the fair-weather wave base. The first observation of a rusty bottom (hereafter referred to as the "rust layer") was documented by Kumar and Sanders, 1976, and may be composed of diatoms, algae, and other microorganisms.

A series of observations taken over a four-week period (August, 1978) during fair-weather conditions indicated a nearly exact correlation between the location of wave-surge cessation and the initiation of the rust layer. This change from debris surging in the ripple troughs to the formation of the rust layer varied across a 50 m band. A series of observations over three consecutive days shows that a thin rust layer can form in 24 h or less. The transition from a yellow-tan to gray-brown colored sediment is also gradational across the 50 m band. However, the color shift may require several months or a season to respond to changes in wave climate as opposed to the rapid response of the rust layer. During low to moderate wave-energy conditions, this color shift is an excellent indicator of fair-weather wave base.

Sedimentary subenvironments of the shoreface

Upper shoreface

The landwardmost shoreface environment is characterized by: (1) a cover of medium sand; (2) a shore-normal sequence of bedforms; (3) three different types of physical structures; and (4) a few biogenic structures. This area extends seaward from MLW to approximately 200 m offshore in water depths ranging from 0 to 6.0 m. The average slope is approximately 2.0° (Fig.4).

The upper shoreface consists of well to moderately sorted medium sand (1.6—2.1 ϕ, 0.45—0.55 σ) that fines seaward and has a white-tan color due to a lack of fine-grained material. The landward progression of bedforms on the upper shoreface exhibits an asymmetric set complete to flat beds as proposed by the Clifton sequence (2b—7 Fig.1A). A typical can core taken from the linear-rippled region on the upper shoreface displays subhorizontal planar lamination overlain by coarser landward-dipping, medium-scale foreset bedding, and capped by small-scale ripple cross-lamination (Fig.5A). Only a few biogenic structures are present despite the great abundance of Haustoriid amphipods. The few short wispy bioturbate structures that are preserved are usually in the top 10 cm of sediment, where amphipods are most commonly found (Howard, 1968).

Longshore trough

This subenvironment seaward of the upper shoreface is distinguished by: (1) a coarse sand size; (2) a sediment surface dominated by large oscillatory ripples; (3) a lack of internal depositional structures; and (4) no biogenic structures. The longshore trough varies from 5.5 to 6.0 m in depth and averages

---

Fig.5. Relief peels of can cores taken from the shoreface: *A.* upper shoreface (depth — 4.6 m). *B.* longshore trough (depth — 6.0 m). *C.* landward slope of the longshore bar (depth — 5.5 m). *D.* crest of the longshore bar (depth — 3.5 m). *E.* seaward slope of the longshore bar (depth — 7.5 m). Some of the ripple cross-lamination at the top of the relief peels have been altered by post-core amphipod bioturbation. Black arrows at top left point landward. Scales at bottom right are 3 cm long.

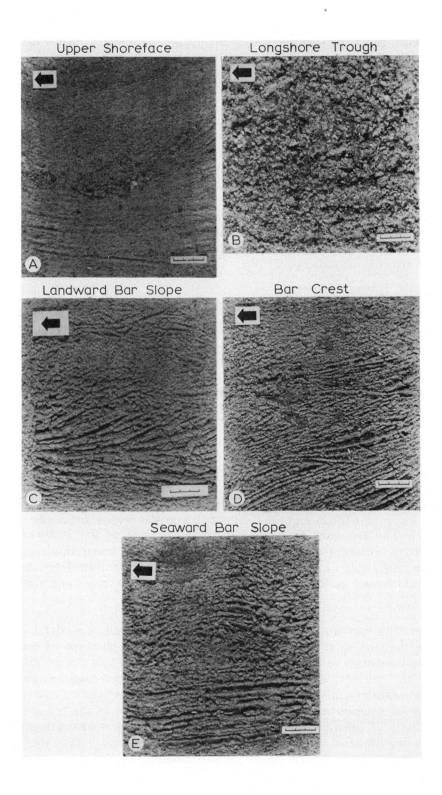

50 m in width. The trough exhibits a concave-upward bathymetric profile with approximately 0.5 m in relief between the center and the landward and seaward boundaries (Fig.4).

The longshore trough is composed of subangular, moderately well-sorted coarse sand (0.5—0.7 ϕ, 0.57—0.65 σ) and at least a 2% component of subrounded gravel. The large oscillatory ripples are continuous in a shore-parallel direction for several hundred meters and have an average 60 cm wavelength and a 15 cm amplitude (2a on Fig.1A). A typical can core taken from the longshore trough displays massively bedded sand and gravel with no discernable physical or biogenic structures (Fig. 5B). The reason for the massive nature of these sediments seems related to the presence of an infaunal fish. During fair weather any disturbance, such as the passage of a large wave overhead, causes the American sandlaunce *Ammodytes americanus* to dart across to an adjacent oscillatory ripple. This in turn is followed by rapid burrowing into the ripple crest, thus disrupting any bedding.

Longshore bar

The third subenvironment is the longshore bar that extends from 250 to 530 m seaward of MLW. It is characterized by: (1) moderately sorted fine to medium sand; (2) several sets of bedforms that follow the Clifton sequence; (3) three distinct vertical sequences of physical structures; and (4) no biogenic structures even though two species of Haustoriid amphipods and the isopod *Cirolana concharum* are in moderate abundance. The three subdivisions associated with the longshore bar are the landward slope, the bar crest, and the seaward slope.

Landward slope. The landward slope of the longshore bar consists of a 50 m shore-parallel band of well- to moderately well-sorted medium sand (1.7—2.1 ϕ, 0.41—0.58 σ) dipping landward at an approximate slope of 3° (Fig.4). The series of bedforms found on this slope follows the asymmetric ripple set from the Clifton sequence except it is in the reverse order (2a—6 on Fig.1A). The sequence from the top to the bottom of the landward slope includes cross ripples (Fig.6), irregular ripples, and linear ripples. A typical can core taken on the middle of the slope displays landward-dipping, medium-scale foreset bedding overlain by small-scale ripple cross-lamination (Fig.5C).

Bar crest. The bar crest is characterized by a narrow band of moderately well-sorted medium sand (1.9—2.0 ϕ, 0.53—0.55 σ). Here, the type of bedform is most sensitive to the incident wave climate. During low wave-energy conditions the bar crest is covered by irregular ripples to cross ripples (3—5 on Fig.1A). Observations during higher wave-energy conditions (1.25 m height and 6.0 s period) revealed a 50 m band centered on of the bar apex that consisted of lunate megaripples having a 60—80 cm wavelength and 14—18 cm amplitude (6 on Fig.1A). A typical can core in this region shows a narrow band of small-scale ripple cross-lamination overlying cosets of

Fig.6. Cross ripples. Arrow points to a secondary ripple crest in a trough between two long-crested ripples. Both sets of ripples are oriented oblique to the oscillatory current. Knife handle is 15 cm long.

landward dipping, truncated, low-angle foreset bedding of megaripple origin (Fig.5D).

Seaward slope. The seaward bar slope is characterized by a 200 m wide band of moderately sorted fine sand (1.8—2.6 ϕ, 0.41—0.67 σ) dipping seaward at approximately 3.0° (Fig.4). During low wave-energy conditions, bedforms of this region vary from symmetric linear ripples at the seaward base of the longshore bar to asymmetric cross ripples (Fig.6) just seaward of the bar crest (2a—5 on Fig.1A). At the time of higher wave-energy, the upper portion of this area shifts to lunate megaripples grading to cross ripples further down the seaward slope (5—6 on Fig.1A). In fair weather, the boundary between symmetric and asymmetric bedforms may be as far as halfway up the seaward slope and then shift downslope in response to increased wave activity. A typical can core in this region reveals a sequence of physical structure quite different from the landward slope. Instead of small-scale ripple cross-laminations overlying medium-scale foreset bedding, the vertical sequence consists of small-scale ripple cross-laminations overlying horizontal to subhorizontal planar lamination (Fig.5E). A similar sequence on the seaward slope of the longshore bar was also reported by Davidson-Arnott and Greenwood (1976) and Greenwood and Mittler (1979).

Sedimentary subenvironments of the inner shelf

Transition

The transition is the landwardmost subenvironment of the inner shelf and is characterized by: (1) moderately well-sorted fine sand; (2) intermittently active, symmetric, linear ripples; (3) a sequence of physical structures similar to the seaward bar slope; and (4) a few biogenic structures. The depth range of this environment varies from 10 m to a depth of 14—16 m. Across the profile, the transition extends from 530 m to a distance of approximately 1.5 km seaward of MLW (Fig.4).

The transition consists of moderately well sorted fine sand (2.4—2.7 ϕ, 0.53—0.67 σ) that is light gray-brown in color. The darker color is due to an increase in the organic constituent of the sediment. The mobility of the linear ripples is a function of the ambient wave conditions and could be precisely determined by observing the onset of the rust layer. A typical can core in this region exhibits the same sequence as the seaward bar slope — small-scale ripple cross-lamination overlying horizontal planar lamination — except the units of the small-scale lamination are thicker (Fig.7A). In addition, at the same level in the core, organic detritus drapes the preserved ripple troughs indicating periods of quiescence when fair-weather wave base was further landward. It is during these quiet periods that the rust layer forms over the sediment (see p. 242). The limited bioturbation (white wispy structures on X-ray radiographs) present in this subenvironment is caused by Lysianassid amphipods.

Offshore

The offshore subenvironment is characterized by: (1) organic-rich, fine sand; (2) inactive symmetric linear to irregular ripples; (3) a lack of physical sedimentary structures; and (4) extensive bioturbation. The landward boundary of this environment begins approximately 1.5 km seaward of MLW in an approximate 15 m depth and continues seaward beyond the 3.25 km distance and the 22 m depth limits of this study area.

The offshore consists of well-sorted, fine sand (2.2—2.5 ϕ, 0.47—0.62 σ) containing a large amount of organic detritus and shell hash. The surface is covered with inactive, symmetric, linear ripples stabilized by the rust layer (1 on Fig. 1A). In addition, a high density of the sand dollar *Echinarachnius parma* covers the surface (Fig.8). Virtually no primary depositional structures are preserved due to the extensive bioturbation by a high density of infaunal organisms. A typical can core for this region exhibits extensive

---

Fig.7. Relief peels (left) and X-ray radiographs (right) of can cores taken from the inner shelf. *A*. Transition (depth — 11.0 m). *OD* on radiograph marks areas draped with organic detritus in the troughs of small-scale ripple cross-lamination. *B*. Offshore (depth — 20.5 m). *IB* marks an infilled burrow and *SD* indicates an in place sand dollar *Echinarachnius parma*. *C*. Coarse-grained deposit (depth — 14.5 m). Black arrows at top left point landward. Scales at bottom right are 3 cm long.

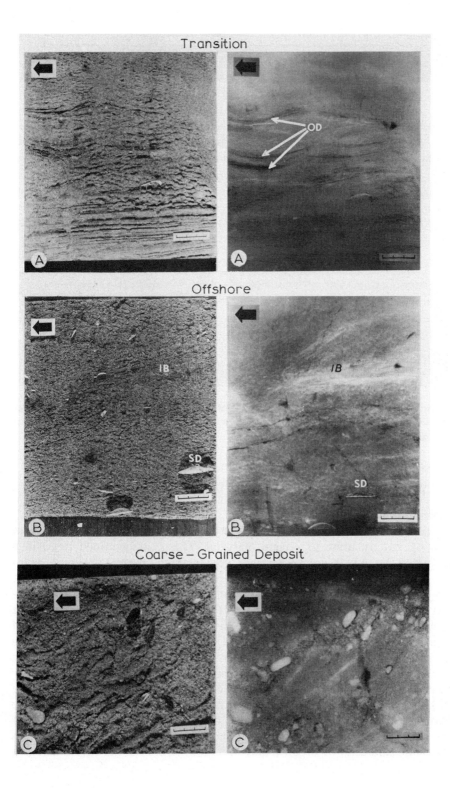

Fig.8. Inactive symmetric linear ripples in the offshore (depth — 20.5 m). *SD* indicates the sand dollar *Echinarachnius parma* and *ST* marks a surface trail of *E. parma*. Scale is 15 cm long.

reworking by *E. parma*. The only bioturbate structures preserved are wispy structures of amphipods and infilled burrows (Fig.7B).

Coarse-grained deposit

The last subenvironment observed in the Long Island nearshore system consists of patchy units of coarse material. This coarse-grained deposit is characterized by: (1) a coarse to very coarse sand texture; (2) large, symmetric, linear ripples; (3) physical structures consisting of medium-scale foreset bedding; and (4) moderate bioturbation. Surficially, the coarse-grained deposit appears similar to the longshore trough. The significant difference between the two areas is the occurrence of cross-bedding and abundance of gravel in the coarse-grained deposit. This subenvironment was observed at numerous locations varying in depth from 10 to 16 m and seems to occur commonly as exposed lenses throughout the transition.

The coarse-grained deposit consists of yellow-tan, moderately sorted coarse sand (0.7—0.9 ϕ; 0.73—0.84 σ) mixed with subrounded gravel. Similar to the longshore trough, the surface is dominated by large symmetric oscillatory ripples having a 60—80 cm wavelength and a 10—20 cm amplitude which contain large amounts of organic and inorganic debris in the ripple troughs.

Unlike the longshore trough, depositional structures are quite abundant. A typical can core reveals steep landward-dipping foreset bedding (Fig.7C). The preservation of bedding in this region is attributed to the absence of *A. americanus*. Wispy bioturbate structures are observed along the bedding planes in the coarse-grained deposit, caused by the presence of Haustoriid amphipods. In addition, several species of bivalves are found in this subenvironment.

DISCUSSION

Bedform sequences

The bedform sequence in the Long Island nearshore system follows the structural model of wave-formed features suggested by Clifton (1976). Even though simultaneous measurements of wave period, wave height, and bedform dimensions were not available, the observations of bedform symmetry and the presence of cross ripples allow qualitative determination of symmetric and asymmetric oscillatory flow for each environment (Fig.9). Unidirectional longshore or rip currents on the shoreface, noted by Clifton (1976) and discussed thoroughly by Davidson-Arnott and Greenwood (1976), were not observed during the present study. Recently, however observations made bi-weekly by aerial reconnaissance have confirmed the existence of seaward flowing rip currents along other parts of the Long Island coast to the west of the study area (G.A. Zarillo, pers. commun., 1983).

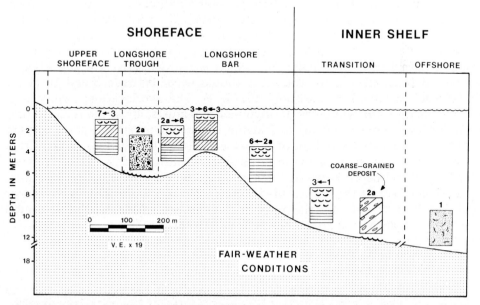

Fig.9. Summary of bedforms and physical sedimentary structures. The numbers above the typical can cores refer to the bedform types in Fig.1A. The arrows associated with the numbers indicate the direction in which the bedform sequence progresses.

The sequence of bedforms across the six subenvironments can be explained as a response to shoaling waves (Fig.1A). The symmetric, linear, to occasionally slightly irregular bedforms of the transition, the coarse-grained deposit, and offshore subenvironments are formed by infrequent large waves during storms or by long period swell. The bedforms in these environments maintain their symmetry, because waves in deeper water depth (>10 m) restricts orbital flow to symmetric oscillations. Waves with periods of 5—8 s on coasts that are affected infrequently by longer period swell (such as Long Island) may generate a symmetrically-rippled bottom to depths of 100 m (Clifton, 1976). The sequence of near-symmetric linear ripples at the seaward base of the longshore bar (depth — 10 m) to lunate megaripple at the bar crest (depth — 4 m) is controlled by wave shoaling and adheres to the order of bedforms proposed by the Clifton sequence. The increased size of the lunate megaripples on the bar crest during higher wave-energy conditions is an expression of increased velocity of asymmetric flows caused by larger waves. A reverse of the Clifton sequence of cross ripples at the bar crest to linear ripples at the bar trough contact (depth — 6 m) is a response to decreased asymmetric flow velocity caused by a rapid increase in depth on the landward slope of the bar. The large symmetric oscillatory ripples in the coarse sand of the longshore trough (depth — 5.5 to 6.0 m) are controlled by two factors — increased depth and larger grain size. The sequence of asymmetric linear ripples (depth — 5.5 m) to flat beds (depth — 0 m) on the upper shoreface is again an expression of an increased asymmetric flow velocity caused by reformed shoaling waves.

In addition to the above qualitative determination of the transition from symmetric to asymmetric bedforms, interpolation of data from Clifton (1976, table 2) provides further support for the empirical model of asymmetric wave-formed structures. Assuming an 8 s wave period and a deep-water wave height of 0.7—0.8 m (which is representative for the Atlantic coast of Long Island), the transition should occur between 5 and 10 m for "typical" fair-weather conditions (Table I). The two locations, where bedforms grade from symmetric to asymmetric features, are at the seaward base of the bar (depth — 10 m) and at the bar trough-upper shoreface contact (depth — 5.5 to 6.0 m). Both of these locations are included in the transition envelope predicted from the interpolation of Clifton's data. The difference in transitional depth between these two locations is caused by the longshore bar. The effect of the bar on the incoming waves is to dissipate wave energy by breaking. Therefore, the reformed waves are smaller and thus orbital velocities near the bed are reduced (Wood, 1971).

Physical and biogenic structures

The sedimentary structures preserved in the upper portion of the can cores are different from those preserved in the lower half (Fig.9). This pattern has been interpreted as low-wave energy reworking of underlying structures deposited during higher energy conditions (Howard and Reineck,

TABLE I

Depth of transition during fair-weather conditions from symmetric to asymmetric ripples in the nearshore zone. T is the wave period in seconds. H_o is the deep-water wave height in meters. Eastern Long Island transition depth is an interpolation of data presented in Clifton (1976) which is verified by observations in the present study

LOCATION	TYPICAL FAIR-WEATHER WAVE		TRANSITION DEPTH
	T(sec)	H_o(m)	
WILLAPA BAY, WASHINGTON	2	0.2	1m
SOUTHEASTERN SPAIN	5	0.5	3-5m
EASTERN LONG ISLAND	8	0.7-0.8	5-10m
SOUTHERN OREGON	10	1.0	10-20m

1972a; Greenwood and Hale, 1980) and provides each core with a record of two phenomena. The top of the cores are the fair-weather phase, whereas the bottom shows the most recent storm event. With only one exception, the Clifton sequence can be used to explain relative flow conditions of the higher wave-energy (deeper) sedimentary structures.

The deeper bedding structures of the upper shoreface consists of planar lamination overlain by landward-dipping medium-scale foreset bedding. The longshore trough would most likely be characterized by landward-dipping, medium-scale foreset bedding similar to the stratification found in the coarse-grained deposit, if it were not for the burrowing activities of *A. americanus*.

The three subdivisions of the longshore bar display separate sequences of physical structures. The landward-dipping, medium-scale foreset bedding on the landward slope indicates a moderate intensity of asymmetric flow, which may be caused by wave energy dissipation on the bar crest. The bar crest is characterized by larger scale foreset bedding (up to 3 ×) signifying larger bedforms and more intense asymmetric flow. The planar lamination of the seaward slope seems inconsistent with the Clifton sequence (i.e., planar lamination found in shallower water depth than foreset bedding). In the study of the barred coast of Kouchibouguac Bay, New Brunswick, Canada, an explanation has been suggested for the absence of foreset bedding on the seaward bar slope. The planar lamination have been interpreted as upper flow-regime structures of a symmetric oscillatory flow (Davidson-Arnott and Greenwood, 1976). This sequence progresses from ripples directly to flat beds with no intermediate formation of megaripples (Fig.1B).

The structures at depth in the cores of the transition subenvironment are the same as those of the seaward bar slope. Since the only difference between both environments is the distance landward or seaward of fair-weather wave base, the major distinction between these two environments is textural.

In the transition the degree of bioturbation increases seaward. This trend is similar to the "parallel laminated-to-burrow sets" observed in cores from the nearshore system off Sapelo Island, Georgia, and viewed in outcrop from the late Cretaceous of east-central Utah (Howard, 1971). The offshore subenvironment is characterized by a lack of physical structures at depth due to the high degree of bioturbation and the apparent infrequency of competent flows. Occasionally, biogenic structures are infilled with organic detritus and preserved. The dominant physical structures throughout the inner shelf coarse-grained deposit are landward-dipping, medium-scale foreset bedding caused by the occasional migration of the steep slipfaces of the oscillatory ripples.

The shoreface—inner shelf contact at the base of the longshore bar (depth − 10 m) marks the most landward indication of major biogenic reworking of nearshore system. Landward of this point, physical sedimentary structures are dominant. The transition subenvironment is characterized by a seaward increase in bioturbation. At the point where few physical structures are present due to bioturbation, the offshore subenvironment begins. In eastern Long Island, the change from the transition to the offshore is approximately at a 14—16 m depth between 1.4—1.6 km seaward of MLW. This study clearly documents the transition as a distinct subenvironment identifiable by discrete sedimentological characteristics. This view challenges the existing convention of a "transition zone" which has been considered as a simple "mixing" of shoreface and offshore components (e.g., Emery, 1960; Reineck and Singh, 1975).

Origin of the coarse-grained deposit

The presence of fields of large, linear bedforms partially composed of coarse-grained sediment has been documented for the inner shelf environment off Atlantic Beach, western Long Island (Swift and Freeland, 1978; Swift et al., 1979). With the aid of side-scan sonar, well-defined sets of features, interpreted as low relief sandwaves, have been observed in water depths between 9 and 14 m. These features exhibit a 10—100 m spacing, 1 m height, and crestlines oriented perpendicular to the shoreline. The sandwaves consist of a smooth, low-angle stoss slope (east-facing), a less smooth higher angle lee slope (west-facing), and a rippled flat trough. The bed roughness of the sandwave troughs is due to large oscillatory ripples generated by infrequent storm waves, and, therefore, are oriented crest-parallel to the shoreline.

Swift and Freeland (1978) and Swift et al. (1979) attributed the origin of the inner shelf sandwaves to unidirectional flow caused by extratropical storms. Measurements by Scott and Csanady (1976) and Lavelle et al. (1978a, b) revealed storm-induced westward bottom currents flowing parallel to the isobaths of the Long Island shelf. The large ripples in the troughs of sandwaves described for the western Long Island shelf seem analogous to the coarse-grained deposits of eastern Long Island observed in the present study.

Stratigraphic implications

A stratigraphic model for progradation of the Long Island nearshore system would likely consists of at least a portion of all six sedimentary subenvironments (Fig.10). The entire sequence would be underlain by organic-rich bioturbated offshore deposits. Conformably overlying the offshore would be the planar-laminated, slightly bioturbated, transition sediments. The contact between the offshore and the transition would be highly gradational and is placed at 15 m in this section. Dispersed throughout the offshore and transition subenvironments would be lenses of inner shelf coarse-grained deposit. Grading upward from the transition are the cleaner, planar-laminated, lower seaward slope sediments of the longshore bar. Above this is a distinct erosional surface indicating the base of the coarse, massive to possibly cross-bedded deposits of the longshore trough. The 2 m unit of bar sands below this erosional surface would be the only

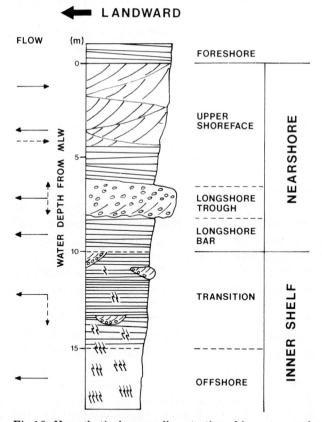

Fig.10. Hypothetical prograding stratigraphic sequence for the eastern Long Island nearshore system. Solid flow arrows indicate observed flow conditions, while dashed arrows indicate postulated flow conditions. The structures in the foreshore were not observed in the present study and are assumed from such sources as Elliott (1978) and Hunter et al. (1979). The lensoidal structures in the transition are the coarse-grained deposits.

portion of the longshore bar preserved. This would be due to erosion of shallow bar crest deposits by the progradation of the longshore trough. Capping the sequence is the planar- to cross-laminated clean sands of the upper shoreface and the proposed planar laminations of the foreshore.

One application of this study to interpretation of ancient environments may be the development of firmer criteria for determining the relative position of fair-weather wave base (or at least position landward or seaward of it). Some of these points have been discussed by earlier workers (e.g., Howard, 1971; Howard and Reineck, 1972a), but consideration of all criteria together may assist future investigations. The three criteria are:

(1) The amount and type of physical and biogenic structures. Preservation of biogenic structures would indicate a position seaward of wave base. Generally, the more bioturbation, the farther seaward the environment.

(2) The presence of drapes (mud or organics) in small-scale ripple cross-lamination (flaser bedding). This criterion is a very sensitive indicator of distance seaward of wave base. The greater the seaward distance, the more opportunity for the finer material to settle out.

(3) An increased darkening of overall sediment color may indicate approximate distance seaward of wave base. This criterion would be subtle and/or poorly preserved, but may be useful when considered with other indicators.

CONCLUSIONS

Four conclusions can be drawn concerning the barred nearshore sedimentary system from this study of Long Island:

(1) Six subenvironments were observed during fair-weather conditions. The shoreface environment is divided into the upper shoreface, the longshore trough, and the longshore bar. The subenvironments of the inner shelf are the transition, the offshore, and the coarse-grained deposits. These subenvironments were defined during fair-weather conditions as distinct combinations of sediment texture and color, bedforms, physical and biogenic sedimentary structures, and infaunal assemblages.

(2) The order of magnitude reduction in slope (3.0—0.3°) at the shoreface-inner shelf boundary coincides with fair-weather wave base. The location of wave base can be characterized at any moment in time by: (a) the cessation of debris surge in ripple troughs; (b) the onset of the rust layer formation over the bedform surface; (c) the darkening of sediments caused by the increased presence of organic detritus; and (4) subtle changes in sediment texture.

(3) Sedimentary structures produced by shoaling waves follow a consistent shoreward progression which is controlled by variations in oscillatory velocity and velocity asymmetry. This oscillatory flow regime concept as proposed by Clifton (1976) and modified by Davidson-Arnott and Greenwood (1976) can be used to qualitatively predict oscillatory flow asymmetry. Specifically, cross ripples are a particularly good indicator of the asymmetric transition between asymmetric ripples and lunate megaripples.

(4) Criteria useful for distinguishing position relative to fair-weather wave base in ancient nearshore environments are: (a) the amount and type of physical and biogenic structures; (b) the occurrence of mud or organic drapes in the small-scale ripple cross-laminations; and (c) the overall possible darkening of sediment color due to the increase of fine-grained material.

ACKNOWLEDGMENTS

This study was carried out as part of the research program of the Coastal Research Division, Department of Geology, University of South Carolina. The project was made possible by a grant from the Morris Companies of Secaucus, New Jersey. The final preparation of the manuscript was undertaken with the support of the facilities at the National Museum of Natural History, Smithsonian Institution. Miles O. Hayes, John C. Horne, Robert Ehrlich, Larry G. Ward, and Gary A. Zarillo read earlier copies of the manuscript. Robin G.D. Davidson-Arnott, Brian Greenwood, and Gary A. Zarillo critically reviewed the final copy. Charlotte Johnson assisted in the drafting of the figures. Appreciation is also extended to Stephanie A. Staples in the preparation of the final manuscript. A special thanks is offered to John H. Barwis and Robert Ehrlich for their initial suggestions and enthusiasm for the undertaking of this study. Their continued support throughout the entire study is also acknowledged.

REFERENCES

Anan, F.S., 1972. Hydraulic equivalent sediment analyzer (HESA). Coastal Research Center, Univ. of Massachusetts, Techn., Rep. 3-CRC, 37 pp.
Bouma, A.H., 1969. Methods for the Study of Sedimentary Structures. Wiley, New York, N.Y., 458 pp.
Clifton, H.E., 1976. Wave-formed sedimentary structures — a conceptual model. In: R.A. Davis and R.L. Ethington (Editors), Beach and Nearshore Sedimentation. Soc. Econ. Paleontol. Mineral., Spec. Publ., 24: 126—148.
Clifton, H.E., Hunter, R.E. and Phillips, R.L., 1971. Depositional structures and processes in the non-barred high energy nearshore. J. Sediment. Petrol., 41: 651—670.
Curray, J.R. and Moore, D.G., 1964. Pleistocene deltaic progradation of continental terrace, Costa de Nayarit, Mexico. In: T.H. van Andel and G.G. Shor (Editors), Marine Geology of the Gulf of California. Am. Assoc. Pet. Geol., Tulsa, Okla., pp.193—215.
Davidson-Arnott, R.G.D. and Greenwood, B., 1974. Bedforms and structures associated with bar topography in the shallow-water wave environment, Kouchibouguac Bay, New Brunswick, Canada. J. Sediment. Petrol., 44: 698—704.
Davidson-Arnott, R.G.D. and Greenwood, B., 1976. Facies relationships on a barred coast, Kouchibouguac Bay, New Brunswick, Canada. In: R.A. Davis and R.L. Ethington (Editors), Beach and Nearshore Sedimentation, Soc. Econ. Paleontol. Mineral., Spec. Publ., 24: 149—168.
Davies, J.L., 1964. A morphogenic approach to world shorelines. Ann. Geomorphol., 8: 127—142.
DeWall, A.E., 1979. Beach Changes at Westhampton Beach, New York, 1962—1973. Coastal Engineering Research Center, Misc. Rep., 79-5, Corps of Engineers, Fort Belvoir, Va., 129 pp.

Dietz, R.S. and Fairbridge, R.W., 1968. Wave base. In: R.W. Fairbridge (Editor), Dictionary of Geomorphology (Vol. III). Reinhold, New York, N.Y.

Elliott, T., 1978. Clastic shorelines. In: H.G. Reading (Editor), Sedimentary Environment and Facies. Elsevier, New York, N.Y., pp.143—175.

Emery, K.O., 1960. The Sea off Southern California. Wiley, New York, N.Y., 366 pp.

Goldsmith, V., Bowman, D. and Krinsley, K., 1982. Sequential stage development of crescentic bars: Hahotherm Beach, southeastern Mediterranean. J. Sediment. Petrol., 52: 233—249.

Greenwood, B. and Hale, P.B., 1980. Depth of activity, sediment flux, and morphological change in a barred nearshore environment. In: S.B. McCann (Editor), The Coastline of Canada: Littoral Processes and Shoreline Morphology. Geol. Surv. Can., Pap. 80-10: 89—109.

Greenwood, B. and Mittler, P.R., 1979. Structural indices of sediment transport in a straight, wave-formed nearshore bar. Mar. Geol., 32: 191—204.

Hamblin, W.K., 1971. X-ray photography. In: R.E. Carver (Editor), Procedures in Sedimentary Petrology. Wiley-Interscience, New York, N.Y., pp.251—284.

Hill, G.W. and Hunter, R.E., 1976. Interaction of biological and geological processes in the beach and nearshore environments, northern Padre Island, Texas. In: R.A. Davis and R.L. Ethington (Editors), Beach and Nearshore Sedimentation. Soc. Econ. Paleontol. Mineral., Spec. Publ., 24: 169—187.

Howard, J.D., 1968. X-ray radiography for examination of burrowing in sediments by marine invertebrate organisms. Sedimentology, 11: 249—258.

Howard, J.D., 1971. Amphipod bioturbate textures in recent and Pleistocene beach sediments. In: J.D. Howard, J.W. Valentine and J.E. Warme (Editors), Recent Advances in Paleoecology and Ichnology. Am. Geol. Inst., Short Course Lecture Notes, pp.213—233.

Howard, J.D. and Frey, R.W., 1975. Estuaries of the Georgia coast, U.S.A. — sedimentology and biology, I: Introduction. Senckenbergiana Mar., 7: 1—31.

Howard, J.D. and Reineck, H-E., 1972a. Georgia coastal region, Sapelo Island, U.S.A. — sedimentology and biology, IV: physical and biogenic sedimentary structures of the nearshore shelf. Senckenbergiana Marit., 4: 81—123.

Howard, J.D. and Reineck, H-E., 1972b. Georgia coastal region, Sapelo Island, U.S.A. — sedimentology and biology, VIII: conclusions. Senckenbergiana Marit., 4: 217—222.

Howard, J.D. and Reineck, H-E., 1981. Depositional facies of high-energy beach-to-offshore sequence: comparison with low-energy sequence. Bull. Am. Assoc. Pet. Geol., 65: 807—830.

Hunter, R.E., Clifton, H.E. and Phillips, R.L., 1979. Depositional processes, sedimentary structures, and predicted vertical sequences in barred nearshore systems, southern Oregon coast. J. Sediment. Petrol., 49: 711—726.

Kent, H.C., 1976. Modern Coastal Sedimentary Environments — Alabama and Northwest Florida. Geological Exploration Associates, Golden, Colo., 96 pp.

Kravitz, J.H., 1966. Using an ultrasonic disruption as an aid to wet sieving. J. Sediment. Petrol., 36: 811—812.

Lavelle, J.W., Swift, D.J.P., Gadd, P.E., Stubblefield, W.L., Case, F.N., Brashear, H.R. and Haff, K.W., 1978a. Fair weather and storm sand transport on the Long Island, New York, inner shelf. Sedimentology, 25: 823—842.

Lavelle, J.W., Young, R.A., Swift, D.J.P. and Clarke, T.L., 1978b. Near-bottom sediment concentration and fluid velocity measurements on the inner continental shelf. J. Geophys. Res., 83: 6052—6062.

Lettau, B., Brower, W.A. and Quayle, R.G., 1976. Marine Climatology. MESA New York Bight Atlas Monograph 7, New York Sea Grant Institute, Albany, New York, 239 pp.

Newton, R.S., 1968. Internal structure of wave-formed ripple marks in the nearshore zone. Sedimentology, 11: 275—292.

Reineck, H-E., 1963. Sedimentgefüge im Bereich der Südlichen Nordsee. Abh. Senckenb. Naturforsch. Ges., 505, 138 pp.

Reineck, H-E. and Singh, I.B., 1971. Der Golf von Gaeta (Tyrrhenisches Meer) — III. Die Gefüge von Vorstrand- und Schelfsedimenten. Senckenbergiana Marit., 3: 185—201.

Reineck, H-E. and Singh, I.B., 1975. Depositional Sedimentary Environments. Springer, New York, N.Y., 439 pp.

Saville, T., 1954. North Atlantic coast wave statistics hindcast by Bretschneider-revised Sverdrup-Munk method. Beach Erosion Board, Techn. Mem. 55, Corps of Engineers, Washington, D.C., 76 pp.

Scott, J.T. and Csanady, G.T., 1976. Nearshore currents off Long Island, J. Geophys. Res., 81: 5403—5407.

Swift, D.J.P. and Freeland, G.L., 1978. Current laminations and sandwaves on the inner shelf, middle Atlantic bight of North America. J. Sediment. Petrol., 48: 1257—1266.

Swift, D.J.P., Freeland, G.L. and Young, R.A., 1979. Time and space distribution of megaripples and associated bedforms. Sedimentology, 26: 389—406.

Taney, N.E., 1961. Geomorphology of the south shore of Long Island, New York. Beach Erosion Board, Techn. Mem. 128, Corps of Engineers, Washington, D.C., 50 pp.

U.S. Army Corps of Engineers District, New York, 1977. Final Environmental Impact Statement for Fire Island Inlet to Montauk Point, New York — Beach Erosion Control and Hurricane Protection Project. Vols. I and II, 375 pp.

U.S. Dept. of Commerce, 1978. Tide Tables — East Coast of North and South America. Natl. Ocean. Atmos. Admin., National Ocean Survey, Rockville, Md., 288 pp.

Wood, W.L., 1971. Transformation of breaking wave parameters over a submarine bar. Unpubl. Ph.D. Diss., Michigan State University, East Lansing, Mich., 224 pp.

BEACH AND NEARSHORE FACIES: SOUTHEAST AUSTRALIA

A.D. SHORT

Coastal Studies Unit, Department of Geography, University of Sydney, Sydney, N.S.W. 2006 (Australia)

(Received December 31, 1982; revised and accepted August 29, 1983)

ABSTRACT

Short, A.D., 1984. Beach and nearshore facies: southeast Australia. In: B. Greenwood and R.A. Davis, Jr. (Editors), Hydrodynamics and Sedimentation in Wave-Dominated Coastal Environments. Mar. Geol., 60: 261—282.

The morphology, texture and facies sequence on seven sand beaches, located in low, moderate and high wave energy, microtidal environments in southern Australia were investigated using box coring and Scuba observations. Systematic variation in facies occur both within and between the beaches. Low-energy reflective beaches are limited in lateral and vertical extent and in facies to beach laminations separated by coarse step deposits from finer nearshore cross-lamination facies. Moderate-energy intermediate beaches characterised by rip circulation possess increasingly wider surfzones with ridge and runnel and bar-trough facies separating the beach and step facies from the more extensive nearshore sequence. High-energy dissipative beaches may have 500 m wide surfzones containing multiple bar-trough topography. Fine beach laminations with backwash structures grade into 4—5 m thick bar-trough sequences then the extensive nearshore facies. As wave energy increases from low ($H_b < 1$ m) to high ($H_b > 2.5$ m) the vertical extent of the beach to nearshore sequence increases from <10 m to approximately 30 m, and the width from 100 m to several kilometres. Consequently one would expect higher-energy paleo-beach sequences to be represented more by diagonal than vertical facies sequences.

INTRODUCTION

The beach and nearshore zone extends from the upper swash limit across the surfzone to modal wave base, the limit to which modal waves actively entrain sediments. The entire zone consists of depositional facies formed by wave-current dynamics and the associated boundary layer flows at the bed. The abundance of paleo-beaches attests to the preservation potential of such deposits during progradational episodes. Understanding the relationship between beach facies and the environmental conditions that produce them is important for several reasons. First, it enables a more complete definition of the morpho-stratigraphic characteristics of a beach; second, in so far as the facies reflect processes at the time of deposition they can indicate present day wave and bed dynamics (e.g., Clifton et al., 1971; Davidson-Arnott and Greenwood, 1976; Davidson-Arnott and Pember, 1980); and third, given the above, if the structures are preserved in the rock record they can equally be used to identify paleo-beaches and to interpret the prevailing environmental conditions at the time of deposition (e.g., Clifton et al., 1971; Reineck and Singh, 1973; McCubbin, 1982; Allen, 1982; Dupré, 1984, this volume).

The pioneering, though recent, investigations in this field recognised the potential range of beach environments in response to varying levels of wave energy and beach configuration. High- and low-energy systems were described by Clifton (1976), and Howard and Reineck (1981). Specific beach configurations such as barred nearshores (Davidson-Arnott and Greenwood, 1974, 1976; Hunter et al., 1979; Greenwood and Mittler, in press) illustrate alternatives to the classic planar nearshore of Clifton et al. (1971). The recently developed models of morphodynamic beach response to low (<1 m) or high (>2.5 m) waves (Fig. 1) by Wright and Short (1983), and Short and Wright (1983), provides a basis for a systematic study of beach and nearshore facies across a range of wave environments. The aim of this paper is to describe the bedforms, textures, structures and resulting facies from a range of beaches, exposed to modally low, moderate and high wave activity. More specifically results are presented from seven contrasting coastal environments in southeast Australia. The characteristic facies sequence of each is presented together with the extent of the various facies within each system.

FIELD SITES AND METHODS

Seven beaches in southeast Australia (Fig.2) that are representative of Wright and Short's (1983) six beach states (Fig. 1) were investigated. The morphodynamic characteristics of each field site are given in Table I. East coast sites are located in a micro-tidal (spring range 1.6 m), east coast swell environment, with highly variable wave regime. The modal deep water wave is 1.5 m in height with a period of 10 s. Goolwa on the south coast has a 1.0 m spring tide range and a modal 3 m, 12 s wave, typical of this west coast swell environment. Sands predominantly consist of quartz with variable percentages of carbonates (shell fragments). All systems were surveyed using the Emery method in shallow water and echosounder in deep water. All systems were box-cored subaerially and subaqueously the later by Scuba divers who also measured bedforms (Short and Wright, in press). Using a modified method of Burger et al. (1969), box-cores were impregnated with Ciba-Geigy araldite (K79 kit) to give a 30 × 20 cm cast of near surface structures. In addition samples of sediment were dry sieved at 0.25 phi intervals to determine grain-size statistics.

BEACH TEXTURES AND STRUCTURES

The results from the seven beach systems are presented briefly to give an indication of the nature, extent and relationship of the facies within each representative beach type. These results are then combined into a more general classification.

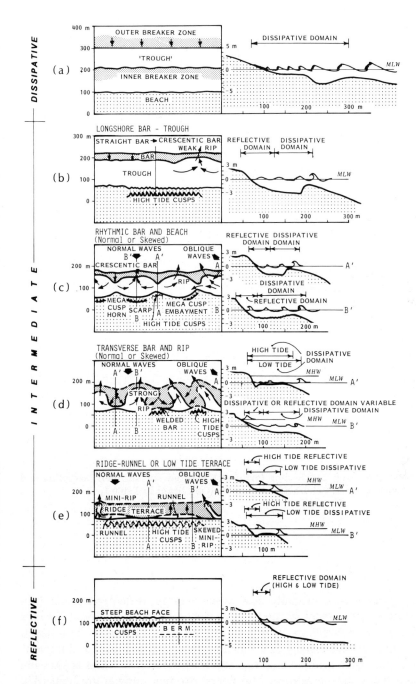

Fig.1. Plan and profile configuration and basic surfzone circulation pattern of the six beach states (from Wright and Short, 1983).

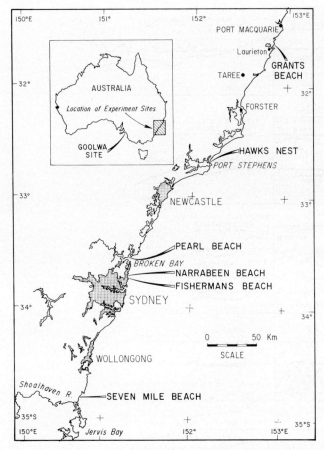

Fig.2. Location of the seven beach sites (arrowed) selected for sediment sampling and box-coring. The wave-sediment characteristics of each are listed in Table I.

LOW-ENERGY REFLECTIVE

Fishermans Beach is a modally low wave energy reflective beach (see Figs. 1f, 3a and 4) composed of medium to very coarse sand. Modal breaker height is 30 cm, with breakers rarely exceeding 1 m (Table I). The beach consists of a moderately steep (10°) beach face capped by an incipient foredune. A very coarse-grained step lies at about mean low water separating the beach face from the nearshore. The nearshore has a lower gradient (3°), and is approximately 50 m wide terminating at a depth of 3 m (modal wave base).

TABLE I

Wave-sediment characteristics of beach sites

LOCATION	FISHERMANS	PEARL	HAWKS NEST	NARRABEEN	GRANTS	M.SEVEN	GOOLWA
BEACH TYPE	REFLECTIVE		INTERMEDIATE				DISSIPATIVE
η	10	17	14	35	21	25	12
Hb(m)	.3	.5	1	1.5	1.6	1.6	3
T	10	10	10	10	10	10	12
Gd(mm)	.35	.5	.26	.3	.3	.27	.2
Ws	.05	.076	.035	.04	.04	.036	.026
Ω	.60	.65	2.85	3.75	4.0	4.4	9.6
Gradient	1:9	1:12	1:13	1:15	1:29	1:37	1:33
B-S	Reflective	Reflective	Ridge Runnel	Transverse Bar-Rip	Rhythmic Bar-Beach	Bar-Trough	Dissipative

η - number of cores; Hb - modal breaker height; Gd - mean grain size; Ws - mean fall velocity (cm sec^{-1}); Ω =Hb/T Ws; Gradient - subaerial beach slope; B-S - beach state (see Figure 1).

Facies (Figs.5 and 6a)

The beach face is composed of thinly bedded sub-parallel, laminations of varying thickness with variable lateral continuity, that dip seaward at $\sim 5°$ (Fig.5a). Sediment varies from very coarse to medium sands. A high degree of variability in texture and laminae thickness reflects the immaturity of the beach face sediments in this low-energy environment.

The step consists of medium-scale sets (~ 10 cm) of cross-stratified gravels (Fig.5b). These gravels are poorly sorted, and rich in carbonate. The step is located at the toe of the beach face at the point of wave surging. It is usually a few decimetres in thickness. It overlies and abruptly grades into nearshore sediments. The nearshore has fine-to-medium, well-sorted, low-carbonate sands (Fig.6); sets of medium-scale, landward-dipping cross-laminations with $0-20°$ dip angles are predominant (Fig.5c). The cross-lamination is produced by asymmetrical oscillation ripples with straight parallel to slightly sinuous crests [length $(L) = 40$ cm, height $(H) = 7$ cm].

HIGHER ENERGY REFLECTIVE

Pearl Beach is a more energetic beach than Fishermans, though still reflective and, apart from its greater extent (Fig. 3b) it has several other distinguishing characteristics resulting from the higher wave energy. The beach is wider consisting of a foredune fronted by a berm runnel and series of beach cusps. The $10°$ beach face grades into a coarse step. The nearshore zone slopes seaward at $5-6°$ from the step, to a depth of $6-8$ m where it levels out into a low gradient ($0.5°$) offshore zone (Fig.4).

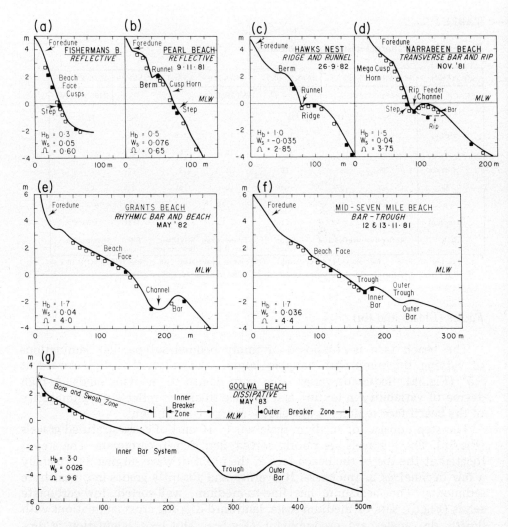

Fig.3. Beach—surfzone profiles across the field sites showing cross-sectional morphology and location of box cores, indicated by squares and solid squares (illustrated in Figs.5 and 7—12). See Table I for definition of H_b, w_s and Ω.

Facies (Figs.6a and 7)

The berm crest consists of 2—10 cm thick units of planar continuous parallel laminae (see Fig.7a). The units alternate between coarse (runnel deposits) and finer sand. The beach face contains seaward dipping (<5°), continuous, planar to curved parallel beds, 10—15 cm thick, and composed of either coarse or finer sand. The lower beach face has steeper (10—20°) more discontinuous planar to curved, non-parallel beds of coarse (shell rich) sand. These grade into the 30 cm thick, partially cross-laminated (seaward dipping 20°), coarse sand beds of the step (Fig.7b). Immediately seaward of the step,

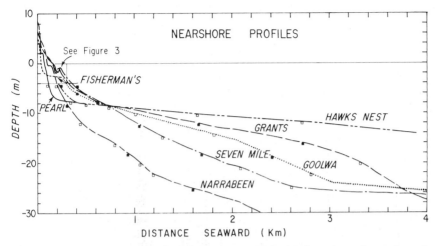

Fig.4. Nearshore profiles across the field sites showing location of nearshore box cores (squares).

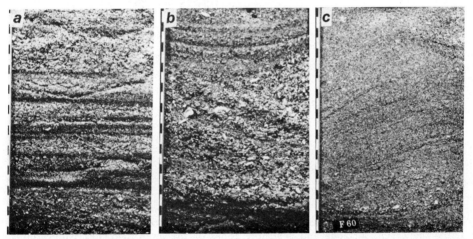

Fig.5. Fisherman's beach box cores. Shore to left. (a) beach face; (b) step; (c) nearshore, depth 2 m. Scale in centimetres.

sediments fine rapidly and sinuous oscillation ripples produce predominantly landward dipping (10°) 5—10 cm thick cross laminations (Fig.7c). These extend offshore to modal wave base at 8 m where bioturbation affected 10—20% of the 30 cm deep core.

RIDGE AND RUNNEL

Hawks Nest Beach is located toward the southern end of the 16 km long Fens embayment. Fens grades from a moderate to high energy rhythmic bar and beach system (Fig.1c) at the northern end to a low-energy, reflective

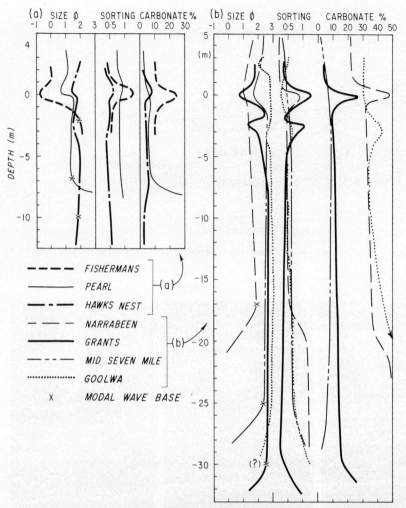

Fig.6. Vertical sequence of sediment characteristics (size, sorting, and percent carbonate) across the beach, surfzone and nearshore zones of the seven field sites. Non-carbonate sediments are predominantly quartz grains.

system at the more protected southern end. Hawks Nest beach experiences a range of wave levels (H_b up to 3 m) and beach types, but modally is a ridge and runnel type (Fig.1e) as it was during the field investigation (Fig.3c). It has a low foredune fronted by a 40 m wide berm-runnel and berm, with a 6° beach face which terminates at a low tide step. The surfzone consists of a shallow narrow runnel and flat ridge. Past the breakpoint, slope increases to 6° until reaching modal wave base and a 0.2° gradient offshore zone at 10 m depth (Fig.4).

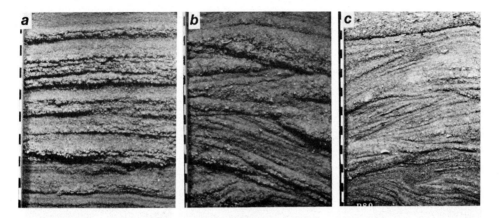

Fig.7. Pearl beach box cores. Shore to left. (a) berm crest; (b) step; (c) nearshore, depth 1 m. Scale in centimetres.

Facies (Figs.6a and 8)

Sediments are predominantly medium to fine sand (Fig.6a), finer than the two reflective beaches described above. The berm consists of continuous parallel laminae, arranged in thin beds, grain size is in the medium to fine sand range. The upper (high tide) beach face contains >30 cm thick beds of seaward dipping (5°) continuous, parallel laminae (up to 3 cm in thickness) of medium-grained sand with occasional coarser grained, shelly laminae (Fig.8a). The lower (low-tide) beach face had shallow scour depressions on the surface with structures alternating between those similar to the upper beach face and 10 cm thick beds of coarse to very coarse seaward dipping (15°) shelly material. The latter represent the step deposits formed at high tide when the beach is more reflective (Fig.8b).

The runnel contained long-crested wave ripples (L = 100 cm, H = 10 cm), which produced steeply dipping (20°) cross-stratification consisting of tabular to trough parallel laminae arranged in 20 cm beds of medium-grained sand. These deposits are overlain by ridge sediments which follow the sequence of Davis et al. (1972). The onshore part of the ridge contained landward dipping (10—30°) cross strata while the crest contained sub-horizontal strata (Fig.8c).

Cross ripples occurred seaward of the break point and were best developed at 4 m depth (L = 20—40 cm, H = 5—8 cm). They produced predominantly landward dipping cross strata (10—30°, Fig.8d) similar to that described by Clifton (1976). Between depths of 4 and 5 m were large megaripples (L = 400 cm, H = 50 cm; which could have been produced by 3 m waves three days previously). The megaripples had tangential, predominantly landward-dipping (10—20°) cross strata on the crest, consisting of fine to medium sand, with shelly cross strata (10—15 cm thick) in the trough. This overlay seaward dipping (10°) slightly bioturbated parallel laminae (pre-high waves?) (Fig. 8e). The zone of megaripples graded into parallel sharp crested, wave

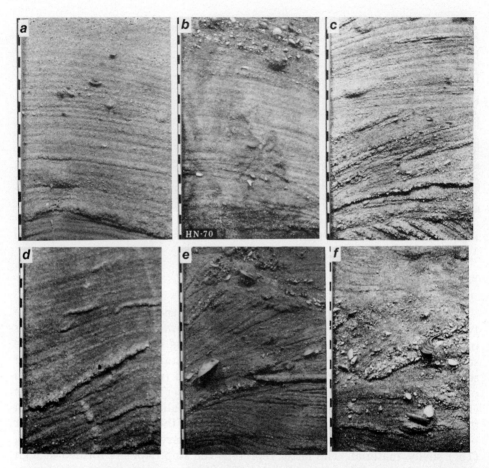

Fig.8. Hawks Nest box cores. Shore to left. (a) upper beach face; (b) lower beach face step; (c) bar crest; (d) nearshore (cross ripples), depth 4 m; (e) nearshore (megaripple trough), depth 5 m; (f) nearshore, depth 6 m. Scale in centimetres.

ripples (L = 30 cm, H = 5 cm) which became sinuous at 8—10 m depth. Internal structures were planar to ripple cross laminations with tangential landward dipping (10—20°) cross beds, of fine to medium sand, and shell rich (*Bankavia*) (20—40°) landward-dipping layers that were bioturbated in the deepest part of this zone (10—15 cm thick). *Bankavia* both living and dead were abundant on the surface and in cores taken between 5 and 9 m water depth (Fig.8f).

TRANSVERSE BAR AND RIP

Narrabeen Beach ranges in morphology from reflective to intermediate type in response to a highly variable wave climate with breakers frequently >3 m and occasionally >5 m; modal wave height is 1.5 m (Short and Wright, 1981). The beach was sampled and cored on three occasions when well-developed transverse bar and rip systems (Fig.1d) dominated the morphology (Fig.3d). The berm and beach face varied in width from 20 to 50 m depending on location relative to the megacusp horns and embayments (Short, 1979). The longshore spacing between megacusps averaged 150 m. The megacusp embayments were fronted by rip feeder and rip channels (0.5—1.5 m deep) which run normal to the shoreline across the bar (see Wright and Short, 1983). The megacusp horns were attached to the transverse bars which continued 50—80 m seaward to the break point. The nearshore zone slopes at 1.5° from the break point to beyond the modal wave base at 18 m (Fig.4). The system is morphologically analogous to the connected inner bar systems described by Greenwood and Davidson-Arnott (1975).

Facies (Figs.6b and 9)

The berm and upper beach face deposits are similar to the previous two beaches. On the megacusp horns the beach face has a 5° gradient with seaward dipping (1—2°) parallel laminae which grade into coarser grained, 10 cm thick cross strata (10—15°) at the junction with the bar. A step is absent. In the embayment coarser grained, 2—5° seaward dipping tangential laminae, grade into a zone of cross strata before a very coarse grained, shelly step with steeply (20°) seaward dipping beds is reached (Fig.9a). The bar facies is similar in sequence to the Hawks Nest ridge deposits; they are however more extensive, slightly coarser and higher in skeletal carbonates. The rip feeder channels, analogous to Hunter et al.'s (1979) 'longshore trough facies' and Davidson-Arnott and Greenwood's (1976) 'trough' facies, contained both wave and current ripple structures. They consisted of very coarse, shell-rich, predominantly seaward dipping (5—40°) cross strata with laminae arranged in alternating 5—10 cm thick sets (Fig.9b). The rip channel contained seaward migrating megaripples (L = 150 cm, H = 25—30 cm). The medium- to coarse-grained megaripple crests produce seaward dipping (10—25°) tangential laminae (Fig.9c). The troughs consist of coarse sand arranged in medium-scale cross strata (10—20° dip). These structures are basically identical to the rip-channel facies described by Davidson-Arnott and Greenwood (1976) and Hunter et al. (1979).

Immediately seaward of the break point and to a depth of 5 m cross ripples dominated (L = 50—100 cm, H = 5—15 cm). They contained 5—10 cm sets of landward dipping (5°) tangential laminations, overlain by sets of steeper (10—40°) predominantly landward dipping cross laminations (Fig.9d). Beyond 6 m depth, sinuous, sharp crested wave ripples (L = 40—80 cm, H = 7—10 cm), pass laterally into parallel-crested forms (L = 40—60 cm, H = 10 cm) which extend to modal wave base at 18 m depth. Cross strata dominate with 5° landward dipping laminae truncating (20°) seaward dipping strata arranged in 5—10 cm thick sets (Fig.9e). Beyond modal wave base sediments rapidly coarsen, becoming shellier with poorer sorting (Fig.6b). However, periodic high waves produce well-developed sharp-crested, parallel wave ripples (L = 50 cm, H = 10 cm) out to a depth of

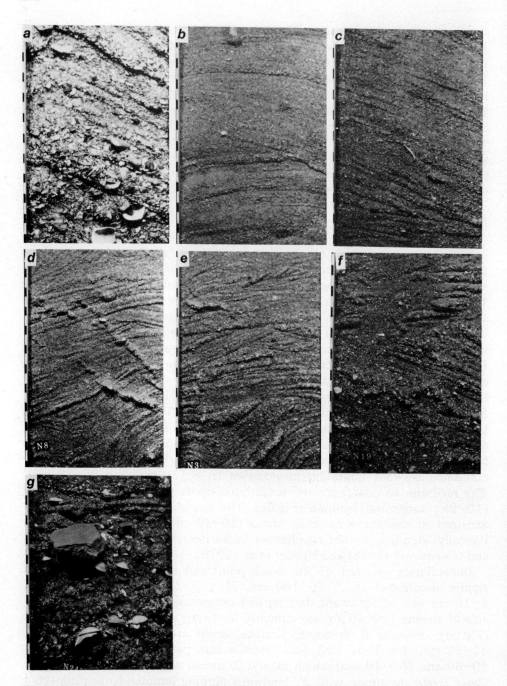

Fig.9. Narrabeen beach box cores. Shore to left. (a) step; (b) longshore trough; (c) rip channel (megaripple crest); (d) nearshore (cross ripples), depth 3 m; (e) nearshore, depth 8 m; (f) nearshore—offshore, depth 19 m; (g) offshore, depth 24 m. Scale in centimetres.

at least 30 m. Crossbedding structures were apparent in the cores. Relatively thick (1—2 cm) landward dipping (5°) laminae overly 10—20° seaward dipping beds (Fig.9f and g). Bioturbation increased markedly seaward of 18 m depth.

RHYTHMIC BAR AND BEACH

Grants Beach is more exposed to deep-water waves and has slightly finer sediments than Narrabeen (Fig.6b). Consequently it is modally more energetic (H_b = 1.6 m) and more often has a rhythmic bar and beach morphology (Fig.1e). It is rarely reflective and under high waves can become dissipative. It was investigated when a well-developed crescentic bar system was present, the bars were not attached to the shoreline. The sampling line crossed the 80 m wide moderate gradient beach, a 60 m wide 3 m deep trough with a 1.5 m deep bar crest lying 70—90 m seaward of the shoreline (Fig.3e). Waves were low (0.5 m) at the time of sampling. The nearshore zone is convex in shape (Fig.4) and perhaps bedrock controlled, though none was observed.

Facies (Figs.6b and 10)

Berm structures consisted of horizontal, parallel, thin laminae composed of medium- and fine-grained sands. The upper beach face contained seaward dipping (5°) parallel and tangential beds of alternating fine and coarse grained laminae (Fig.10a). These graded into predominantly seaward dipping (10—15°) cross strata on the lower beach face (Fig.10b) and a coarse-grained shelly 10—20 cm thick step sequence at low water. The beach face sequence was similar in gross form to Hawks Nest and the Narrabeen rip embayment. The distinctive characteristic of this beach type is however the extensive bar and trough sequence. The deep rip trough contained straight to sinuous parallel wave ripples (L = 50—60 cm, H = 10—20 cm) in coarse-grained shelly sediments. Structures were similar to the Narrabeen rip channel with seaward dipping (20°) laminae, arranged in 10—20 cm thick sets of alternately coarse and fine sand (Fig.10c). The trough facies contrasts with the finer-grained better sorted bar sediments which contained predominantly landward dipping (5—20°) parallel laminae overlain with ripple cross strata on the crest (Fig.10d), produced by sinuous wave ripples (L = 50 cm, H = 15 cm) present at the time of coring.

Seaward of the breaker zone lunate megaripples (L = 50—60 cm, H = 10 cm) produced 10—15 cm thick sets of predominantly landward dipping (10—30°) parallel laminae. Between 4 and 12 m depth, sharp-crested parallel, wave ripples (L = 6—8 cm, H = 1—2 cm) were encountered. They produced horizontal to slightly landward dipping wavy laminae (Fig.10e). These appear analogous to the 'inner offshore' facies of Hunter et al. (1979). At Grants Beach the small scale structures and increasing bioturbation, which occurred seaward of 8 m depth (Fig.10f and g), reflected the prevailing

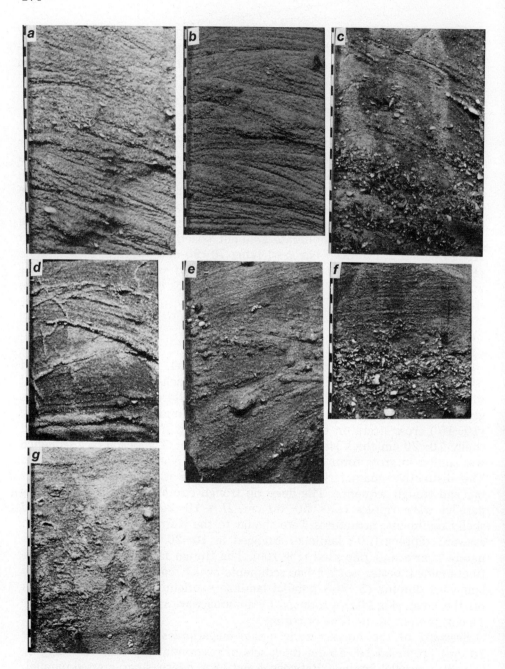

Fig.10. Grants Beach box cores. Shore to left. (a) mid beach face; (b) lower beach face, above step; (c) longshore trough; (d) bar crest; (e) nearshore, depth 8 m; (f) nearshore, depth 12 m; (g) nearshore, depth 16 m. Scale in centimetres.

low swell conditions. Under more normal wave conditions a sequence such as the Narrabeen nearshore would be expected. The small wave ripples became sinuous beyond 16 m depth. *Bankavia* were prominent both on the surface and in the shallow cores between 12 and 21 m depth.

BAR TROUGH

Mid-Seven Mile beach is fully exposed to the regional deep-water wave regime (H_b = 1.6 m) which, combined with predominantly fine sand, results in a low gradient beach and a double bar-trough surfzone (Figs.1b and 3f). The inner bar varies from ridge and runnel to bar-trough in response to varying wave conditions, while the outer bar-trough, apart from on-offshore movement of the bar crest maintains its form year round (Short and Wright, 1983). The outer bar commonly lies over 100 m seaward of low water (Fig.3f). The beach, inner bar-trough and nearshore (Fig.4) were cored, while breakers prevented coring of the outer bar-trough system.

Facies (Figs.6b and 11)

The wide, low gradient (2°) generally featureless beach face, exhibits characteristics of the high-energy dissipative beach face. On the upper beach-face parallel, horizontal to slightly seaward dipping laminae are arranged in uniformly fine-grained sets in thick to very thick beds. On the lower beach face low-frequency backwash associated with surfbeat set-down (see Wright et al., 1982) produces a slowly seaward-migrating, hydraulic jump and leaves antidunes (L = 70—100 cm, H = 1.5 cm) on the surface. They result in landward dipping truncation of horizontal beds and landward dipping (10°) even laminae (Fig.11a). Similar structures are described by Panin and Panin (1967) and Reineck and Singh (1973, p. 303).

The step characteristic of lower-energy beach systems and common in coarser-grained bar-trough systems is usually absent on the fine-grained more dissipative beach faces. The coarser-grained lower beach face sediments grade into the inner trough sequence. The troughs and associated rip channels are dominated by both wave and current ripples and in gross form follow the Narrabeen sequence. Generally coarser-grained sediments and medium-scale, cross bedding underly seaward migrating lunate megaripples, with 10—15 cm thick co-sets of cross lamination dipping both seaward (20—30°) and landward (10—20°) (Fig.11b). The fine-grained, relatively thin (<1 cm) laminae and absence of coarse-grained beds distinguishes these trough deposits from those in the previous beach systems (i.e. Hawks Nest, Narrabeen, Grants). The inner bar deposits followed the usual bar pattern with steeply landward dipping (30°) parallel laminae on the inner section, and lower angle (10°) laminae on the crest (Fig.11c). The outer bar-trough was not cored. However, it would be expected to follow sequence described by Davidson-Arnott and Greenwood (1976) and Greenwood and Mittler (in press) with cross bedding in the trough and a mixture of cross

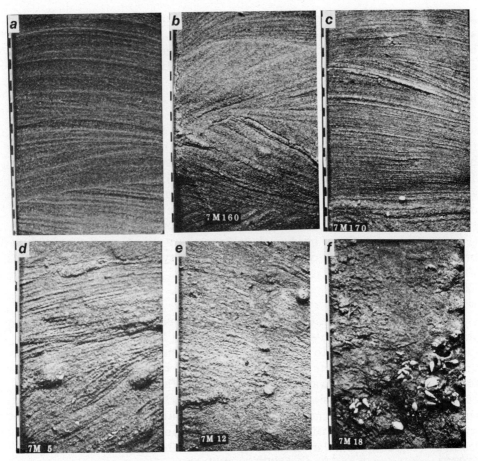

Fig.11. Seven Mile Beach box cores. Shore to left. (a) lower beach face; (b) longshore trough; (c) bar crest; (d) nearshore, depth 5 m; (e) nearshore, depth 12 m; (f) nearshore, depth 18 m. Scale in centimetres.

bedding or horizontal laminae on the crest, possibly with larger scale bedforms owing to the more energetic wave conditions. The nearshore sequence at greater than 5 m depth consisted of sinuous-crested wave ripples, initially of large scale (L = 150 cm, H = 20 cm) which produced medium beds of landward dipping (10—30°) tangential laminae (Fig. 11d). Ripple size decreased (L = 20—50 cm, H = 5—10 cm) between 12 and 21 m depth. Medium scale cross-bedding and increasing bioturbation by *Bankavia* dominated (Fig.11e) with few structures apparent below a depth of 20 m (Fig.11f). Beyond the modal wave base at 25 m, large well-developed, parallel sharp-crested, wave ripples (L = 100 cm, H = 25 cm) were present. These forms, produced by 2 m waves five days previously, were composed of very coarse, shelly, poorly sorted offshore sediments, which combined with bioturbation masked any structures.

DISSIPATIVE

Goolwa Beach in South Australia is a modally high energy, (H_b = 3 m, with H_b > 2 m 70% of the year), fine-grained dissipative beach, with a wide, low gradient beach face (1.5°), and 500 m wide surfzone. The surfzone consists of an inner and outer breaker zone separated by a 4 m deep trough region (Fig.3g). Sediment and bedform observations were made across the region during a period of low (~1 m) waves, however cores could only be obtained from the beach face and depths greater than 6 m. The sediment characteristics are given in Fig.6b.

Facies (Figs.6b and 12)

The 100 m wide beach face was essentially similar to the previous beach (Mid-Seven Mile). However the upper beach face did contain coarser sediments arranged in parallel laminations (Fig.12a). The coarseness perhaps represents swash limit deposition of coarser particles and is equivalent to the coarser berm and cusp deposits of lower-energy beaches. The lower beach face contained thin parallel-to-tangential horizontal-to-low-angle seaward and landward dips (Fig.12b); the latter are due to the backwash processes described for Seven Mile beach.

In the 4 m deep trough and over the 3 m deep bar crest wave oscillations maintained a plane bed. Small ephemeral parallel wave ripples (L = 5 cm, H = 1 cm) began at 4 m depth on the seaward slope and dominated from 6 to 10 m depth. These produced slightly landward dipping thin parallel laminations, in 20 cm sets over an erosional contact (Fig.12c). The erosion was probably due to 4—5 m high, 12 s waves two days previously. At 14 m depth low parallel ripples (L = 30—40 cm, H = 2—3 cm) were underlain by similar structures and a shell rich erosion contact, (Fig.12e). Given the previous high waves these cores (Fig.12c, d and e) resemble the "shoreface storm layers" with upper laminated tempesites over an erosion contact, described by Aigner and Reineck (1982). If so the lower convex curved laminations observed in the 10 m depth core may represent hummocky cross stratification.

A coarsening in grain size below 18 m (due to inner shelf lag deposits) produced large (L = 150 cm, H = 30 cm), sharp crested, parallel wave oscillations ripples, with predominantly steeply landward dipping cross strata (Fig.12f). Smaller active ripples (L = 40 cm, H = 10 cm) with more sinuous crests were observed at 22 and 25 m depth.

DISCUSSION

The foregoing description of beach morphology, texture, bedforms and structures provide a 'representative' facies sequence for each of the six beach states presented in Fig.1. In Fig.13 the observed facies are ranked according to their beach type. The figure illustrates two important features of the nature and extent of facies relative to wave energy.

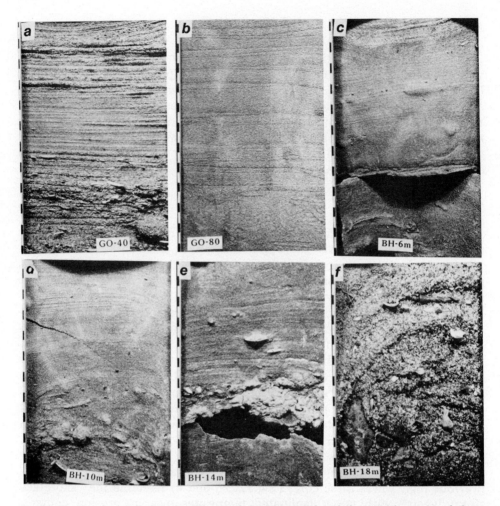

Fig.12. Goolwa beach box cores. Shore to left. (a) upper beach face; (b) lower beach face; (c) nearshore, depth 6 m; (d) nearshore, depth 10 m; (e) nearshore, depth 14 m; (f) nearshore, depth 18 m. Scale in centimetres.

First, the thickness or depth (beach to nearshore) of each facies sequence increases with increasing wave energy from 5 to 10 m in low-energy reflective beaches, to 10 to 30 m in intermediate beaches and 30 m or more in high-energy dissipative beaches (Fig.13). At the same time the horizontal extent of the active sequence increases from 100 m to several kilometres (Fig. 4). On a prograding shoreline this means that a complete vertical sequence of a reflective beach may be preserved after 100 m of shoreline progradation, whereas several kilometres of progradation could be required to produce a similar vertical sequence for higher-energy intermediate and dissipative sequences. In other words the higher-energy sequences will be spread over a greater horizontal distance, as indicated by Fig.4. On stationary or regressive shorelines a diagonal sequence would at best be preserved.

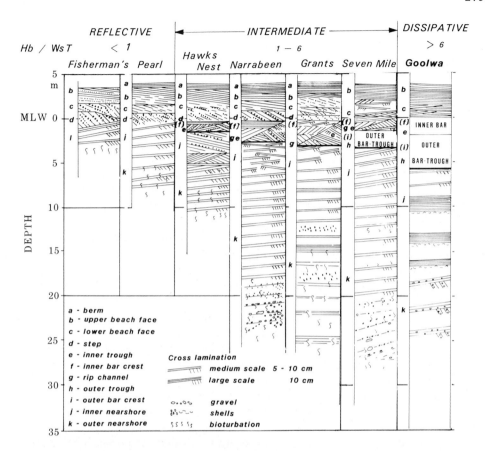

Fig.13. Idealised vertical sequence of all possible facies for each beach system. Bracketed facies (*f* and *i*) have a low preservation potential. Higher-energy intermediate and dissipative sequences are more likely to occur in diagonal sequences due to massive progradation required to produce vertical sequences.

Second, the occurrence of individual facies (a—k in Fig.13) is dependent of the prerequisite morphodynamic coupling. Consequently, higher-energy beaches will have features and structures not found on lower energy beaches and vice versa. The lowest-energy reflective beaches consist solely of a beach face, step and nearshore (b, c, d, k sequence, Fig.13). Higher-energy reflective beaches have a berm crest, upper and lower beach face, step, and deeper nearshore (a, b, c, d, j, k). The berm persists through the intermediate beaches becoming wider and lower in gradient. The berm is usually absent on dissipative beach faces which widen and have more extensive upper and lower beach-face deposits. Steps at first prominent become discontinuous longshore on rhythmic beach shorelines (absent on horns) and absent on finer-grained, high-energy intermediate and dissipative beaches. Mackaness (1981) used discriminant analysis of beach face textures and structures to statistically distinguish reflective, intermediate and dissipative beaches. The bar-trough facies is initiated on a small scale in the low- to moderate-energy ridge and runnel beach state; it is prominent in the bar and rip state, with the trough increasing in depth to 3 m below MLW in the bar-trough state. Davidson-Arnott and Greenwood (1976) and Hunter et al. (1979) suggest that in the progradation of such systems only the swash (beach face), swash-trough transition, longshore trough (rip feeder channel), rip channel and nearshore facies would be preserved with little preservation of the bar sequence.

The inner nearshore regions of intermediate and dissipative beaches are dominated by what Clifton et al. (1971) termed the 'outer rough facies' containing megaripples. The outer nearshore to modal wave base, and nearshore of reflective beaches, is dominated by Clifton et al. (1972) 'asymmetric ripple facies'. This sequence has also been described by Shipp (1984, this volume).

Moderate-energy intermediate beaches will therefore have an a-b-d-e-(f)-g-j-k sequence, with higher-energy, intermediate and dissipative beaches a-c-e-(f)-g-h-(i)-j-k sequence. The bars (f and i) have a low preservation potential, and the nature of the outer nearshore facies (k) is highly dependent on grain size.

The overall preservation potential of beach-nearshore systems has been well documented in the literature (Clifton et al., 1971; McCubbin, 1982). In southeast Australia, Thom et al. (1981) have completed extensive augering of numerous Holocene and Pleistocene barrier systems. Using grain size, colour and percent carbonate they have been able to discriminate between dune, beach-nearshore, and shelly nearshore (offshore) facies. The size and extent of these systems, which included intact buried Pleistocene barriers, suggest an overall high preservation potential. In southern and western Australia formation of calcrete and consequent partial lithification of the barriers increases preservation potential enabling them to survive sealevel transgressions (Short and Hesp, in press).

The gradation in shoreface facies between low- and high-energy systems, first proposed by Clifton et al. (1971) and elaborated by Davidson-Arnott and Greenwood (1976) has been both confirmed and extended. This study of seven beaches located in low, moderate and high wave environments has provided additional information on the beach morphodynamics and associated texture, bedforms and structures. While Figs.6 and 13 illustrate the vertical sequence of the systems, the increasing width of the higher energy systems would dispose them to a more diagonal sequence of preservation. The high energy dissipative Goolwa system would require several kilometres of shoreline progradation to produce a straight vertical sequence of all beach-nearshore-offshore facies.

The occurrence, sequence and extent of individual facies (a—k) in Fig.13 may assist identification of paleo-beach type and thereby levels of wave energy. The arrangement of the facies sequence, vertical to diagonal, will be an indication of degree of shoreline stability, with vertical high-energy sequences indicative of massive shoreline progradation, and diagonal sequences of stable and/or regressive shorelines.

ACKNOWLEDGEMENTS

This study was supported in part by the Australian Research Grants Committee and Australian Marine Science and Technologies Committee. In the field G. Lloyd was essential, assisting in all SCUBA operations along with the excellent assistance of P. Cowell, J.M. Short, N.L. Trenaman, and L.D. Wright. Reviews by J.R. Dingler and B. Greenwood greatly assisted the revision of this manuscript. Figures were drafted by J. de Roder, cores photographed by A. Pritchard and manuscript typed by J.M. Martin.

REFERENCES

Aigner, T. and Reineck, H-E., 1982. Proximality trends in modern storm sands from the Helgoland Bight (North Sea) and their implications for basin analysis. Senckenbergiana Marit., 14: 183—215.
Allen, J.R.L., 1982. Sedimentary Structures: Their Character and Physical Basis. (Dev. Sedimentol., 30A, 593 pp; 30B, 663 pp) Elsevier, Amsterdam.
Burger, J.A., Klein, G. deV. and Sanders, J.E., 1969. A field technique for making epoxy relief-peels in sandy sediments saturated with saltwater. J. Sediment. Petrol., 39: 338—341.
Clifton, H.E., 1976. Wave-formed sedimentary structures — a conceptual model. In: R.A. Davis, Jr. and R.L. Ethington (Editors), Beach and Nearshore Sedimentation. Soc. Econ. Paleontol. Mineral., Spec. Publ., 24: 125—148.
Clifton, H.E., Hunter, R.E. and Phillips, L., 1971. Depositional structures and processes in the non-barred high-energy nearshore. J. Sediment Petrol., 41: 651—670.
Davidson-Arnott, R.G.D. and Greenwood, B., 1974. Bedforms and structures associated with bar topography in the shallow-water wave environment, Kouchibouguac Bay, New Brunswick, Canada. J. Sediment. Petrol., 44: 698—704.
Davidson-Arnott, R.G.D. and Greenwood, B., 1976. Facies relationships on a barred coast, Kouchibouguac Bay, New Brunswick, Canada. In: R.A. Davies, Jr. and R.L. Ethington (Editors), Beach and Nearshore Sedimentation. Soc. Econ. Paleontol. Mineral., Spec. Publ., 24: 149—168.
Davidson-Arnott, R.G.D. and Pember, G.F., 1980. Morphology and sedimentology of multiple parallel bar systems, southern Georgian Bay, Ontario. In: S.B. McCann (Editor), The Coastline of Canada. Geol. Surv. Can., Pap. 80-10, pp. 417—428.
Davis Jr., R.A., Fox, W.T., Hayes, M.O. and Boothroyd, J.C., 1972. Comparison of ridge and runnel systems in tidal and non-tidal environments. J. Sediment. Petrol., 42: 413—421.
Dupré, W.R., 1984. Reconstruction of paleo-wave conditions from Pleistocene marine terrace deposits, Monterey Bay, California. In: B. Greenwood and R.J. Davis, Jr., (Editors), Hydrodynamics and Sedimentation in Wave-Dominated Coastal Environments. Mar. Geol., 60: 435—454 (this volume).
Greenwood, B. and Davidson-Arnott, R.G.D., 1975. Marine bars and nearshore sedimentary processes Kouchibouguac Bay, New Brunswick. In: J. Hails and A. Carr (Editors), Nearshore Sediment Dynamics and Sedimentation. Wiley-Interscience, London, pp. 123—150.
Greenwood, B. and Mittler, P.R., in press. Vertical sequence and lateral transitions in the facies of a barred nearshore. J. Sediment. Petrol.
Howard, J.D. and Reineck, H.-E., 1981. Depositional facies of high energy beach to offshore sequence: comparison with low-energy sequence. Am. Assoc. Pet. Geol., 65: 807—830.
Hunter, R.E., Clifton, H.E. and Phillips, R.L., 1979. Depositional processes, sedimentary structures, and predicted vertical sequences in barred nearshore systems, southern Oregon coast. J. Sediment. Petrol., 49: 711—726.

Mackaness, J., 1981. Microsedimentary structures of intertidal sand beaches. B.A. Hons Thesis, Department of Geography, University of Sydney, Sydney, 111 pp.

McCubbin, D.G., 1982. Barrier-island and strand-flat facies. In: P.A. Schoole and D. Spearing (Editors), Sandstone Depositional Environments. Am. Assoc. Pet. Geol., Tulsa, Okla., pp. 247—279.

Panin, N. and Panin, St., 1967. Regressive sand waves on the Black Sea Shore. Mar. Geol., 5: 221—226.

Reineck, H-E. and Singh, I.B., 1973. Depositional Sedimentary Environments. Springer, Berlin, 439 pp.

Shipp, R.C., 1982. Nearshore depositional facies of Long Island, New York, U.S.A. In: B. Greenwood and R.J. Davis, Jr. (Editors), Hydrodynamics and Sedimentation in Wave-Dominated Coastal Environments. Mar. Geol., 60: 235—259 (this volume).

Short, A.D., 1979. Three dimensional beach stage model. J. Geol., 87: 553—571.

Short, A.D. and Wright, L.D., 1981. Beach Systems of the Sydney Region. Aust. Geogr., 15: 8—16.

Short, A.D. and Hesp, P.A., in press. Coastal morphodynamics of the South East Coast of South Australia. Coastal Studies Unit Technical Report, 84/1, Coastal Studies Unit, Department of Geography, University of Sydney, Sydney, N.S.W.

Short, A.D. and Wright, L.D., 1983. Physical variability of sandy beaches. In: A. McLachlan and H. Erasmus (Editors), Sandy Beaches as Ecosystems. Junk, The Hague, pp. 133—144.

Short, A.D. and Wright, L.D., in press. Field methods in wave dominated surfzone and nearshore environments. Occasional Papers, Department of Biology, Memorial University of Newfoundland.

Thom, B.G., Bowman, G.M., Gillispie, R., Temple, R. and Barbetti, M., 1981. Radiocarbon dating of Holocene beach-ridge sequences in south-east Australia. Monogr. 11, Department of Geography, University of N.S.W., R.M.C., Duntroon, A.C.T., 36 pp.

Wright, L.D. and Short, A.D., 1983. Morphodynamics of beaches and surfzones in Australia. In: P.D. Komar (Editor), Handbook of Coastal Processes and Erosion. CRC Press, pp. 35—64.

Wright, L.D., Guza, R.T. and Short, A.D., 1982. Dynamics of a high energy dissipative surfzone. Mar. Geol., 45: 41—62.

STRUCTURES IN DEPOSITS FROM BEACH RECOVERY, AFTER EROSION BY SWELL WAVES AROUND THE SOUTHWESTERN COAST OF ARUBA (NETHERLANDS ANTILLES)

J.H.J. TERWINDT[1], C.H. HULSBERGEN[2] and L.H.M. KOHSIEK[1] *

[1] *Department of Physical Geography, State University of Utrecht, P.O. Box 80.115, 3508 TC Utrecht (The Netherlands)*
[2] *Delft Hydraulics Laboratory, P.O. Box 152, 8300 AD Emmeloord (The Netherlands)*

(Received February 21, 1983; revised and accepted July 14, 1983)

ABSTRACT

Terwindt, J.H.J., Hulsbergen, C.H. and Kohsiek, L.H.M., 1984. Structures in deposits from beach recovery, after erosion by swell waves around the southwestern coast of Aruba (Netherlands Antilles). In: B. Greenwood and R.A. Davis, Jr. (Editors), Hydrodynamics and Sedimentation in Wave-Dominated Coastal Environments. Mar. Geol., 60: 283—311.

Hurricane- or storm-generated swell waves may cause erosion and deposition along coasts which are situated thousands of kilometers outside the generating wind field. Marked beach erosion, caused by such swell waves, was observed along the micro-tidal west coast of Aruba. During the process of erosion a swash bar was formed, which moved up-beach during the waxing part of the swell event. The swash bar welded to the beach during the waning part of the event. Rapid sedimentation occurred on the upper beach. Finally, recovery of the beach was observed. The formation of a swash bar was attributed to an erosive, dissipative interval of a normally accretionary reflective beach. The sedimentary structures, although generally in line with observations on other beaches, show several peculiar characteristics: (1) the great thickness of the laminae in these calcareous sands; (2) the succession of low-angle sigmoidal and tangential sets in the swash bar; (3) the relatively steep erosional lower set boundaries and the wedge-shaped lamination in the successive stages of beach recovery; and (4) the several types of deformation structures.

INTRODUCTION

The destructive effect of swell waves generated by hurricanes or severe storms which pass or hit coastal zones has been long recognized in different parts of the world (Howard, 1939; McKee, 1959; Hayes, 1967; Stoddart, 1971; Hopley, 1974; Kumar and Sanders, 1976). It has generally been observed that in most cases the recovery of the beaches takes place rather rapidly after the swell has ceased (King and Williams, 1949; Reineck, 1963; Hayes, 1967; Hayes and Boothroyd, 1969; Davis et al., 1972; Owens and Frobel,

---

*Present address: Rijkswaterstaat, Delta Service, Van Alkemadelaan 400, 2597 AT The Hague, The Netherlands.

0025-3227/84/$03.00 © 1984 Elsevier Science Publishers B.V.

1977). However, the effects of hurricanes and storms are not restricted to areas in the vicinity of the track of the eye. The generated swell waves may influence coastal processes in areas, thousands of kilometers outside of the wind field.

Along the west coast of the island of Aruba (Netherlands Antilles), the existence of such swell waves, having travelled a considerable distance, was demonstrated by Wilson (1968, 1969) and Wilson et al. (1973). The low, but long waves, reaching shallow water produce high breakers, which cause havoc and damage to the recreational resort areas (Kohsiek et al., in prep.).

The sequence of events in the surf zone during erosion and initial recovery of the beach is difficult to establish because profiling and diving is almost impossible during heavy surf. Another way to analyse these events is to study the sedimentary succession. Three large, 2 m deep trenches were excavated across the entire beach after an erosional event. This paper describes the observed characteristics of the exposed sediments.

GENERAL SETTING OF THE AREA OF INVESTIGATION

Aruba is situated in the Caribbean Sea in the zone swept by the east trade winds (Fig.1). The mean wind velocity is about 7.7 m s^{-1}. There are minor seasonal variations in wind direction and speed. The diurnal tides are low around Aruba: the spring tidal range is 0.43 m and the neap range is 0.13 m.

The wave climate is almost exclusively dominated by the trade winds. In the Caribbean Sea the wind waves are heading to the west for 67% of the time, for 18% to the southwest and for 11% to the northwest. The average wave height is about 1.5 m and the average period is 7 s.

The calcareous sandy beaches are situated at the leeward side of the island between Malmok and Oranjestad (Fig.2) and are unprotected by barrier reefs. The latter are present to the southeast of Oranjestad.

The westernmost part of the island is called Manshebu and the present study is focused on Pelican Beach between Manshebu and Oranjestad and on Eagle beach between Manshebu and Pos Chikitu (a bluff of an old lower terrace barrier reef deposit).

HURRICANES AND TROPICAL STORMS AFFECTING THE AREA

The effect of tropical storms (8—11 B) or hurricanes (>11 B) on the Aruban beaches differs whether the track is within or without the Caribbean Island Arch.

Within the Arch the storms are generated in the east and track toward the west. As a result, important variations in wind velocity and wind direction may occur and this affects the direction of the swell waves. Around Aruba the swell starts heading to a westerly direction merging in time toward the south and finally toward the southeast. An example is presented in Fig. 3, illustrating the track of the Hurricane David (Aug.—Sept. 1979)

Fig.1. Location map of the eastern part of the Caribbean.

and the successive wave fields. Since 1970, a total of 19 tropical storms or cyclones has passed the 70°W meridian in the vicinity of Aruba.

Aruba may also be reached by waves generated by storms or hurricanes tracking over the Atlantic, north of the Caribbean Island Arch. These storms move from east to west, and if powerful enough, may generate wind waves of sufficient height to become swell waves. Only swell waves travelling in the direction of the corridors, the Mona and Anegada Passage, and sometimes even the Guadeloupe and Martinique Passage, can enter the Caribbean Sea. This means that in the Caribbean, the travelling direction is more uniform, while, due to the greater distance, the wave height is smaller and the period longer in comparison with the Inner Arch swell waves under similar boundary conditions. In Table I some recent data are gathered for events of

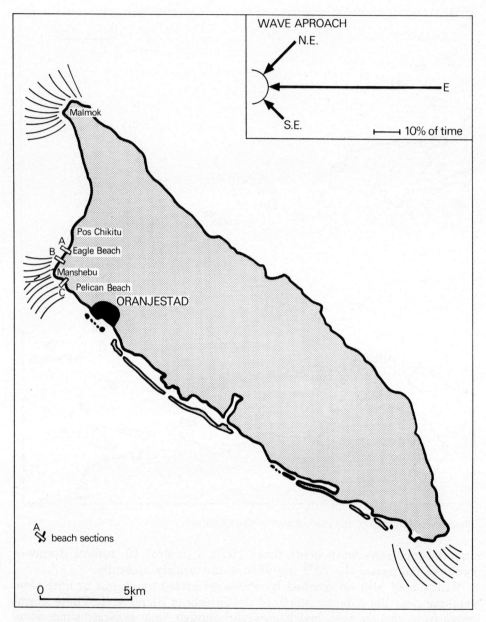

Fig.2. Island of Aruba. Sandy beaches are situated between Malmok and Oranjestad. Dominant wave approach is from the east; wave refraction takes place around the north and south capes of the island; the meeting area of refracted waves is near Manshebu.

severe erosion along the west coast of Aruba and the inferred tracks of tropical cyclones on the Atlantic.

In conclusion, Aruba may be reached by different types of swell waves either being multi-directional with variable travelling distances or uni-directional with long travelling distances.

TABLE I

Recent data on erosional events along the west coast of Aruba and the hindcast causes and swell wave tracks

Date of severe erosion on Aruban coast	Meteorological cause; L = low H = high	Inferred passage of the Caribbean Island Arch
Nov. 24, 1977	L, Cuba to Puerto Rico	Anagada
Dec. 1, 1977	H, Atlantic 50°W, 35°N	Guadeloupe or Martinique
Feb. 16, 1978	L, north of Cuba	Mona
Aug. 26, 1978	L, near Cuba	Mona
Nov. 18, 1978	L, northeast of Puerto Rico	Mona
Feb. 20, 1979	L, east of Florida	Anagada or Guadeloupe
May 21, 1979	L, north of Hispaniola	Mona

THE EFFECT OF WIND AND SWELL WAVES ON THE SAND MOVEMENT

Wind and swell waves approaching the east coast of Aruba are refracted around the north and south capes of the island (Fig.2). These refracted waves meet each other at the west point of the island near Manshebu, generating a complicated cross pattern (Fig.4). The location of this meeting area and the dominancy of one pattern over the other depends on the direction of the incoming wave trains.

Under normal wind wave conditions the meeting area of the refracted waves is situated just west of Manshebu. There is a net longshore drift along Pelican as well as Eagle Beach towards Manshebu (Kohsiek et al., in prep.) resulting in an accumulation of sediment at the west point (Fig.5).

A different refraction pattern occurs during the presence of swell waves which originate from outside the Caribbean Arch and come from a more northerly direction. The southward travelling waves dominate along Eagle and Pelican Beach, resulting in a southward littoral drift (Fig.5).

If the swell waves are derived from a storm within the Caribbean Arch, the refracted waves approach the west point dominantly from the northwest, gradually shifting toward the west. Thus, at first, there is an increasing littoral drift and erosion along Eagle Beach and, at Manshebu, a decreasing drift and even sedimentation along Pelican Beach. Swell, approaching from the west causes erosion at Manshebu but ultimate accretion along Eagle and Pelican Beach (Fig.5).

It appears that the littoral drift during a swell event is partly opposite to that of the normal wind wave conditions. As an example Fig. 10 compares the net sediment movement during normal conditions with that during the passage of the hurricane David (Aug.—Sept. 1979). The sediment discharge estimates are based on beach profiling and measurements of wave characteristics, angle of incidence and littoral currents. The breaking swell waves created high sediment transport during a short time as compared with the wind wave transport. Furthermore, these high breakers caused a rapid erosion of the foreshore and backshore. The eroded sediment is partly transported offshore below the breaker zone, but a larger part is transferred to

288

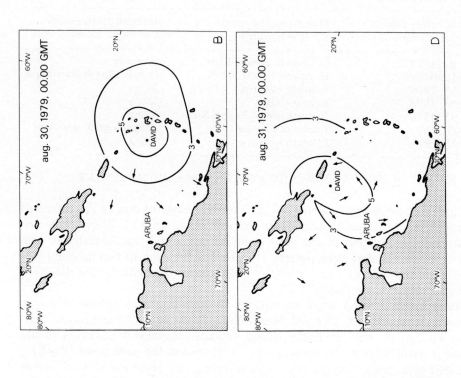

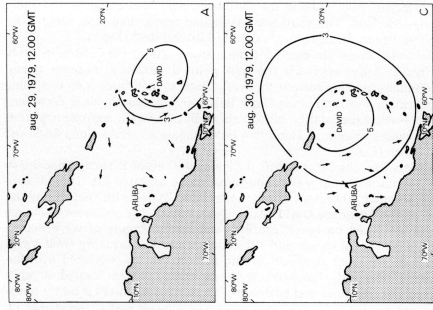

289

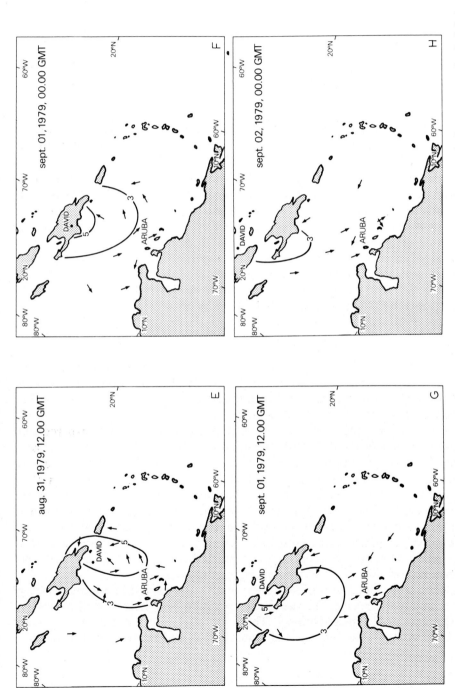

Fig.3. Track of hurricane David, August 29–September 1, 1979 and the calculated wave fields, based on meteorological data computations (courtesy Royal Dutch Meteorological Institute). Arrows indicate direction of wave travel.

adjacent stretches of the coast, as is evidenced by the changes in the beach profiles. When the swell ceases the sediments gradually return by the action of the smaller wind waves.

DESCRIPTION OF SECTIONS

The main purpose of this study was the ultimate fate of the beach sediments and the way in which erosion changes into accretion and the restoration of the beach during recovery.

Three sections, 2 m deep, were excavated crossing the entire beach. Two were located on Eagle Beach and one on Pelican Beach. The sections showed almost similar features and therefore only one (A on Fig.2) from Eagle Beach will be treated here in detail.

All sections comprise the deposits of coastal recovery after a heavy erosion by swell waves which occurred in the beginning of April 1979. The profile of the ultimate erosion could be established by a deposit of rocks, broken asphalt and coral rubble which was dumped by the local authorities in order to stop or delay the retreat of the beach. It appeared that soon after the dumping of the rocks, etc., the beach erosion stopped, the wave climate calmed down and the beach recovery started. The sections were made on 9—10 May 1979, about 4 weeks after the erosion. During this time the beach recovery was 20—25 m; a mean of about 0.8 m per day.

Eleven sedimentary units can be distinguished in the section under consideration (Fig. 6). The boundaries between the units are erosional surfaces. The units are numbered according to their inferred order in the succession of the recovery. The section shows a landward dipping part (unit 1a—1e), an almost horizontal part (unit 2, 4—6) lying on top and landward of unit 1c and several seaward dipping units (3, 5—11). A similar distinction could be made in the other sections.

In the sections from Eagle Beach and Pelican Beach, dumped rock and pieces of asphalt originating from the erosion prevention measures are found on top of the landward dipping unit and are incorporated in the first seaward dipping unit, comparable to unit 3 (Fig. 6). The horizontal and seaward dipping units lie above this rubble. This indicates that unit 1 was deposited before the dumping of the fill. As it was impossible to perform beach profiling during the period of heavy swell we cannot be absolutely sure that the landward dipping unit was deposited during the swell period and that it is not a relict. However, we think that unit 1 fits very well in the deduced succession of events, and the accompanying deposits.

Furthermore, a similar unit is present in all sections and in a similar setting (approximate distance from the shore, position and height in the profile, the character of adjacent units). Thus, although not certain, we consider unit 1 to belong to the sequence of swell erosion and recovery of the beach.

Fig.4. Aerial photograph of the cross-pattern of refracted waves near Manshebu.

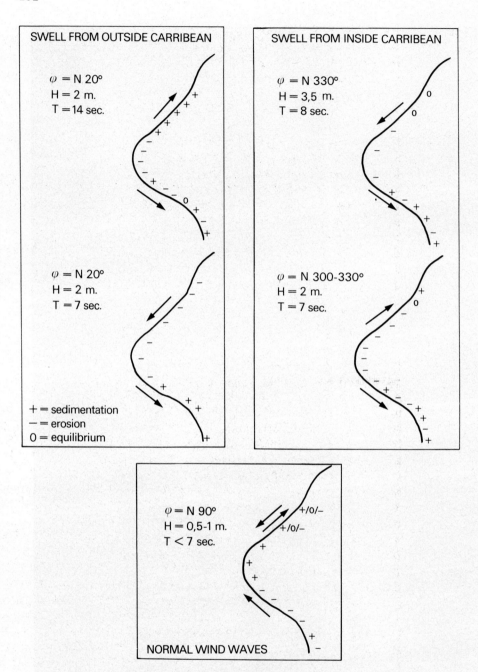

Fig.5. Erosion and sedimentation on the west coast of Aruba (Manshebu) in relation to swell and wind-wave conditions.

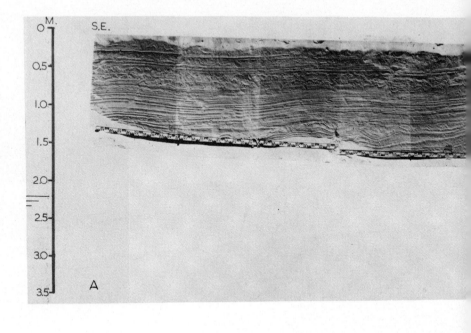

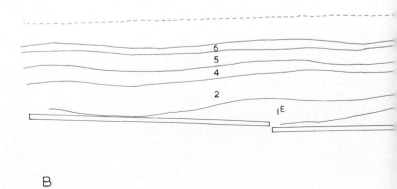

Fig. 6. Section showing sedimentary structures across Eagle Beach.

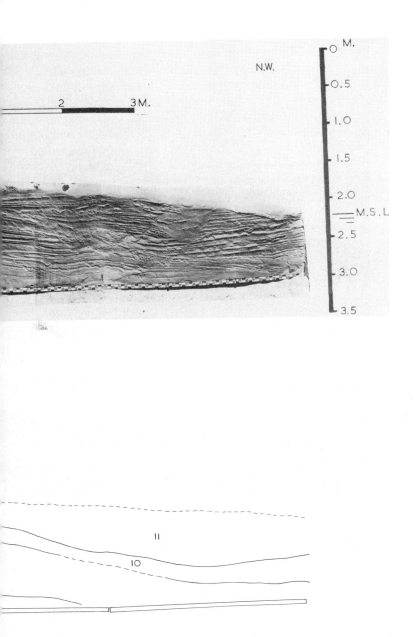

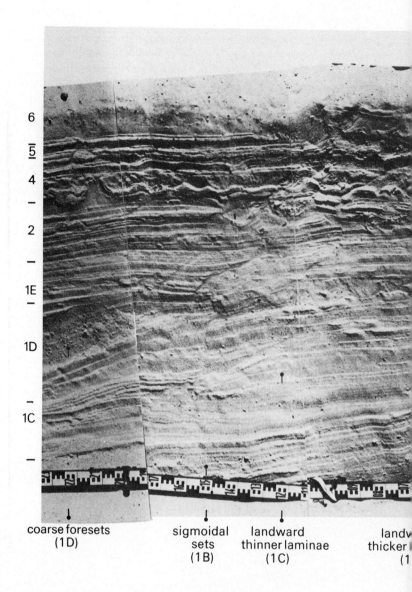

Fig. 7. Detail of section, showing the landward dipping unit 1.

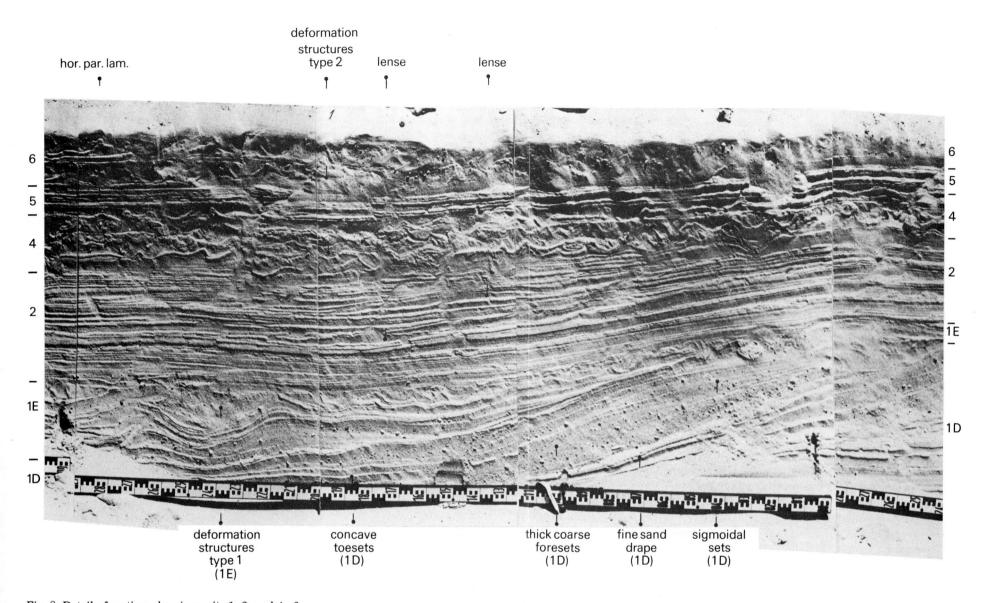

Fig. 8. Detail of section, showing units 1, 2, and 4—6.

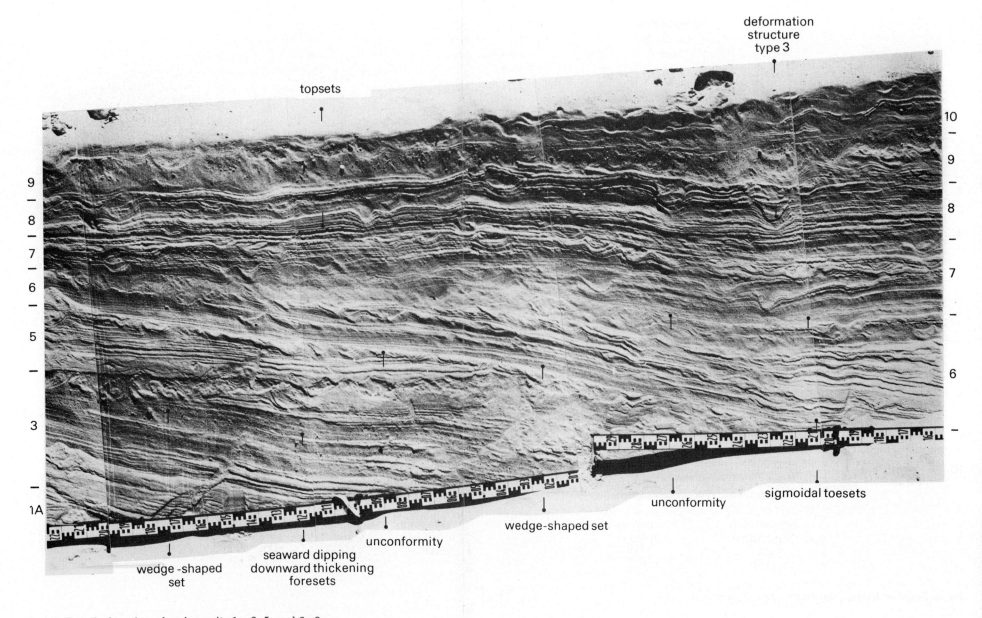

Fig. 9. Detail of section, showing units 1a, 3, 5, and 6—9.

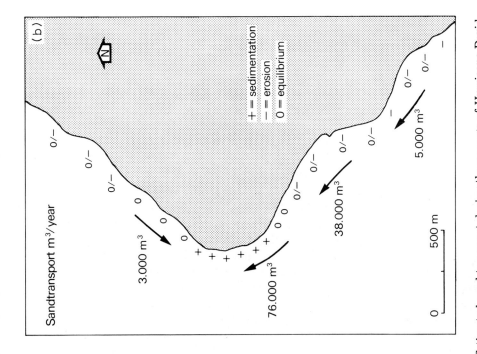
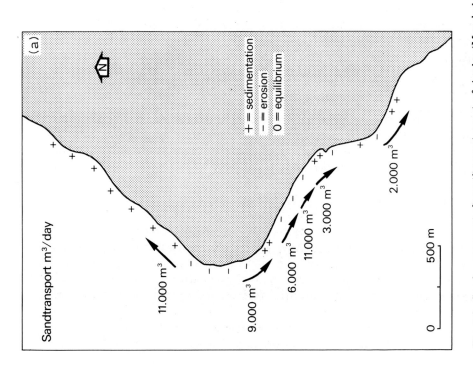

Fig. 10. Sand movement along the west coast of Aruba (Manshebu). *a*. Estimated sand transport during the passage of Hurricane David (in m³ per day). *b*. Estimated net yearly sand transport during normal wind wave conditions (in m³ yr⁻¹).

DESCRIPTION OF THE UNITS

Unit 1

This unit is made up of low to moderate dipping landward, cross-strata (Figs. 7 and 8). The unit is subdivided into five subunits (Fig. 6). The boundaries are unconformities, either erosion surfaces or sudden and marked changes in textures and/or structural properties (foreset angles, nature of toesets, specific changes in thickness of the laminae).

Unit 1a is coarse at the base (d_{50} = 1.0—1.5 mm), including small coral rubble and fines upward to medium/coarse sand (d_{50} = 0.4—0.8 mm). The basal horizontal lamination merges upward and landward into landward dipping (15—20°) foresets and toesets.

Unit 1b shows a much finer sediment (0.4—0.6 mm) and the structures consist of very thinly laminated almost horizontal lamination, in landward direction gradually merging into low-angle (10°) foresets and sigmoidal toesets. The laminae become progressively thicker going landward.

Unit 1c differs from 1b in that the foresets are better developed, are somewhat steeper (15°), have thicker laminae, but the toesets are concave and the laminae become progressively thinner.

Unit 1d is a continuation of unit 1c, but there are differences; the subhorizontal topsets are very thin and there is almost no vertical accretion. There are thick, coarse-grained (0.8—1.2 mm) foreset laminae, which become thinner toward the concave toesets. Two of such coarse-grained foreset beds can be distinguished, separated by a thin intercalation of finer material consisting of concave foresets and sigmoidal fore- toesets.

Unit 1e shows concave topsets merging into increasingly thicker foresets and sigmoidal toesets, in rather coarse material (d_{50} = 1.2—1.5 mm).

Unit 2

This unit consists primarily of thin, horizontal or slightly landward dipping parallel lamination in medium-sand (d_{50} = 0.3—0.5 mm; Fig. 8). Most laminae can be traced over 2—5 m and landward they become thicker. Some very flat lenses may be observed.

Units 3—11

Each unit is bounded at the base by an erosional surface, truncating the underlying strata. The units show almost similar characteristics throughout. Going from the beach toward the sea, the units consist of almost horizontal, parallel lamination which merge into seaward-dipping downward-thickening foresets and sigmoidal, almost horizontal, parallel-laminated toe and bottomsets (Fig. 9).

The increasing thickness of the laminae results in a wedge-shaped set with gradually flattening foreset angles. This would finally lead to very low dipping strata. Occasionally, however, there were erosional conditions which resulted in erosion surfaces having a steeper slope than the underlying strata. In most cases coarser sediment gradually fines upward from these erosion surfaces.

The almost horizontal nature of slightly landward-dipping topsets is nearly concordant with unit 2. However, the material is much coarser and the laminae become progressively thicker in landward direction.

The deposition of units 3—11 resulted in an overall aggradation.

INTERPRETATION

Initially the interpretation of the sedimentary units does not seem to be very problematic.

Many authors have provided evidence that the landward-dipping stratification like that of unit 1 may be attributed to a shoreward migrating bar (Reineck, 1963; Panin, 1967; Psuty, 1967; Hayes and Boothroyd, 1969; Clifton et al., 1971; Davis et al., 1972; Davis and Fox, 1972; Wunderlich, 1972; Davidson-Arnott and Greenwood, 1976; Fraser and Hester, 1977; Owens and Frobel, 1977; Van den Berg, 1977; Hine, 1979; Hunter et al., 1979; Greenwood and Davidson-Arnott, 1979).

Unit 2 may be interpreted as an infilling of a depression landward of the bar. This infilling finally results in a welding of the bar to the beach, analogous to the description by Sonu and Van Beek (1971), Davis et al. (1972), Wunderlich (1972), Sonu (1973), Owens and Frobel (1977), Van den Berg (1977), Hine (1979), Short (1979) and Wright et al. (1979).

Units 3—11 may be referred to as the seaward outbuilding (recovery) of the beach showing similar deposits as described by Thompson (1937), Panin (1967), Clifton et al. (1971), Davis et al. (1972), Van den Berg (1977), Hine (1979) and Hunter et al. (1979).

Although these general interpretations are supported by previous observations of many investigators, several peculiar points may be recognized when looking in greater detail.

First, a decision has to be made as to whether unit 1 belongs to the swell event or is a relict of a former beach deposit. During the heavy swell the beach was strongly eroded and the strand line shifted landward. The bottom sets of units 3—7 were not exposed and were below the base of the section. The same holds for unit 2. This seems to indicate that the lowermost beach profile, which developed during the April swell event lies below the present section. Unfortunately this lowermost profile could not be excavated because it is situated below the ground water table. Unit 1 in this conception lies above the lowermost erosion profile. Furthermore, if unit 1 was a relict bar then an overall truncation plane should be present on top of the unit. There are truncations on the upper side, but not on the landward side. Here unit 1 gradually merges into unit 2, which may indicate a genetic relationship. Thus, we think that it is justified to incorporate unit 1 in the features connected with the swell event of April 1979.

Under normal conditions of refracted wind waves, the Aruban beaches have an almost straight, uniform beach profile, dipping at about 6—8° with no bars and a small 0.1 m high step at the seaward side of the swash zone (Fig.11). The step is composed of very coarse sand and coral rubble. Below the step there is a linear, low-gradient nearshore profile consisting of fine sand. There are no runnels, rips, circulation cells, etc.; however, beach cusps of varying dimensions were frequently observed on the beach. These characteristics have some resemblance to those described by Wright et al. (1979) for reflective beaches. It should be noted that the Australian beaches described by Wright et al. are pocket beaches, while the Aruba beaches occur along an open coast. Nevertheless, the overall characteristics are quite similar.

According to Guza and Inman (1975) strong reflection of incident waves may occur if the reflectivity parameter $\epsilon < 2.0$—2.5. The reflectivity parameter is defined as:

$$\epsilon = a_i \, \omega^2 / g \, \tan^2 \beta \tag{1}$$

in which a_i = incident wave amplitude near the breakpoint, $\omega = 2\pi/T$, where T = wave period, g = acceleration of gravity and β = beach slope.

Under the normal conditions along the western part of the Aruban coast (a_i = 0.5—1.0 m, β = 8—15°, T = 5—8 s), ϵ has values below 2.5, and the beaches may be considered reflective. However, during a swell event the beach gradient becomes much flatter. As $\tan \beta$ decreases and a_1 increases ϵ will be above 2.5 despite the fact that T increases and consequently ω decreases somewhat. Then a dissipative type of beach may occur. It is characterized by the presence of bars, rip cells, etc. (Wright et al., 1979).

Thus the swell event may be considered as an *erosional dissipative* interval in the normal *accretionary, reflective* beach conditions although small erosional fluctuations may be present. This interval is of a short duration (a few days) and although sand movement is extensive, a complete adaptation of the beach characteristics to the dissipative condition is hardly to be expected in such a short time. Apparently, during this dissipative interval an initial swash bar is formed. This is postulated from the fact that unit 1 is covered

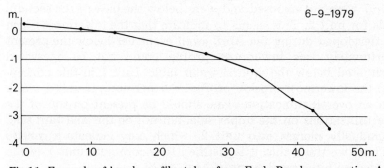

Fig.11. Example of beach profile, taken from Eagle Beach near section A.

by the rocks and asphalt, dumped during the erosion of the beach. Thus the swash bar was already present in the swash zone, *when the upper part of the beach was still eroding*. The conditions during the swell event are comparable to a very rapid transition from beach stage 1—2 and reverse to 1 as described by Short (1979) and a sequence of beach profile types put forward by Sonu and Van Beek (1971) and Sonu (1973).

The continuous erosion of the upper beach during the dissipative interval resulted in a landward migration of the swash zone and hence the swash bar. This was accomplished by erosion of the seaward flank of the bar and deposition at the leeward side where a slight depression exists. In the depression at the leeward side of the bar the suspension outfall of the coarse material, produced the low-angle sigmoidal foresets and bottom sets of unit 1a, as described by Thompson (1937), Psuty (1967) and Panin (1967). In some instances, perhaps during low water the orbital velocities over the bar decreased somewhat. Then the steeper dipping tangential foresets of unit 1b may be deposited. This is analogous to many observations on ridge and runnel beaches where steep slip faces of the bar are encountered during rather low orbital motion when water sweeps over the bar under limited water depth (Wunderlich, 1972; Owens and Frobel, 1977). Finally, an equilibrium developed between the flattened beach profile and the swash and backwash produced by interacting swell and wind waves, thus stopping migration of the bar.

Then the period of the gradual decrease of the swell waves started. This had a direct effect on the intensity of the swash and backwash. The mean wave run-up decreased. Still, occasional high swash ran far up the more or less dry beach but the backwash was reduced due to percolation. As a result, much material moved upslope and remained in the upper part of the profile, especially in the depression landward of the swash bar. In this way unit 2 was formed which has a similar nearly horizontal, parallel, slightly outwedging stratification as described by Thompson (1937), Panin (1967), Psuty (1967) and Hine (1979), although the laminae are remarkably thick. This suggests a high deposition rate.

As the swell waves were further reduced the relative importance of the wind waves increased. Low, wind-wave-generated swash did not overtop the swash bar anymore and a seaward outbuilding occurred on the sea facing slope of the bar. However, high (wind and swell wave) swash occasionally still swept over the bar, but on the landward side there was a very flat beach. As a result of percolation the backwash was very ineffective and almost all material moved forward by the swash, remained in the uprush zone producing an almost horizontal lamination and a heightening of the upper beach (units 3, 4).

Further decrease of the swell waves resulted in a gradual transition toward the normal steeper reflective wind-wave profiles and a seaward outbuilding of the beach (units 5—11). However, during this outbuilding, rather important erosional events took place as evidenced by truncated lamination. Over the erosive lower boundaries of the units much coarser material is found than is in the flatter upper parts of the units. The greater coarseness may be attributed to stronger wave action (Thompson, 1937; Hine, 1979). Apparently, increased wave energy results in erosion, steepening and selection of coarser grains in the swash zone. Waning wave energy produces accretion, less steep profiles and deposition of finer material in upward pointing wedge-shaped laminae. Such type of lamination was also reported by Thompson (1937) and Van den Berg (1977) although a conclusive explanation is still lacking.

DEFORMATION STRUCTURES

Three types of deformation structures were observed in the section: Type 1 in unit 1e and the lower part of unit 2; type 2 in unit 3, 4 and 6, landward of the swash bar; and type 3 in several places in the seaward dipping parts of units 5—11 (Figs.8, 9, and 10).

The deformations of type 1 in unit 1e and lower part of unit 2 are typically encountered in the trough of the swash bar and in the last (upper) deposited sediments in this trough (Fig.9). Water escape structures and convolutions (Lowe, 1975; Reineck and Singh, 1980) were observed at several places. At other places the lamination was completely absent indicating post-depositional rearrangement of the fabric. Noteworthy are the dimensions of the undulations in the trough which have amplitudes of some decimeters. These deformation structures seem to point to conditions of "momentary failure" due to flow, induced within the porous bed (Madsen, 1974). As indicated by Madsen (1974) and by Sleath (1970), a horizontal pressure gradient may be initiated of sufficient magnitude under breaking waves to create a momentary bed failure. Structural indications of this failure are generally observed in the trough of the swash bar and not at the crest. This may indicate, that wave induced liquefaction, as described by Dalrymple (1979, 1980), was not the generating process of the deformation, because this type of liquefaction works out preferentially in the crestal zone of major bedforms. The position of the deformation all over the trough seems to indicate that not only horizontal but also vertical pressure gradients play an important role.

The deformations of type 2 in units 5 and 6 (Figs. 8 and 9) consist of water escape structures (pillars), convolutions, concavities, and structureless sand patches. At some places the stratification is completely disturbed. No cavernous sand is observed in the deformed layers. This excludes the formation of the convolution as a result of entrapped air as described by De Boer (1979). Furthermore, there is almost never a kind of stratification visible in the lows between the updomed parts of the convolute layers, a feature considered representative by De Boer for an air-trapped origin of the convolutions. The deformed layers much more resemble the water escape structures (dish structures, pillars) as described by Lowe (1975, figs.4 and 5). These structures are attributed to liquified behaviour of the sediment. According to Lowe, after resedimentation liquified sands may show: (1) undeformed primary structures; (2) deformed primary structures; (3) nearly complete homogenization; and (4) water escape structures (pillar and dish structures). The deformed sediments of units 5 and 6 do show these features. Loose packing may be expected in these swash-produced layers in this part of the units 4 and 5. The great extension of these deformed layers and the rather sharp lower and upper boundary excludes a local origin. Generating conditions appear to have been operative over the whole top part of the upper beach. After the liquefaction the normal parallel beach lamination is restored. The reason for this alternation of liquified and non-liquified sedimentation is not clear.

The third type of deformation structure (Figs. 8 and 9) is especially visible at some places in units 7—11. It consists of small concavities filled in with concave-upward stratified laminae. This structure shows a resemblance of hoofprints or footprints (Van der Lingen and Andrews, 1969; Lewis and Titheridge, 1978), made by horses or human beings, walking on an almost water-saturated beach or moist dune sands. This type 3 deformation structure also resembles those shown by Lindström (1979, fig.4).

CONCLUSIONS

(1) Under normal trade wind-wave conditions the calcareous-sandy microtidal beach along the west point of Aruba has a reflective character with steep beach profiles and no bars. The beach is mostly accretionary, although with small erosional fluctuations.

(2) Occasional swell events produce strong erosion and flatter beach profiles of a dissipative nature. During this condition a swash bar may be formed moving shoreward and welding to the upper beach, showing a succession of steeper and flatter landward-dipping tangential and sigmoidal sets.

(3) The gradual transition from dissipative to the normal reflective conditions creates rapid deposition in the upper part of the swash zone with substantial sub-horizontal parallel-laminated sets. In the lower part of the swash zone the beach recovery occurs by a succession of several units having a steep erosive lower boundary over which is deposited coarse material. Upward the material becomes finer and is incorporated in upward-pointing wedge-shaped sets, resulting in a gradual and slight decrease in the steepness of the beach profile.

(4) As to the preservation potential in a regressive sequence we may anticipate deposits resembling units 3—11 in the lower part overlain by deposits related to the swell wave events. Such events, although of rare occurrence, may dominate the sedimentary sequence.

(5) Three types of deformation structures could be observed in the beach section.

ACKNOWLEDGEMENTS

This study was a part of a larger investigation requested by the Government of the Island of Aruba. We gratefully acknowledge the approval by this Government to publish these results. Furthermore, we benefitted much from the kind and effective cooperation of the Public Service Department (D.O.W.). Our special gratitude goes to Mr. Camminga of D.O.W. We further acknowledge the Royal Dutch Meteorological Institute for their computation of the wave fields of Hurricane David.

REFERENCES

Clifton, H.E., Hunter, R.E. and Phillips, R.L., 1971. Depositional structures and processes in the non-barred high-energy nearshore. J. Sediment. Petrol., 41: 651—670.
Dalrymple, R.W., 1979. Wave-induced liquefaction: a modern example from the Bay of Fundy. Sedimentology, 26: 835—844.
Dalrymple, R.W., 1980. Wave-induced liquefaction: an addendum. Sedimentology, 27, p. 461.
Davidson-Arnott, R.G.D. and Greenwood, B., 1976. Facies relationships on a barred coast, Kouchibouguac Bay, New Brunswick, Canada. In: R.A. Davis and R.L. Ethington (Editors), Beach and Nearshore Sedimentation. Soc. Econ. Paleontol. Mineral., Spec. Publ., 24: 140—168.
Davis, R.A. and Fox, W.T., 1972. Coastal processes and nearshore sand bars. J. Sediment. Petrol., 42: 401—412.
Davis, R.A., Fox, W.T., Hayes, M.O. and Boothroyd, J.C., 1972. Comparison of ridge and runnel systems in tidal and non-tidal environments. J. Sediment. Petrol., 42: 413—421.
De Boer, P.L., 1979. Convolute lamination in modern sands of the estuary of the Oosterschelde, the Netherlands, formed as a result of entrapped air. Sedimentology, 26: 283—294.
Fraser, G.S. and Hester, N.C., 1977. Sediments and sedimentary structures of a beach-ridge complex, southwest shore of Lake Michigan. J. Sediment. Petrol., 47: 1187—1200.
Greenwood, B. and Davidson-Arnott, R.G.D., 1979. Sedimentation and equilibrium in wave formed bars: a review and case study. Can. J. Earth Sci., 16: 312—332.
Guza, R.T. and Inman, D.L., 1975. Edge waves and beach cusps. J. Geophys. Res., 80 (21): 2997—3012.
Hayes, M.O., 1967. Hurricanes as geological agents, south Texas coast. Bull. Am. Assoc. Pet. Geol., 51: 937—956.
Hayes, M.O. and Boothroyd, J.C., 1969. Storms as modifying agents in the coastal environment. In: M.O. Hayes (Editor), Coastal Environments, N.E. Massachusetts and New Hampshire. Coastal Res. Group, Contrib. 1: 245—265.
Hine, A.C., 1979. Mechanisms of berm development and resulting beach growth along a barrier spit complex. Sedimentology, 26: 333—351.
Hopley, D., 1974. Coastal changes produced by cyclone Althea in Queensland, December 1971. Aust. Geogr., 12: 446—456.
Howard, A.D., 1939. Hurricane modification of the offshore bar of Long Island, New York. Geogr. Rev., 29: 400—415.
Hunter, R.E., Clifton, H.E. and Phillips, R.L., 1979. Depositional processes, sedimentary structures and predicted vertical sequences in barred nearshore systems, southern Oregon Coast. J. Sediment. Petrol., 49: 711—726.
King, C.A.M. and Williams, W.W., 1949. The formation and movement of sand bars by wave action. Geogr. J., 112: 70—85.
Kumar, N. and Sanders, J.E., 1976. Characteristics of shoreface storm deposits: modern and ancient examples. J. Sediment. Petrol., 46: 145—162.
Lewis, D.W. and Titheridge, D.G., 1978. Small scale sedimentary structures resulting from foot impressions in dune sands. J. Sediment. Petrol., 48: 835—838.
Lindström, M., 1979. Storm surge turbation. Sedimentology, 26: 115—124.

Lowe, D.R., 1975. Water escape structures in coarse grained sediments. Sedimentology, 22: 157—204.
Madsen, O.S., 1974. Stability of a sand bed under breaking waves. Proc. 13th. Conf. Coastal Engineering, 2: 776—794.
McKee, E.D., 1959. Storm sediments on a Pacific atoll. J. Sediment. Petrol., 29: 354—364.
Owens, E.H. and Frobel, D.H., 1977. Ridge and runnel systems in the Magdalen Islands Quebec. J. Sediment. Petrol., 47: 191—198.
Panin, N., 1967. Structure de dépôts de plage sur la côte de la mer noire. Mar. Geol., 5: 207—219.
Psuty, N.P., 1967. The geomorphology of beach ridges in Tabasco, Mexico. Louisiana State Univ. Studies, Coastal Studies Ser., 18, 51 pp.
Reineck, H.E., 1963. Sedimentgefüge im Bereich der südlichen Nordsee. Abh. Senckenbergiana Naturforsch. Ges., 505, 138 pp.
Reineck, H.E. and Singh, I.B., 1980. Depositional Sedimentary Environments. Springer, Berlin, 549 pp.
Short, A.D., 1979. Three dimensional beach-stage model. J. Geol., 87: 553—571.
Sleath, J.F.A., 1970. Wave-induced pressures in beds of sand. J. Hydraul. Div. ASCE, 96: 367—378.
Sonu, C.J., 1973. Three dimensional beach changes. J. Geol., 81: 42—64.
Sonu, C.J. and Van Beek, J.L., 1971. Systematic beach changes on the outer banks, North Carolina. J. Geol., 79(4): 416—425.
Stoddart, D.R., 1971. Coral reefs and Islands and Catastrophic Storms. In: J.A. Steers (Editor), Applied Coastal Geomorphology. MacMillan, London, pp. 155—197.
Thompson, W.O., 1937. Original structures of beaches, bars and dunes. Geol. Soc. Am. Bull., 48: 723—752.
Van den Berg, J.H., 1977. Morphodynamic development and preservation of physical sedimentary structures in two prograding recent ridge and runnel beaches along the Dutch coast. Geol. Mijnbouw, 56(3): 185—202.
Van der Lingen, G.J. and Andrews, P.B., 1969. Hoof-print structures in beach sand. J. Sediment. Petrol., 39: 350—357.
Wilson, W.S., 1968. On the origin of certain breakers off the Island off Aruba. Techn. Rep. Chesapeake Bay Inst. No. 43, 27 pp.
Wilson, W.S., 1969. Field measurements of swell off the Island Aruba. Techn. Rep. Chesapeake Bay Inst. No. 56, 64 pp.
Wilson, W.S., Wilson, D.G. and Michael, J.A., 1973. Analysis of swell near the Island of Aruba. J. Geophys. Res., 78(33): 7834—7844.
Wright, L.D., Chappell, J., Thorn, B.G., Bradshaw, M.P. and Cowell, P., 1979. Morphodynamics of reflective and dissipative beach and inshore systems: south-eastern Australia. Mar. Geol., 32: 105—140.
Wunderlich, F., 1972. Beach dynamics and beach development. Senckenbergiana Marit., 4: 47—79.

WHAT IS A WAVE-DOMINATED COAST?

RICHARD A. DAVIS, Jr. and MILES O. HAYES

Department of Geology, University of South Florida, Tampa, FL 33620 (U.S.A.)
Research Planning Institute, 925 Gervais St., Columbia, SC 29201 (U.S.A.)

(Received February 19, 1983; revised and accepted July 29, 1983)

ABSTRACT

Davis, Jr., R.A. and Hayes, M.O., 1984. What is a wave-dominated coast? In: B. Greenwood and R.A. Davis, Jr. (Editors), Hydrodynamics and Sedimentation in Wave-Dominated Coastal Environments. Mar. Geol., 60: 313—329.

During the past decade or so, various coasts have been designated as wave-dominated or tide-dominated. Typically there is an association made between coastal morphology and the dominant process that operates on the coast in question. Most authors consider long, smooth, barrier coasts with few inlets and poorly developed ebb deltas as "wave-dominated". These coasts are associated with microtidal ranges. Conversely, mesotidal coasts tend to develop short, drumstick-shaped barriers with well-developed ebb deltas. They are considered as tide-dominated barriers. Such generalizations may be restricted to coasts with moderate wave energy although this is commonly not stated.

Exceptions to these stated generalizations are so numerous that wave energy and tidal prism must also be included in characterizing coasts. The *relative* effects of waves and tides are of extreme importance. It is possible to have wave-dominated coasts with virtually any tidal range and it is likewise possible to have tide-dominated coasts even with very small ranges. The overprint of tidal prism will also produce tide-dominated morphology on coasts with microtidal ranges.

INTRODUCTION

The title of this volume and the symposium from which it originated implies that there are some coasts where the physical processes to which they are subjected are dominated by waves. There is the implication that some coasts are dominated by tides, the other major physical process acting upon coastal environments. Although a continuum is present between domination by one process as contrasted to the other, there is a portion within this continuum where there is subequal impact by both waves and tides. It should be possible therefore to identify three types of coasts from the standpoint of their influencing processes: (1) those dominated by waves; (2) those dominated by tides; and (3) those with a balance between waves and tides.

One would expect that the dominant physical process or processes would leave an imprint on the coastal morphology in the form of the geometry and type of sediment bodies that accumulate in the coastal zone. It is also

0025-3227/84/$03.00 © 1984 Elsevier Science Publishers B.V.

likely that the combination of coastal morphology and sediment bodies would produce a recognizeable stratigraphic sequence in the geologic record. The recognition of sequences as wave-dominated or tide-dominated would be of considerable value in reconstruction of the depositional history and paleogeography of a coastal setting.

The following discussion will attempt to answer the question posed by the title using modern coasts as examples. It will be apparent to the reader that the oversimplified and overgeneralized approaches to this problem which have been stated in the literature are unwarranted (e.g. Price, 1955; Hayes, 1975, 1979). It is now time to look at the numerous exceptions to these general "rules" and see if it is possible to refine and restate them in light of a more comprehensive look at the variables involved.

PREVIOUS APPROACHES TO THE PROBLEM

A recent and comprehensive review of wave-dominated coastal environments by Heward (1981) states that "wave-dominated shorelines are those where wave action causes significant sediment transport and predominates over the effects of tides" (Heward, 1981, p. 223). This is certainly a reasonable definition and will be used as the basis for this discussion. It is appropriate to assume that the same definition can be used for tide-dominated shorelines with appropriate substitutions in the wording.

Although coastal morphology has been examined and classified for many years, it was not until the general classification of coasts according to tidal range by Davies (1964, 1980) and by Hayes (1975, 1979) that the application of tide-dominated and wave-dominated coasts became prominent. Davies (1964) considered three major categories of coasts; microtidal (range <2 m), mesotidal (2—4 m) and macrotidal (>4 m). Although the boundaries are somewhat artificial, a general morphotype can be used to characterize each tidal range category.

Hayes (1975) applied Davies' terminology to general types of shoreline morphology. The microtidal coast is characterized by long, narrow and rather straight barriers with widely spaced inlets. Washover features are prominent and flood tidal deltas are well developed but ebb deltas are small or nonexistent (Fig.1a). Waves are the dominant physical process (Hayes, 1975, 1979).

Mesotidal barrier coasts display short, rather wide barriers with closely spaced inlets. The barriers have a drumstick configuration (Hayes, 1975) with one end much wider than the other due to shoreline progradation. Inlets have well-developed ebb deltas (Fig.1b) testifying to a relatively pronounced influence by tidal currents.

Macrotidal coasts do not develop barriers due to the predominance of strong tidal currents in a shore-normal orientation. Funnel-shaped embayments may be present containing linear sand bodies oriented parallel to tidal currents (Hayes, 1975, 1979).

Hayes (1979) has further sub-divided these three categories into five as a refinement of Davies' (1964) original classification. In this classification, the microtidal coast has a tidal range of <1 m with the same characteristics as mentioned above. The low-mesotidal coast has a tidal range of 1—2 m and is characterized by increasing numbers of tidal inlets with washovers diminishing. It is a coast of mixed tidal and wave energy but with waves dominant. The New Jersey coast is an example.

The high-mesotidal coast has tidal ranges of 2—3.5 m with abundant tidal inlets, large ebb-tidal deltas and drumstick barriers (Hayes, 1979). Plum Island, Massachusetts and the Georgia—South Carolina coasts serve as excellent examples. This is also a mixed-energy coast but tides are dominant. The low-macrotidal coast has tidal ranges of 3.5—5 m. Some wave-built bars may be present but barriers are not developed. The German Bight of the North Sea is an example of this type of coast. The last category is the macrotidal coast which has tidal ranges of >5 m with numerous tidal-current formed ridges and extensive tidal flats and marshes. Bristol Bay in Alaska and the Bay of Fundy are examples (Hayes, 1979).

This approach to coastal classification by Hayes (1979) uses tidal ranges which do tend to develop similar morphologies through a spectrum of wave climates. The classification is based on a broad range of coastal areas throughout the world but generally is restricted to those having moderate wave energy.

SOME NOTABLE EXCEPTIONS

The above described classifications and generalizations about the interactions of waves and tides with the coasts and the resulting morphology are based on numerous examples and they make good sense. Like most generalizations they are somewhat simplified and there are many exceptions.

High wave energy

As stated previously, Hayes' (1975, 1979) classification is based largely on coasts with low to moderate wave energy. It is also intended to be applied to trailing edge, depositional coasts which would have high preservation potential. Along such coasts there is commonly a demonstrable relationship between tidal range and morphology. The coast of Oregon and Washington is a leading edge coast but it displays well developed beaches and contains a rather large sediment prism along the coast and shoreface. This area is within the zone of highest wave energy in Davies' (1980) classification. Tidal ranges along this coast reach 4 m thus placing it in Hayes' (1975, 1979) low-macrotidal type. The morphology of the coast is however, distinctly wave-dominated with shore-parallel spits developed across estuary mouths and essentially no ebb tidal deltas present (Fig.2). Bays in this area which have large tidal prisms (e.g. Willapa Bay, Washington) may develop some ebb delta morphology (H.E. Clifton, pers. commun., 1983).

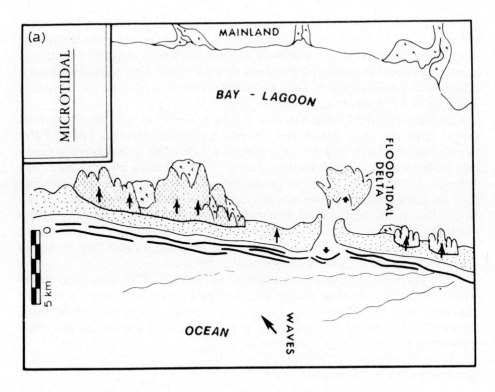

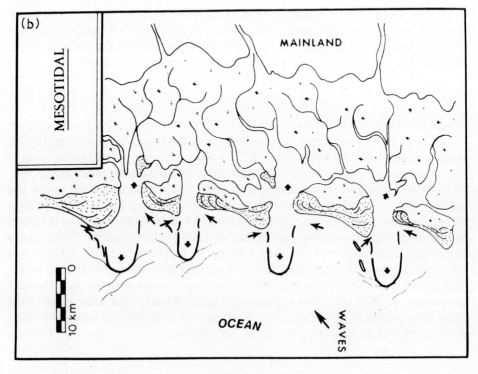

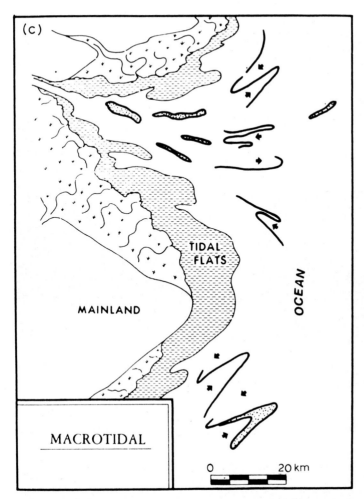

Fig.1. Diagrams of Hayes' coastal morphology types. (a) Microtidal, showing long narrow barriers with numerous washovers and few inlets; (b) mesotidal, showing short, wide barriers with numerous inlets; and (c) macrotidal, on which barriers are absent (after Hayes, 1979).

Another good example is in the Bay of Fundy, an area well-known for its extreme tidal ranges. The coast along the east side of Scot's Bay, just west of the entrance to the Minas Basin experiences a tidal range of ~10 m. This coast is characterized by a smooth shoreline with a well-developed cobble beach exhibiting a large storm berm (Fig.3). This coast, although only 9 km long, is a classic example of a wave-dominated coast yet it is in an area of the largest tidal ranges in the world. The Scot's Bay coast is exposed to the long fetch of the Bay of Fundy and the storm waves that are generated by southwesterly winds.

A similar situation exists along the northwest coast of France although sediment supply is limited in this area. Such a coast would not fall under Hayes' (1975, 1979) classification because it is not a trailing edge, deposi-

Fig.2. Oblique aerial photo at Siletz Spit, Oregon. No ebb delta is present along this wave-dominated coast although spring tides are 4 m (photo by W.T. Fox).

tional coast. It is apparent however, that in coastal regions of high wave energy, the tidal range is not a significant factor in determining coastal morphology.

Low wave energy

Whereas the above section considers problems associated with high wave energy and high tidal range, it is also important to comment on exceptions at the other end of the spectrum. Coasts that are characterized by low wave energy and low tidal range also display numerous exceptions to the aforestated generalizations.

Among the most notable of these is the west peninsular coast of Florida along the Gulf of Mexico. This coast is characterized by barriers for a distance of more than 250 km. The barriers give way to mangrove swamps to the south and to coastal salt marshes to the north. Tidal range is less than 1 m throughout the barrier coast placing it in Hayes' (1979) microtidal category. Wave energy is low with mean annual wave height of 30 cm (Tanner, 1960; Hayes, 1979). Although this coast is within the belt of tropical storms, the impact of these storms is infrequent; the most recent being in 1960.

The general morphology of the west-peninsular Florida barrier system is very similar to that proposed by Hayes (1975, 1979) for the high-mesotidal

Fig.3. Cobble beach with washovers on marsh at Scot's Bay, Bay of Fundy. The tidal range on this wave-dominated coast is ~10 m.

coast of mixed energy but with tides-dominant. There are abundant tidal inlets, large ebb deltas (Fig.4) and drumstick-shaped barriers (Fig.5). Downdrift offset characterizes most of the inlets; another feature common to mesotidal coasts. Although tidal range is low in this area, the wave energy is also very low. This coupled with rather large tidal prism permits large ebb deltas to develop. The infrequent occurrence of extreme storms allows these ebb deltas to persist. Their presence causes the appropriate set of process-response conditions that gives rise to drumstick barriers.

The coastal areas at both ends of the west-peninsular Florida barrier system experience somewhat higher tidal ranges (1—1.2 m to the north and 1—1.4 m to the south) and lower wave energy (Tanner, 1960). As a consequence the coast assumes a morphology that is tide-dominated (Price, 1955) with extensive salt marshes or mangrove swamps, tidal current ridges and tidal flats, although the latter two features are not well-developed on most of the Florida coast.

Another example of tide-dominated morphology in a region of low tidal range is present on the west side of Andros Island in the Bahamas. This area experiences tidal ranges of slightly less than 1 m yet it exhibits a morphology that is quite similar to the German Bight coast of the North Sea (Fig.6), an example of the typical low-macrotidal coast (Hayes, 1979). The major reason for the situation on Andros Island is the very low wave energy

Fig.4. Oblique aerial photo of Redfish Pass on the Florida Gulf coast. This large ebb delta develops with a spring tidal range of 0.7 m.

coupled with the extensive low relief coastal zone. It is much like the "zero-energy" coast of Florida (Tanner, 1960) described above.

ADDITIONAL IMPORTANT VARIABLES

This discussion has only considered wave energy and tidal range as the variables which control the morphology and configuration of the coast. Although these parameters are obviously quite important, consideration must also be given to other factors which may influence a coast and which may determine if it is or is not wave-dominated. Among these are coastal physiography, tidal prism, availability of sediment and influence of riverine input; more or less in order of decreasing importance. The availability of sediment and coastal physiography are commonly related. Coasts bounded by resistant bedrock typically limit the amount of sediment especially on high wave energy coasts where sediment is swept out of the shoreline area.

Physiography may be a limiting factor in that high relief coasts provide little space for sediment to accumulate either above or below mean sea level. Much of the west coast of North and South America serves as an example. These are high relief areas and are also high wave energy coasts. As a result sediment collects only in small pockets or reentrants along the coast. The shelf is narrow to non-existant so that there is no place for

Fig.5. Oblique aerial photo of Caladesi Island, Florida, an excellent example of a drumstick barrier developed with a spring tidal range of 0.8 m.

large sediment accumulations to develop. Such coasts are not likely to accumulate sequences which would be preserved in the stratigraphic record.

Coastal plain areas provide abundant sediment as well as extensive sites for sediment accumulation. Most of the world's barrier systems are developed along these physiographic provinces (Glaeser, 1978). It is this type of coast that was the basis for Hayes' (1975, 1979) original classification.

Tidal prism represents one of the most important but commonly overlooked factors in determining the morphology of barrier island type of coasts. The amount of water that passes through any inlet is determined by the product of the tidal range and the area of the bay landward of the barrier which is serviced by this tidal inlet. Obviously in a situation where the bay area is unchanged, a change in tidal range will effect the prism and the reverse will also hold true.

Generally there will be little variation in the tidal range along a particular reach of barrier coast although there are exceptions. For example, virtually the entire Gulf of Mexico experiences tides of less than 1 m. When changes occur they are of a regional nature such as along the south Atlantic coast of the United States. From the Outer Banks there is a general and rather pronounced increase in tidal range from 1 m toward the southern South Carolina coast where it reaches 2.5 m, then a decrease to the south into

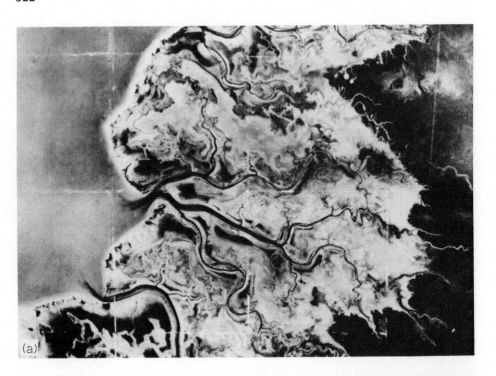

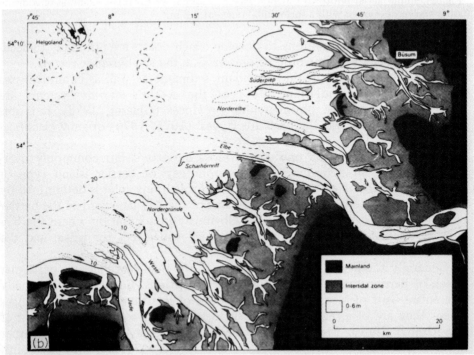

Fig.6. *a*. Vertical aerial photo of tide-dominated coast on the west side of Andros Island, Bahamas. Tidal range is 0.9 m (photo courtesy of E.A. Shinn).

b. General map of coastal morphology in the German Bight area along the North Sea, a macrotidal to high-mesotidal coast (after Hayes, 1979).

Florida where it is again about 1 m. There are many inlets along both of these coasts and the tidal prisms of these vary greatly both from one to the next and also from one area to another. The reason for this is the variety of sizes of estuaries, lagoons and bays which the inlets service. It is possible therefore to have adjacent inlets with markedly different tidal prisms or to have areas where tidal prisms are great even though the tidal range is small.

Data of this type are shown for several inlets on the Atlantic and Gulf coasts and for one on the Pacific Coast (Fig.7). Only those inlets where well-documented data on tidal range and prism are included. It can be seen from this plot that there is no relationship between tidal range and tidal prism. Note that the highest and the lowest prism values are on the Gulf coast where all tidal ranges are less than 1 m.

The reason that this lengthy discussion of tidal prism is important to the question at hand is because large tidal prisms, especially in areas of low wave energy, can explain large, well-developed ebb tidal deltas and development of drumstick barriers.

Riverine input to the coast in the form of both sediment and the physical energy of the river itself has an important impact upon the coast. It is obvious that much sediment is provided by rivers and that this sediment may be dispersed in a variety of manners and in various amounts once it reaches the coast. Great quantities of sediment may accumulate in the form of deltas but the morphology of these deltas covers a broad spectrum depending

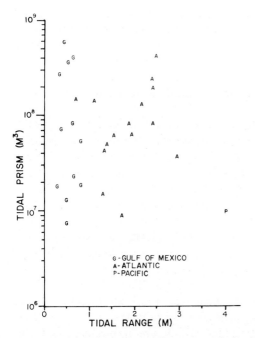

Fig.7. Plot of inlets along the coast of the United States showing relationship between tidal range of tidal prism. Data from NOAA tables.

largely on the relative roles of the river, tides and waves. Such interactions and their resulting morphology have been classified by Galloway (1975) and by Scott (1969). In general, riverine dominated deltaic coasts are characterized by digitate lobes on the delta with the Mississippi Delta being a good example. Wave-dominated deltas achieve an accurate or cuspate outline with a smooth outer boundary. Well-developed beaches and beach ridges are common. The São Francisco River of Brazil is a good example (Wright, 1978). Tide-dominated deltas are similar in appearance to tide-dominated estuaries. There is generally little protrusion beyond the regional shoreline and sediment bodies are linear, paralleling the tidal currents. The Ord River on the northern coast of Australia is an excellent example (Wright, 1978).

DISCUSSION

The data and examples described above and the numerous examples described in the literature do enable some useful generalizations to be made about the morphology of wave-dominated coasts and these in turn permit related generalizations about stratigraphic sequences which accumulate on wave-dominated coasts. The latter can serve as valuable data sets in reconstructing paleoenvironmental conditions in the coastal regime.

Modern wave-dominated coasts

There can be no disagreement with the basic premise that rather straight and smooth coasts, characterized by well-developed beaches are the result of physical conditions dominated by waves and by wave-generated currents. If barriers are present, they are typically long and smoothly accurate or straight. Storm generated washovers may be present but are not *a required characteristic.* Pocket beaches along bedrock coasts also fall into the wave-dominated category.

The above characteristics are similar to those described by Hayes (1975, 1979). The only important deviation from Hayes' classification is the association of a particular tidal range (e.g. microtidal) with the wave-dominated coast. Although this is a common association, there is no need to relate tidal range to coastal morphotypes. The important relationship is that wave-energy overwhelms tidal energy and in so doing, a characteristic morphology is produced.

The relationships between tidal processes and wave-generated processes have been shown by Hayes (1979) who presented a generalized diagram based on many geographic areas (Fig.8). The five fields presented range from tide-dominated to wave-dominated. An approximate limit of barrier island development would be within the field labeled tide-dominated (low). Notice that this field, as do all, covers a spectrum of tidal ranges and wave heights. It is the *relative* effects of these processes which are important, not the absolute values.

It is also important to be aware of the rather delicate balance between

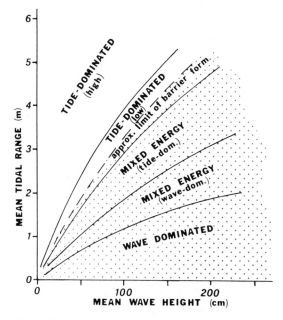

Fig.8. General relationships between tidal range and wave height as it relates to coastal morphology. A particular coastal region may span several fields (modified after Hayes, 1979).

tide and wave processes as lower and lower values are approached. All of the five fields converge at the low end of the spectrum (Fig.8). The consequences of these relationships are that tide-dominated, wave-dominated or mixed energy morphologies may develop with very little difference in tide and wave parameters. Such relationships further emphasize the importance of excluding absolute tidal range or wave height values from coastal morphotypes.

The shoreline in wave-dominated environments is characterized by elongate shore-parallel sediment bodies. These include longshore bars, beaches, beach ridges, foredunes and even outer lobes of ebb-tidal deltas which are wave-formed. Such features may be present along all types of coasts including river deltas but are generally best developed along coastal plain shore zones. They may be present on prograding coasts or on transgressive coasts.

The only sediment body type which commonly occurs on wave-dominated coasts that is not shore-parallel is washover or blowover features. These lobate or fan-shaped features are storm generated phenomena which may be developed on all coastal types but are most common on wave-dominated coasts.

Sediment which accumulates on wave-dominated coasts generally shows moderate to very good sorting due to continual reworking by waves but it may be of virtually any grain size. Sand is by far the most common size as it is in shore zones overall. Mud does occur, such as along the coast of

Surinam (Wells and Coleman, 1981) and on chenier coasts. Gravel and cobble sediment is also common especially along many wave-dominated beaches in moderate to high latitudes where glacial deposits are reworked along the coasts and in some low latitudes where coral reef debris is abundant.

To summarize therefore, modern wave-dominated coasts are characterized by shore-parallel sediment bodies which may span the entire grain size spectrum. Tidal range itself is not a criterion in determining tidal- or wave-dominance.

Stratigraphic sequences from wave-dominated coasts

A fundamental reason for considering the distinctions between wave-dominated and tide-dominated coasts is the ability of making these distinctions in the stratigraphic record. In order to consider this problem it is necessary to consider both the individual stratigraphic column such as might be seen in a core and also the geometry and interrelationships between lithosomes.

Transgressive, wave-dominated sequences typically display a coarsening upward texture. Considering a barrier island sequence, there would be a sequence of estuarine, marsh or tidal flat, washover and beach and nearshore deposits from bottom to top (Fig.9). Such a sequence could be resting comformably on almost any lithosome type. Many Holocene barrier systems on the present Atlantic and Gulf coasts of the United States exhibit such sequences (e.g. Fisk, 1959; Kraft, 1971).

A progradational (regressive) sequence of a wave-dominated system would also display a generally coarsening upward texture with beach and dune ridge lithosomes overlying nearshore and shoreface units. Thin units of muddy and organic-rich sediment representing swales or cat's eye ponds would be expected to be incorporated into the well-sorted, cross-stratified sands of the prograding beach ridge lithosome (e.g. Barwis, 1976).

By way of contrast, the prograding tide-dominated coast would be expected to accumulate a fining-upward or at least a shoaling-upward sequence with sandy subtidal shoal deposits, intertidal shoals, tidal flats and marsh deposits from bottom to top. Good examples are those from the Wash in England (Evans, 1965, 1975) and the Bay of Fundy (Knight and Dalrymple, 1975). A transgressive tide-dominated system would display a stratigraphic sequence that is similar to that of the transgressive wave-dominated system. It would contain a generally coarsening-upward textural trend with salt marsh, tidal flat, and tidal sand bodies from the bottom to top. Subtle differences would include nature and orientation of cross-stratification, extensive reactivation surfaces in the tide-dominated sequence and the contained fauna.

The most demonstrable method of distinguishing between these two types of coastal sequences is by considering the overall geometry and orientation of the sand bodies. The wave-dominated system displays shore-parallel sand lithosomes whereas the tide-dominated system contains shore-normal sand

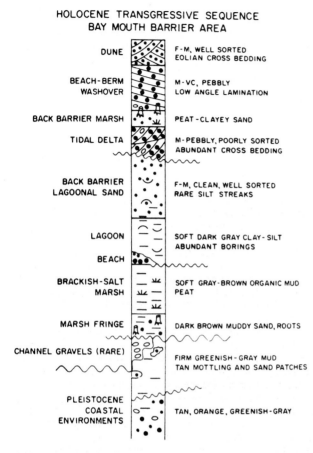

Fig.9. Stratigraphic sequence of Holocene transgressive barrier system along the Delaware coast (after Kraft, 1971).

lithosomes. Various directional and orientational structures such as cross-strata, reactivation surfaces, parting lineation, scour marks and other indicators of current direction must be used to confirm these distinctions.

SUMMARY

A wave-dominated coast is simply a coast which is subjected to physical processes that are dominated by wave energy. This dominance is strictly relative; it is not based on any absolute wave parameters. The same is true for tide-dominated coasts. To relate the dominant process to a particular tidal range or wave parameter is incorrect.

Modern coastal morphology confirms this by the wide range of wave and tidal conditions which produce similar appearing coastal configurations. Stratigraphic sequences in the ancient record may be recognized as tide- or wave-dominated but the association with specific levels of tidal range is at best, difficult. Paleotidal range determinations have been discussed by Klein (1971; 1972) but such determinations are rarely attempted from the stratigraphic record. Paleo-wave climate has been investigated by Komar (1974) and Allen (1981) by relating bedform characteristics to wave parameters. Dupré (1984, this volume) has interpreted Pleistocene wave climate on the basis of sediment and bedform features.

REFERENCES

Allen, P.A., 1981. Wave-dominated structures in the Devonian lacustrine sediments of south-east Shetland and ancient wave conditions. Sedimentology, 28: 369—379.
Barwis, J.H., 1976. Internal geometry of Kiawah Island beach ridges. In: M.O. Hayes and T.W. Kana (Editors), Terrigenous Clastic Depositional Environments. Univ. South Carolina, Coastal Res. Div., Columbia, S.C., pp.II-115—II—125.
Davies, J.L., 1964. A morphogenic approach to world shorelines. Z. Geomorphol., 8: 27—42.
Davies, J.L., 1980. Geographical Variation in Coastal Development (2nd ed.). Longman, New York, N.Y., 212 pp.
Dupré, W.R., 1984. Reconstruction of paleo-wave conditions from Pleistocene marine terrace deposits, Monterrey Bay, California. In: B. Greenwood and R.A. Davis, Jr. (Editors), Hydrodynamics and Sedimentation in Wave-Dominated Coastal Environments. Mar. Geol., 60: 435—454 (this volume).
Evans, G., 1965. Intertidal flat sediments and their environments of deposition in the Wash. Q. J. Geol. Soc. London, 121: 209—245.
Evans, G., 1975. Intertidal flat deposits of the Wash, western margin of the North Sea. In: R.N. Ginsburg (Editor), Tidal Deposits. Springer, New York, N.Y., pp.13—20.
Fisk, H.N., 1959. Padre Island and the Laguna Madre flats, coastal south Texas. Natl. Acad. Sci., Natl. Res. Council, 2nd Geogr. Conf., pp.103—151.
Galloway, W.E., 1975. Process framework for describing the morphologic and stratigraphic evolution of the deltaic depositional systems. In: M.L. Broussard (Editor), Deltas, Models for Exploration. Houston Geol. Soc., Houston, Texas, pp.87—98.
Glaeser, J.D., 1978. Global distribution of barrier islands in terms of tectonic setting. J. Geol., 86: 283—297.
Hayes, M.O., 1975. Morphology of sand accumulations in estuaries: an introduction to the symposium. In: L.E. Cronin (Editor), Estuarine Research, vol. 2. Academic Press, New York, N.Y., pp.3—22.
Hayes, M.O., 1979. Barrier island morphology as a function of tidal and wave regime. In: S.P. Leatherman (Editor), Barrier Islands. Academic Press, New York, N.Y., pp.1—27.
Heward, A.P., 1981. A review of wave-dominated clastic shoreline deposits. Earth Sci. Rev., 17: 223—276.
Klein, G. deV., 1971. A sedimentary model for determining paleotidal range. Geol. Soc. Am. Bull., 82: 2585—2592.
Klein, G. deV., 1972. Determination of paleotidal range in clastic sedimentary rocks. 24th Intl. Geol. Congr. Proc., Sec. 6: 397—405.
Knight, R.J. and Dalrymple, R.W., 1975. Intertidal sediments from the south shore of Cobequid Bay, Bay of Fundy, Nova Scotia, Canada. In R.N. Ginsburg (Editors), Tidal Deposits. Springer, New York, N.Y., pp.47—55.
Komar, P.D., 1974. Oscillatory ripple marks and the evaluation of ancient wave conditions and environments. J. Sediment Petrol., 44: 169—180.
Kraft, J.C., 1971. Sedimentary facies patterns and geologic history of a Holocene marine transgression. Geol. Soc. Am. Bull., 82: 2131—2158.
Price, W.A., 1955. Development of shoreline and coasts. Dept. Oceanography, Texas A & M Univ., Project 63, 9 pp.

Scott, A.J., 1969. Figure 7, p. 82. In: W.L. Fisher, L.F. Brown Jr., A.J. Scott and J.H. McGowen, 1969. Delta Systems in the Exploration for Oil and Gas — A Research Colloquium. Univ. Texas, Bur. Econ. Geol., Austin, Texas, 102 pp.

Tanner, W.F., 1960. Florida coastal classification. Trans. Gulf Coast Assoc. Geol. Soc., 10: 259—266.

Wells, J.T. and Coleman, J.M., 1981. Physical processes and fine-grained sediment dynamics, coast of Surinam, South America. J. Sediment Petrol., 51: 1053—1068.

Wright, L.D., 1978. River deltas. In: R.A. Davis, Jr. (Editor), Coastal Sedimentary Environments. Springer, New York, N.Y., pp.5—68.

SHOREFACE MORPHODYNAMICS ON WAVE-DOMINATED COASTS

ALAN WILLIAM NIEDORODA[1], DONALD J.P. SWIFT[2], THOMAS SAWYER HOPKINS[3] and CHEN-MEAN MA[4]

[1]R.J. Brown and Assoc., Suite 200, 2010 North Loop West, Houston, TX 77018 (U.S.A.)
[2]Exploration and Production Research, PGR-G130, ARCO Oil and Gas Company, P.O. Box 2819, Dallas, TX 75221 (U.S.A.)
[3]Oceanographic Sciences Division, Department of Energy and Environment, Brookhaven National Laboratory, Upton, NY 11973 (U.S.A.)
[4]Woodward-Clyde-Oceaneering, 7330 Westview Drive, Houston, TX 77055 (U.S.A.)

(Received January 17, 1983; revised and accepted July 29, 1983)

ABSTRACT

Niedoroda, A.W., Swift, D.J.P., Hopkins, T.S. and Chen-Mean Ma, 1984. Shoreface morphodynamics on wave-dominated coasts. In: B. Greenwood and R.A. Davis, Jr. (Editors), Hydrodynamics and Sedimentation in Wave-Dominated Coastal Environments. Mar. Geol., 60: 331—354.

An open ocean shoreface typical of long, wave-dominated sandy coasts has been examined through a combination of extensive field measurements of wave and current patterns with computations of marine bedload transport and sedimentation. Sand transport on the upper shoreface is dominantly controlled by waves with only secondary transport by currents. Sand on the middle and lower shoreface, as well as the inner continental shelf is entrained by storm waves and transported by a complex pattern of bottom boundary layer currents.

Storm events have been studied and modeled for the shoreface off Tiana Beach, Long Island. The dominant effect of coastal frontal storms is to cause significant shore-parallel bedload transport with important shore-normal secondary components. These storms tend to result in net offshore transport of sand removed from the beach and surf zone systems. The bedload transport during a storm is convergent on the shoreface leading to accretion. Most accretion occurs on the upper shoreface with lesser deposits covering the middle and lower shoreface as well as the inner continental shelf. Longer-term equilibrium can be maintained by slow return of sand up the shoreface during non-storm conditions.

Annual and geologic time-scale budgets of shoreface sand transport and sedimentation yield equilibrium, net accretion or net deposition. The annual balance results from an integration of the event-scale bedload transport patterns and morphologic responses. These processes and responses have feedback mechanisms which stabilize the system over longer, but not geologic, time scales. Geologic time scale balances are controlled by relative sea level changes and relative availability of sediment supply with the event-scale shoreface sand transporting processes providing the mechanism to produce the changes in long-term morphology and sedimentation patterns. In the area of study, the long-term pattern is one of net shoreface erosion, and the permanent loss of sand to the shelf floor.

0025-3227/84/$03.00 © 1984 Elsevier Science Publishers B.V.

INTRODUCTION

A major morphological feature of many continuous, sandy, wave-dominated coasts is the concave-upward bottom slope lying between the outer reaches of the surf zone and the inner continental shelf. This feature is called the shoreface. Although the recognition of the prevalence of this feature was demonstrated in the early portion of the 20th Century, there have been remarkedly few studies conducted to understand the nature of the sediment transport processes which occur in this zone and, in fact, result in controlling its morphology. This zone is difficult to reach with instruments supported from the beach and is generally considered too close to the shore for classic oceanographic examinations. In addition, the waves, currents, and sediment transport patterns in this zone are characterized by strong spatial and temporal gradients. Unsteady flow conditions make study of the region difficult without exhaustive theoretical or field measurement programs.

In treating the overall morphodynamics of the shoreface, three time scales must be considered. These are the event scale (order of 2—7 days), the annual scale (order of a year), and the geologic time scale (order of centuries or longer). The bulk of the following discussion will be directed at event scale processes as these must be understood before longer time scale effects can be defined. Only long, straight sandy coasts are considered. Alterations in shoreface processes related to tidal inlets, estuary mouths, or rocky headlands are not considered. Much of the experimental data presented resulted from a series of field measurements off Tiana Beach, Long Island, N.Y. These data are assumed to be reasonably representative of typical shoreface wave, current, and sediment transport conditions because the shoreface in this area has a relatively simple shape and constant exposure to open-ocean conditions.

DEFINITION OF THE SHOREFACE

From the days of classical geomorphology, the shoreface has been recognized as a significant zone along much of the world's coastline. Early discussion of this zone can be found in Fenneman (1902), Barrell (1912), and Johnson (1919). On long sandy wave-dominated coasts the shoreface is the region lying immediately seaward of the surf zone and terminating at the inner continental shelf. In general, it has a concave-upward shape with the steepest slope near its top (Swift, 1976). The landward portion has steepnesses on the order of 1:20. This slope diminishes seaward until the shoreface merges with the inner continental shelf where the slope is on the order of 1:2000. The depths of the upper and lower boundaries of the shoreface are variable depending upon local sediment supply and on wave and current conditions. On the long sandy coasts of the mid-Atlantic and New York bights, the upper shoreface joins the surf zone at a depth of approximately 4 m and the lower shoreface joins the inner continental shelf at a depth of approximately 25 m.

Many authors have concluded that the shape and morphology of the shoreface is an equilibrium response of an unconsolidated coast to the typical local wave and current regime (Johnson, 1919; Bruun, 1962; Moody, 1964; Swift, 1976; Niedoroda and Swift, 1981). However, quantitative relationships between process and response in this zone have been generally lacking.

Limited information concerning the nature of dominant shoreface processes has resulted from geological interpretation of the character and distribution of shoreface sediments. Relatively coarse sand in the surf zone grades to fine sand on the upper portion of the shoreface and is abruptly replaced by coarser sand at the middle and base of the shoreface. Cook (1970), Langford-Smith and Thom (1969), and Cook and Gorsline (1972) have concluded that this pattern can be partially explained by the winnowing of fine sand from deposits within the surf zone and deposition of this fine sand by rip currents which diffuse over the upper shoreface. Variations in the slope and curvature of shorefaces in widely separate regions have been qualitatively attributed to differential sorting of sand deposits through the action of wave-induced currents and to differences in the amount and type of sediment available to local shoreface processes (Wells, 1967; Wright and Coleman, 1972). More recent work by Niedoroda (1980) and Niedoroda and Swift (1981) has outlined many of the basic sediment dynamic processes on the shoreface.

DYNAMIC ZONES OF THE COASTAL OCEAN

Niedoroda (1980) and Niedoroda and Swift (1981) have shown that the marine bedload transport and morphodynamics of the shoreface are primarily controlled by the net local wave and current environment. The dynamics controlling these two phenomena are independent of each other as are their associated length scales, neither of which is necessarily coincident with the width of the shoreface. All length scales are, however, ultimately related to the mean bottom slope, the sediment supply, and the amount of energy available to the system. Wave processes dominate the upper reaches of the shoreface while current processes dominate the middle and lower reaches of the shoreface.

For typical open ocean wave heights and periods non-linear wave behavior becomes important in defining near bottom wave orbital kinematics in depth ranges from about 8 to 18 m (approximately 1.5 km offshore). Thus, non-linear wave behavior characterizes the upper portion of the shoreface zone. The waves typically begin to break at depths of 3—5 m (approximately 0.5 km offshore). Under open-ocean storm conditions breaking waves may extend offshore to a depth of approximately 10 m. Thus, the inner edge of the shoreface is occasionally occupied by the outer reaches of the surf zone.

The shore-normal or diabathic length scale for coastal ocean currents is considerably wider than that for nonlinear behaviour of coastal ocean waves. The coastal boundary layer represents the region of the ocean where the presence of shoaling depths and the shoreline strongly affect the dynamics of currents. Coastal ocean currents result from tidal forcing, wind forcing, and horizontal pressure gradients related to slopes of the mean sea surface and the isobars of the internal field of mass.

Non-tidal coastal ocean currents develop as a result of a complex relationship between surface wind stresses, diabathic slopes of the mean sea surface and/or the pycnocline, as well as flow in the upper and lower frictional boundary layers. In general, the significant baroclinic diabathic length scale is on the order of 5—10 km and the corresponding barotropic length scale is on the order of 25—50 km for typical mid-Atlantic summer conditions (Csanady, 1978; Winant and Beardsley, 1979; Hopkins, 1982; Hopkins and Dieterle, 1983).

The characteristic time scales of coastal ocean waves and currents serve to define the durations for which forcing mechanisms must remain nearly constant to yield coherent and steady flow conditions. The steady coastal ocean currents tend to occur in distinguishable patterns, whereas unsteady or transient currents tend to develop in quite variable patterns. Waves occur in a relatively narrow range of frequencies and adjust to the wind within durations of hours to one day. Coastal ocean currents have a wide range of characteristic time scales which depend on the relative magnitude of the forcing mechanisms and the nature of the local water column stratification.

Surface wind stresses are a principal forcing mechanism for coastal currents and their temporal dependency. Over deep water the wind must blow with a nearly constant speed and direction for more than the inertial period for steady currents to result. Near coastal boundaries, wind transport causes sea-level distortion and accompanying barotropic flow. Once initiated these flows can remain coupled to the wind forcing, provided no sudden changes in the wind occur. The greater the change the more uncoupled the flow becomes until the lag is again at the inertial time scale (Hopkins, 1974). At mid-latitudes this time scale falls between the semidiurnal and diurnal frequencies (about 19 h). Over shallower water columns of the shoreface, the response to wind forcing (including the barotrophic effects) is reduced by the effects of increased turbulence and shallower depths. If the coastal ocean is stratified, energy is diverted to the baroclinic mode which opposes and has longer time scales than the barotrophic mode. However, the baroclinic mode is reduced over the shoreface because stratification there is often weak due to wave mixing. Well-organized coastal flows which are coupled to the surface wind stresses require at least a day of nearly steady conditions to develop. This often complicates the procedure of defining typical flow patterns from measured data.

CURRENT PATTERNS IN THE COASTAL OCEAN

The purpose of this section is to define a series of typical current patterns over an open ocean shoreface. In subsequent subsections this information will be used in combination with other measured wave and current data to compute patterns of shoreface bedload transport.

The principal source of data concerning characteristic coastal ocean currents over open ocean shorefaces is from a series of field experiments conducted during 1976, 1977 and 1978. A complete description of these experiments is given in Niedoroda (1980). Therefore, only a brief description is provided here.

The field experiments were part of two interrelated projects. These were project INSTEP (conducted jointly by the Coastal Research Center of the University of Massachusetts and AOML/NOAA) and project COBOLT (conducted jointly by Brookhaven National Laboratory and Woods Hole Oceanographic Institute). A diabathic transect was established offshore of Tiana Beach, Long Island (see Fig.1). Boat stations were established at 1 km intervals over a 12 km line across the coastal boundary layer in 1976. In 1977, these boat stations were established along a 3.5 km line across the shoreface at approximately 500 m intervals. Nearly synoptic measurements of currents, temperature, and salinity were made at 1 and 2 m depth intervals at these boat stations. The locations of these sampling points are given in Figs.2 and 3. During 1976, two Shelton Spars were located on this transect at 3 and 9 km offshore (see Fig.1). Each spar supported four 2-axes electromagnetic current meters and eight temperature and conductivity probes. A

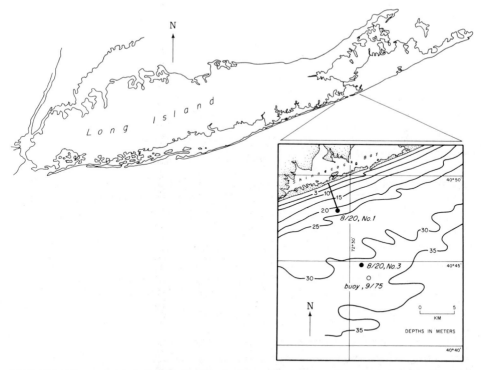

Fig.1. Location map of the Tiana Beach, Long Island shoreface experiments (Projects INSTEP and COBOLT). Solid line shows location of the boat station transect during 1976 and 1977. Dots are locations of the Project COBOLT Shelton Spars during 1976.

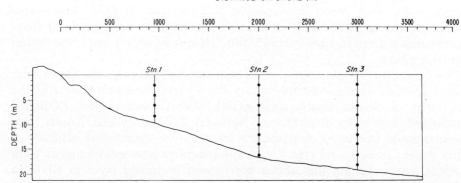

Fig.2. Location of daily current meter, conductivity and temperature measurements along the 1976 Tiana Beach Shoreface Transect.

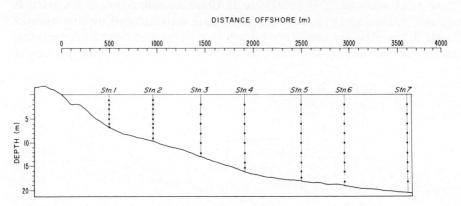

Fig.3. Location of daily current meter, conductivity and temperature measurements along the 1977 Tiana Beach Shoreface Transect.

complete description of the project COBOLT Shelton Spars is given in Dimmler et al. (1976), Scott et al. (1978) and Tucker (1974a, b, 1975). A complete set of the data resulting from the boat station measurements is given in a report by Niedoroda (1980).

This relatively comprehensive set of oceanographic data from a typical open ocean shoreface has been processed to show characteristic flow patterns. Measurement of the currents at the boat stations required a total of approximately two hours. Therefore the data are nearly synoptic. In order to compare data taken on different times and days it is necessary to remove the relatively large tidal components.

Time series of the currents measured at the 3 and 9 km Shelton Spars of the COBOLT project were processed using rotary spectra and complex demodulation (Mooers, 1973). The principal semidiurnal tidal current amplitudes were determined. These are plotted on Fig.4. The regression lines shown on Fig.4 indicate that the distribution of tidal current velocities can

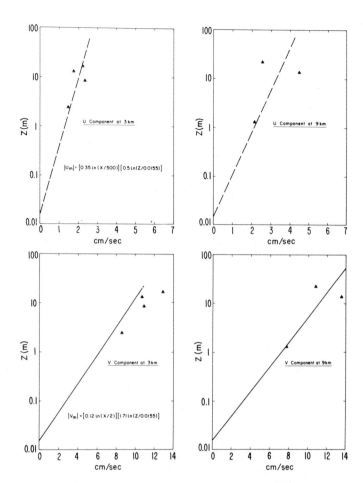

Fig.4. Average semidiurnal tidal current components for various depths at the 3 and 9 km Project COBOLT Shelton Spars. Approximate semilogarithmic relationships are shown.

be approximated by a semilogarithmic function whose equation is given on the figure. Using these empirically derived semilogarithmic functions for the nearshore tidal components a series of curves defining the maximum tidal amplitude at each depth for each boat station of both the 1976 and 1977 measurements was determined. This is shown in Fig.5.

Data from Fig.5 were used to generate a series of masks representing estimated tidal current components at the time that each data set was collected. Maximum tidal amplitude was adjusted according to a cosine function whose argument was the phase of the semidiurnal tide. These values were then subtracted from the measured current values. Representative results are shown in Figs.6 and 7.

The occurrence of coastal upwelling or downwelling has long been recognized but few data are available on the characteristic velocities and actual

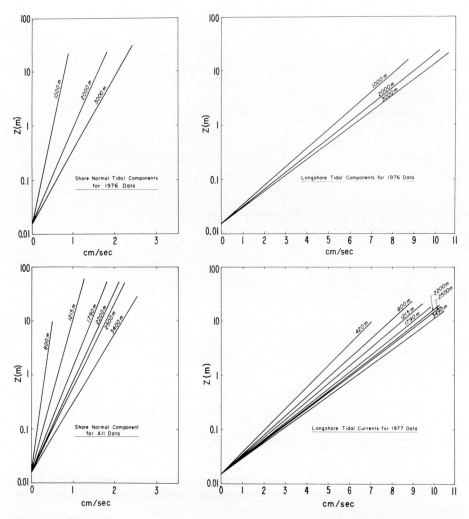

Fig.5. Average tidal component curves used to generate synoptic tidal component masks. These values were corrected for the instantaneous tidal phase via a cosine function.

flow patterns in shoreface regions. Figure 6 shows a typical flow pattern off Tiana Beach under a moderate southwest wind. Inside of the 12 m isobath the shore-parallel current component dominates. The surface Ekman transport, which has an offshore component, causes a horizontal divergence of the upper layer over the shoreface. The divergence results in an upwelling of bottom waters over the shoreface. This flow situation is represented diagramatically in the upper panel of Fig.8. The opposite situation is given by Fig.7 which shows flow over the Tiana Beach shoreface resulting from a moderately strong easterly wind. A horizontal convergence in the upper layer over the shoreface is caused by onshore surface Ekman transport. These data are shown diagramatically in the lower panel of Fig.8.

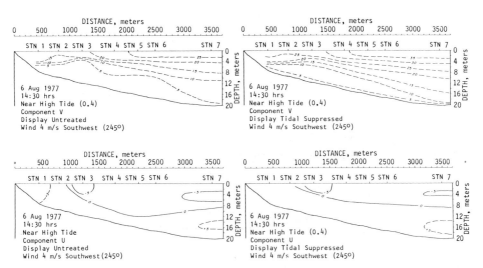

Fig.6. Shore parallel (parabathic, V) and shore normal (diabathic, U) components of shoreface currents on 6 Aug. 1977. Contours are in cm/s. Dashed contours and positive V-values show eastward parabathic flow. Dashed contours and positive U-values show onshore flow. The left-hand panels are raw data and the right-hand panels show flow with the tidal components removed. An eastward/upwelling flow event is shown.

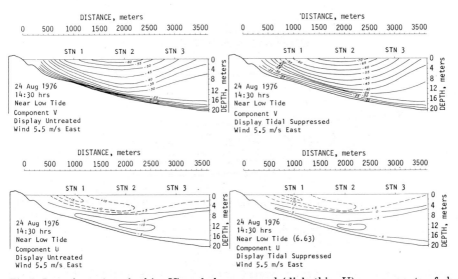

Fig.7. Longshore (parabathic, V) and shore normal (diabathic, U) components of shoreface currents on 24 August 1976. Contours are in cm/s. Solid contours and negative V-values show westward parabathic flow. Solid contours and negative U-values show offshore flow. The left-hand panels are raw data and the right-hand panels show flow with the tidal components removed. A downwelling westward flowing coastal jet is shown.

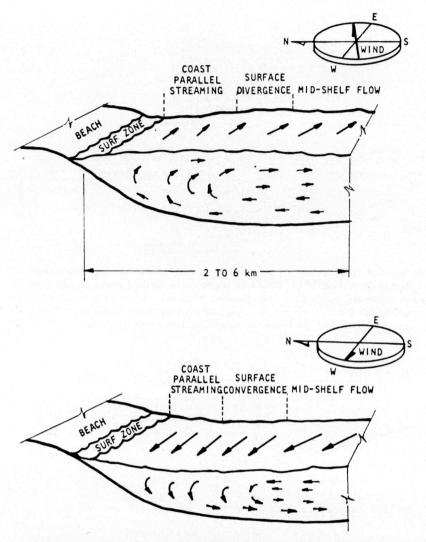

Fig. 8. Diagrams of two of the important classes of coastal ocean flows over the shoreface. The upper panel shows an eastward upwelling event driven by a southwest wind along an east—west coast. The lower panel shows a westward downwelling event driven by an east wind along an east—west coast.

Similar characteristic flow patterns (not shown) for various wind stresses were assembled from the original data. These patterns show representative flows for westward downwelling, westward upwelling, eastward downwelling, and eastward upwelling flow over the Tiana Beach shoreface. These data were subsequently used to model flow and resulting marine bedload transport on the shoreface during storm events.

SHOREFACE BEDLOAD SEDIMENT TRANSPORT

The subject of shoreface marine bedload transport has been discussed by Niedoroda and Swift (1981). It was shown that during non-storm conditions the sand residing on most of the shoreface was not transported even under the combined effects of strong mean- and tidal-currents. Only at the upper portions of the shoreface is there a tendency for sand to be driven upslope by the asymmetrical near bottom wave orbital velocities. During storm events, however, the larger waves cause large near-bottom orbital velocities which entrain sand over the entire shoreface. This sand is then transported both parallel and normal to the shore by the currents. Transport extends across the shoreface onto the inner continental shelf. In this paper we extend the results of our earlier work to show typical shoreface bedload transport patterns and morphological responses due to coastal storms.

Event scale bedload transport

The relative rates and patterns of shoreface sediment transport were examined by combining data on waves and currents during storms to estimate the spatial distribution of both high- and low-frequency fluid shear stresses acting on the sea floor and computing the corresponding bedload. Wave and current measurements from a profiling concentration velocity probe (hereafter PCV-probe) were combined with current data from the boat stations located across the shoreface to produce a synthetic time series of wave and current conditions across the shoreface during the storms.

The PCV-probe consists of an electromagnetic two-axes current meter supported 1.1 m above the sea floor. This probe also has a pressure-type wave gauge and a suspended sediment concentration transducer (Huff and Friske, 1980; Young et al., 1982). This instrument was deployed at a 10 m depth on the Tiana Beach shoreface during the fall of 1977 through the spring of 1978.

Figure 9 shows a record of data measured by the PCV-probe during a coastal storm which occurred between the 26th and the 29th of March 1978. As the storm developed, the waves built rapidly to a maximum during the afternoon and evening of March 27 (Julian Day 86). As the storm center crossed the area, the winds backed sharply and blew offshore. As a result, the wave heights diminished rapidly during March 28 (Julian Day 87).

This storm was a classic Northeaster. The initial winds came from the east to southeast and developed the heavy seas. During the first phase of the storm the coastal flow was directed strongly to the west with the surface current convergence over the shoreface producing a downwelling. The wind veered slowly to the south and southwest before backing to the northwest as the storm center passed. As the wind suddenly began to blow offshore, the wave heights were rapidly reduced while the currents reversed to an easterly direction with first a downwelling and then an upwelling component. These phases of the storm are noted on Fig.9.

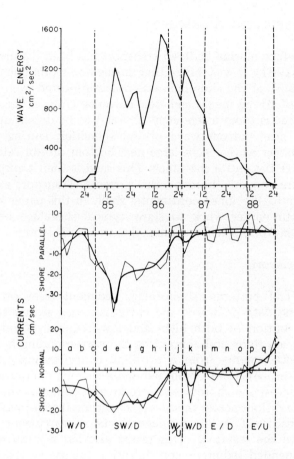
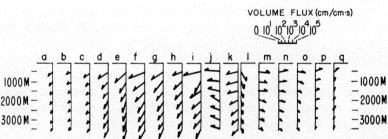

Fig. 9. Upper portion — Time series of total wave energy (upper panel), and current component (middle panels) as measured by the NOAA PCV-probe at a depth of 10 m (current meter 1.1 m above the sea floor) on the Tiana Beach shoreface during the storm of 26 March (day 85) to 29 March (day 88) 1978. The symbols W/D, SW/D, W/U, E/D and E/U mean westward/downwelling and eastward/upwelling, respectively. The letters a through q represent 6-h time intervals used to display shoreface bedload transport in the lower panel of this figure.

Lower portion — The lower panel shows computed bedload transport at 500 m intervals across the shoreface. Each subfigure (a through q) shows bedload fluxes for a particular time interval as defined by the corresponding letter's position in the time series of the upper panel. The bedload flux vectors are on a logarithmic scale. The short horizontal lines beneath the letters represent the shoreline. The vertical lines show distances offshore.

The data from the storm of 26—29 March 1978 shown on Fig.9 were used in conjunction with the previously described catalog of characteristic shoreface flow patterns to estimate bedload transport over the shoreface during the event. The tidal oscillations were subjectively removed by smoothing the shore-parallel and shore-normal current records. The smooth lines shown in the middle and lower panel of Fig.9 represent the smooth flow patterns used. To compute shoreface bedload transport the tidal oscillations were added to the corresponding characteristic flow patterns to yield estimates of the near bottom current over the duration of the storm. The single-point measurement of current direction and intensity was thus expanded to estimate bottom boundary layer flow over the whole shoreface using the characteristic flow patterns for each of the different stages of the storm (e.g., westward downwelling flow, westward upwelling flow, etc.).

A computer program based on the method for computing marine bedload sediment transport due to the combined effects of waves and currents (Madsen and Grant, 1976) was utilized. This method applies the Einstein-Brown bedload transport relationship to marine conditions through the use of a combined wave-current friction factor (Jonsson, 1966) and a modified Shields parameter. Points at 500 m intervals across the Tiana Beach shoreface were represented in the program. Wave and near-bottom current conditions at each of these points were estimated from measured data at two-hour time intervals over the duration of 26—29 March, 1978, storm. The time-averaged bedload transport for each point and time interval was computed. An "instantaneous" bedload volume flux was computed for time sub-intervals equal to one sixteenth the wave period. The sum of these values normalized by the wave period and multiplied by the time interval was used to represent the bedload transport of each two-hour interval. This method and computer program has recently been calibrated with field data (Niedoroda et al., 1982) and shown to yield calculated results within 20 percent of measured transport values. The relative magnitudes of sediment transports are less sensitive to calculation inaccuracies and are more important in understanding the response of shoreface sediments to storm events.

Bedload transport was calculated for each two-hour interval of this storm at seven locations at 500 m intervals across the shoreface. Transport rates and directions for each of these points across the shoreface were summed for 17 intervals (labeled a through q on Fig.9) over the duration of the storm to compute the net transport during the storm.

The lower panel of Fig.9 shows vectors of the bedload sediment flux whose magnitude is related to a logarithmitic scale given on the figure. In each part of this figure (labeled a through q) the short horizontal line represents the shoreline. The vertical line represents a transect extending 3500 m offshore. At each 500 m interval, a bedload sediment transport flux vector is shown indicating the magnitude and direction of the bedload transport at each location during that period of the storm. The succession of figures shows the time series of this transport.

These results indicate that during the early portion of the storm, sediment was driven westward and offshore. As the storm intensified and the waves became larger, more sediment was entrained and transported westward and offshore. As the storm center passed through the area (between time period k and time period l), the direction of shore parallel transport reversed to the east but diabathic (shore-normal) transport remained offshore. Only during the last portions of the storm did an upwelling flow develop. However, by this time the offshore winds had significantly reduced the wave heights resulting in only a minor pulse of upslope transport in the very final stages of the storm.

Figure 10 shows the result of summing the diabathic component of shoreface bedload transport during the entire storm. At all points, the net volume flux is offshore. The greatest volume flux occurs at the calculation point highest on the shoreface (500 m). The net volume flux reduced sharply to a distance of approximately 1500 m offshore and then fell off less rapidly to the last calculation point near the base of the shoreface. The second curve shown on this figure represents the difference in sediment transport between each adjacent calculation point. Because of the convergence of the offshore transport there is a deposition of sediment on the shoreface. The greatest deposition occurs at the upper shoreface. The relative accretion of the shoreface during the storm decreased over a distance of approximately 1700 m offshore. Seaward of this distance, the rate of deposition is approximately constant. Extrapolation beyond the last calculation point suggests that this rate of deposition would continue out onto the inner continental shelf.

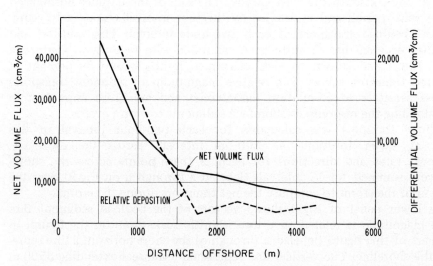

Fig.10. The net bedload sediment flux (solid line) and net differential transport or deposition on the shoreface (dashed line) as computed for the 26 through 29 March 1978 northeastern storm off Tiana Beach, Long Island. Note that the largest offshore sand transport and largest spatial gradients of offshore sand transport (hence deposition) occurred on the upper shoreface. However, offshore sand transport and deposition extends across the entire shoreface to the inner continental shelf.

The data shown on Fig.10 indicate that sediment removed from the surf zone by northeast storms is deposited across the shoreface with the majority coming to rest on the upper shoreface. Transport and deposition of this sediment extends across the entire shoreface and onto the adjoining inner continental shelf. That sediment which arrived at the lower shoreface or inner portion of the continental shelf may, or may not, be returned to the beach system. The sediment which was deposited during the storm on the upper shoreface constitutes a reservoir which is then depleted by the slow return of sand up the upper portions of the shoreface and into the surf zone and beach during periods of non-storm conditions.

An obvious question is whether the one storm studied is indicative of shoreface sediment transport and morphological response during other coastal storms. This point is addressed with the following considerations. If the storm follows an offshore track relative to say the New Jersey coastline, then the first winds characterizing this storm will develop and strengthen from the northeast. As the storm center migrates past the coastal point the winds rapidly back and blow offshore from the northwest. With progressive movement of the storm along the offshore track these northwesterly winds diminish in intensity. If the storm follows an onshore track the winds will develop from the southeast and strengthen as the storm center moves adjacent to the coastal point. As the storm center progresses past the coastal point of interest the winds rapidly veer to the southwest and diminish in intensity. These conditions are illustrated in Fig.11.

In either case, the occurrence of the low-pressure storm causes the initial winds to have an onshore component leading to the rapid generation of heavy seas. The passage of the storm center results in a rapid shift of the wind so that it has an offshore component which substantially reduces the wave heights.

Figure 12 shows an idealization of this pattern. The upper stick diagram in this figure indicates the time history of winds during the passage of a northeastern storm for a typical location on the Long Island shoreline. The second panel indicates rapid growth in local wave height during the early phases of the storm. This corresponds to the generation of an organized wind-driven circulation over the shoreface. The onshore and southward surface currents are characterized by a strong horizontal convergence over the shoreface. This results in a downwelling. Combination of high waves and downwelling current results in offshore sediment transport along the shoreface. As the storm center passes and the winds back rapidly to the northwest, the surface currents change and are characterized by a horizontal divergence over the shoreface. This results in upwelling. However, the offshore winds rapidly decrease the local wave heights so that the return of sediment up the shoreface during the latter portion of the storm does not compensate for the offshore transport in the early phase of the storm.

Figure 13 illustrates a similar pattern for a low-pressure storm following an inland track. In this case the shore-parallel currents and resulting sediment transport during the initial downwelling stage are directed in the same

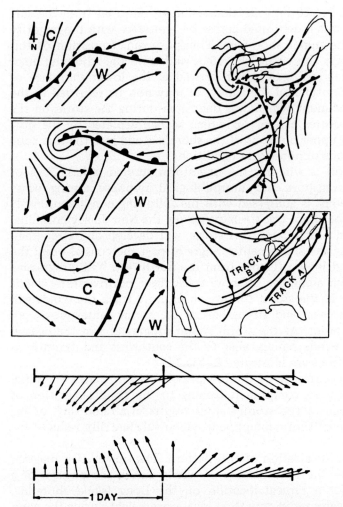

Fig.11. Typical coastal storm wind patterns for the U.S. Northeast coast. The upper left three panels show typical fronts and wind streamlines of an evolving low pressure storm. Upper right panel shows a typical storm superimposed on a map of the eastern U.S. Typical storm tracks are shown beneath. An inland track (B) and an offshore track (A) have been identified. The two wind stick diagrams show typical time histories of the storm winds (track A, upper and track B, lower) for a particular point on the northeast coast as the storm moves past. The beginning of the storm appears to the left and the end appears to the right in the stick plots of the winds.

direction but are weaker than that of the previously discussed case. The waves during the first stage of the storm will be larger than in the previous example because of a more direct exposure of the coast to the wind. The pattern of convergence of surface currents over the shoreface with the resulting of downwelling during the initial period of the storm when the waves are high, followed by a divergence of surface currents over the shoreface and upwelling when the waves are low, is maintained.

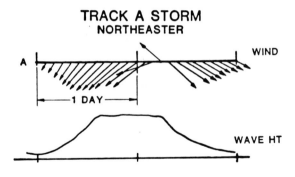

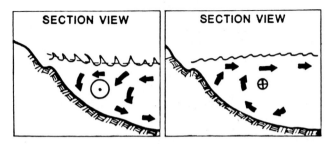

Fig.12. Typical winds, wave heights and coastal ocean (shoreface) currents for a northeastern storm.

Although the above examples of storm induced shoreface currents show only two of the complex set of strongly forced shoreface circulation patterns, they do demonstrate a general effect. Storm winds with strong onshore components cause high locally generated waves, convergence of surface currents and downwelling over the shoreface. Most frontal storms produce a rapid shift of the wind as the front or storm center propagates past the coastal site. These later winds have offshore components which rapidly decrease the height of the waves and cause surface current divergence and upwelling over the shoreface. A hysteresis develops in the diabathic bedload transport over the shoreface resulting in net offshore transport during storms. More sediment is moved offshore in the first part of the storm, when downwelling and high waves prevail, then is moved onshore when lower waves and upwelling prevail. The processes causing the bedload transport vary significantly over the shoreface. Most typically, this transport decreases offshore leading to a bedload convergence and accretion on the shoreface.

Geographic generalization of this pattern of storm induced diabathic shoreface bedload transport hysteresis is possible. Along the northern Gulf Coast frontal storms occur in winter. They are called "Blue Northers" and

TRACK B STORM
SOUTHEASTER

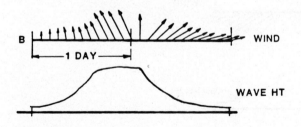

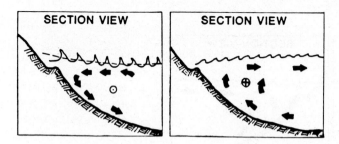

Fig.13. Typical winds, wave heights and coastal ocean (shoreface) currents for a frontal storm following track B on Fig. 11.

result from cold fronts which sweep across the coast and propagate eastward. As the front (which normally extends at a high angle or perpendicular to the coastline) approaches a given coastal site the winds freshen from the south or southeast. As the wind continues to increase they tend to swing to the southwest and then veer suddenly to the northwest as the front passes. The southeast winds begin to build the waves while causing surface current convergence and downwelling over the shoreface. As the wind strengthens and veers to the south the waves increase in height while the convergence in surface currents over the shoreface amplifies the downwelling. If the wind continues to strengthen and veer to the southwest the shoreface shore-parallel flow may reverse and the zone of surface current convergence will move closer to shore. Downwelling may be confined to the mid- and upper-shoreface with a second upwelling circulation developing further offshore. As the front passes the wind veers rapidly to the northwest. This strong offshore wind quickly reduces the wave heights while causing divergence of the surface currents and upwelling over the shoreface. The dominant near-bottom current over the shoreface is offshore when the waves are high and onshore when the waves are diminishing in height.

A somewhat similar pattern is exhibited for storms incident on U.S. Pacific northwest coast (i.e., north of Cape Mendocino). As winter low pressure storms move eastward out of the Gulf of Alaska into Canada, the early winds peak early out of the south or southwest and then veer and become weaker northwesterlies as the low moves inland. Although the winds remain onshore through the cycle the initial downwelling forcing is stronger than the final upwelling forcing and thus the sense of the net shoreface bedload transport hysteresis is preserved.

Many effects have not yet been considered. The prevalence of swell on the Pacific coast must reduce the contrast in shoreface bedload transport between storm and non-storm conditions. Hurricanes and tropical storms have not been considered. The relatively small size and rapid propagation of these storms can cause complex coastal currents which have marked longshore gradients and may not be well coupled with the local winds.

Annual and geologic scale shoreface processes

The preceeding discussion of event-scale shoreface processes has shown that the upper shoreface forms a reservoir for sand removed from the beach and surf zone during coastal storms. It has been shown in an earlier analysis (Niedoroda, 1980; Niedoroda and Swift, 1981), that fair-weather wave processes in the area of study will tend to return this material to the surf zone and beach. Thus, this study, which extends further seaward than have most studies of the coastal sediment budgets, leads to a somewhat more complete and spatially comprehensive view of the annual cycle of coastal sedimentation. Many earlier studies have described a cycle in which sand tends to be stored in breakpoint bars during the stormy winter months, but returns to the beach prism during the quiescent summer months, as the bars migrate landward and weld to the berm. In this study, the bar-berm cycle is seen to be part of a larger-scale cycle which extends down the shoreface.

The study also carries important implications for coastal processes at geologic time scales, since it demonstrates that a significant amount of sand may be transported entirely across the shoreface, and on to the adjacent inner shelf floor. Companion studies indicate that this sand is not necessarily lost; the asymmetrical orbital currents of shoaling waves cease to be effective in moving sand landward at about 15 m water depth (Niedoroda, 1980; Niedoroda and Swift, 1981), but wave-current interactions continue to move sand landward on the lower shoreface below 15 m, and on the adjacent inner shelf (Vincent et al., 1983). However, there is clearly the potential for permanent loss, if downwelling storm currents transport sand out onto the inner shelf faster than wave-current interactions can return it. In fact, two basic coastal sand budgets can be defined in terms of the three basic parameters of coastal evolution; the rate of sediment supply, the rate of relative sea level change, and the rate of fluid power expenditure (Fig.14).

In an erosional regime, such as that of the Atlantic shelf of North America, sea level is rising sufficiently rapidly with respect to the rate of river sediment input that the river mouths have become estuaries. They trap not only

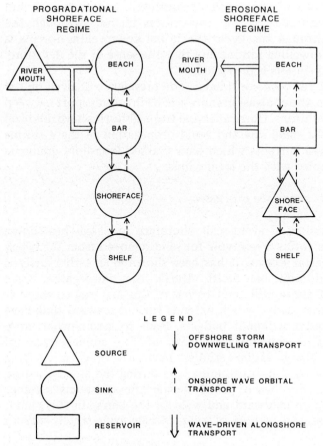

Fig.14. Schematic representation of general coastal sand budgets in prograding and eroding shoreface regimes.

all fluvial sand, but also littoral drift (Meade, 1969). Storm downwelling currents must move sand seaward faster than wave orbital currents can return it, because the coast is retreating almost everywhere, and cores of most shorefaces reveal a thin veneer of modern sand (several decimeters thick) over older back barrier strata (Swift et al., 1985). In this kind of coastal system, estuaries and the inner shelf floor become the ultimate sinks for sand. The primary source is the eroding shoreface. Storms may strip off the entire beach prism and back-barrier. Tree stumps, clays, and peats become briefly exposed along hundreds of kilometers of beach before fair weather waves return the sand (Harrison and Wagner, 1964). This denudation extends down the entire shoreface, during major events (Charlesworth, 1968), with ancient back-barrier deposits exposed at the sea floor to 10 or 15 m water depth. The Holocene sand sheet on the Atlantic shelf surface appears to have been generated in this fashion by shoreface erosion, as the shoreface retreated back across the shelf surface in response to post glacial sea-level rise (Swift, 1976; Swift et al., 1985).

In a second coastal system described by Curray et al. (1969), the rate of river sand input overwhelms the effects of sea-level rise, and river mouths become deltas which inject sand directly into the surf zone. The beach, bar and shoreface all receive sand more rapidly than they can exchange it with adjacent environments. During neap tides, bars are captured as new berms and the coast progrades seaward as a series of beach ridges, to form a strand plain.

CONCLUSIONS

The data, analyses, and results presented in this paper lead to the following conclusions.

The shoreface region of wave-dominated sandy coasts is a zone of active bedload sediment transport. The processes which cause and control this sediment transport are distinct from those which control sediment transport in the surf zone. The rate of sediment transport on the shoreface is strongly controlled by the height of waves which serve to entrain the bottom sediments. Sediment on the upper shoreface is entrained by relatively small waves and generally shifted landward during non-storm conditions as a result of the asymmetrical wave orbital fluid stresses caused by non-linear nearshore waves. This landward shift of sediment results in erosion of the shoreface sand reservoir. The pattern of sediment transport on the shoreface during storm events is controlled by the fluid shear stresses and transport of water in the bottom boundary layer of the coastal ocean.

Frontal storms produce a relatively predictable pattern of waves, currents, and bedload sediment transport on wave-dominated open-ocean shorefaces. At the event-scale, the result of coastal storms tends to be an offshore transport of sediment over the entire shoreface. The offshore transport of bedload during these storm events is convergent on the shoreface yielding accretion. It has marked spatial gradients such that the majority of the deposition is concentrated on the upper shoreface. A small percentage of the sediment can be transported offshore, across the shoreface and deposited on the inner continental shelf. Thus, the upper shoreface forms a reservoir for sand removed from the surf zone during storm events. This reservoir provides a source of the sand which generally returns to the surf zone and beach during non-storm intervals.

A shoreface bedload transport hysteresis commonly results during storms. On most of the U.S. coasts the initial stages of storms generally yield high locally generated waves and bottom currents with an offshore component due to downwelling over the shoreface. Strong offshore winds rapidly reduce the wave heights and cause onshore bottom currents due to upwelling over shoreface. The net bedload transport is offshore because the lower waves associated with upwelling cannot transport as much bedload onshore as was moved offshore by downwelling flow.

Annual scale sediment budgets and morphological balances result from the integration of storm- and non-storm shoreface bedload transport patterns. Such annual averaging tends to yield imbalances related to the average number and intensity of storms during the year. Relatively sudden shifts in shoreface position can result but its shape and morphology is preserved.

Over longer time spans the sediment transport processes on the shorefaces provide a major avenue for the removal of sediment from coastal areas and its consequent deposition at the base of the shoreface. Shorefaces can be either erosional or depositional features depending upon the local supply of coastal sediments from longshore processes and slow changes in relative sea level.

ACKNOWLEDGEMENTS

Research which resulted in this publication was in part funded by contracts from the National Oceanic and Atmospheric Administration (Contract COM/NOAA 03-6-022-3511), the Office of Naval Research (Contract N00014-76-C-0145), and the Department of Energy (DE-AC-2-76CH00016) as well as a subcontract to Brookhaven National Laboratories from NOAA/AMOL. Additional funding came from the University of Massachusetts, Woodward-Clyde Consultants, and Dames & Moore Consultants. Portions of the data were supplied by the Sediments Dynamics Group at the Atlantic Oceanographic and Meteorologic Laboratories, NOAA, Miami, from the INSTEP program which was funded by the Northeast Office, Office of Marine Pollution Assessment, NOAA.

Many people assisted in this research. Particular appreciation is owed to Mr. David Battisti for assistance with the field work and in the analysis of tidal currents. Additional thanks is given for field assistance to H. Peg, S. Wall, L. Wall, J. Collins, J. Lenard, G. Bowers, Dr. B. Brennickmeyer, E. Divit and D. Carlson. Assistance in logistical support, data collection, and data exchange obtained from Southampton College, Boston College, and Woods Hole Oceanographic Institute is also acknowledged.

REFERENCES

Barrell, J., 1912. Criteria for the recognition of ancient delta deposits. Geol. Soc. Am. Bull., 23: 377—446.
Bruun, P., 1962. Sea-level rise as a cause of shore erosion. J. Waterways Harbors Div., A.S.C.E., 88: 117—130.
Charlesworth, L.J., 1968. Bay, inlet and nearshore marine sedimentation: Beach Haven—Little Egg inlet region, New Jersey. Doct. Diss., Univ. Michigan, Dept. Geol., 23 pp.
Cook, D.O., 1970. Occurrence and geologic work of rip-currents in Southern California. Mar. Geol., 9: 173—186.
Cook, D.O. and Gorsline, D.S., 1972. Field observations of sand transport by shoaling waves. Mar. Geol., 13: 31—55.
Csanady, G.T., 1978. The arrested topographic wave. J. Phys. Oceanogr., 8: 47—62.
Curray, J.R., Emmel, F.J. and Crampton, P.J.S., 1969. Lagunas costercas, in simposio. In: Mem. Simp. Int. Lagunas Costercas, UNAM—UNESCO, Nov. 28—30, 1967, Mexico, pp.63—100.
Dimmler, D.G., Rankowitz, S., Huszagh, D.W. and Scott, J.T., 1976. A controllable real-time data collection system for coastal oceanography. Proc. Oceans '76, 256: 1—13.
Fenneman, N.M., 1902. Development of the profile of equilibrium of the subaqueous shore terrace. J. Geol., 10: 9—32.

Harrison, W. and Wagner, K.A., 1964. Beach changes at Virginia Beach, Virginia Beach, Virginia. U.S. Army Coastal Eng. Res. Center, Misc. Pap. No. 6-64, 25 pp.

Hopkins, T.S., 1974. On time dependent wind induced motions. Rapp. P.V. Réun. Cons. Int. Explor. Mer, 16: 721—736.

Hopkins, T.S., 1982. On the sea level forcing of the Mid-Atlantic Bight. J. Geophys. Res., 87: 1997—2006.

Hopkins, T.S. and Dieterle, D.A., 1983. An Externally Forced Barotropic Circulation Model for the New York Bight. Continental Shelf Res., 2: 49—73.

Huff, L.C. and Friske, D.A., 1980. Development of two sediment transport instrument systems. Proc. 17th Int. Conf. Coastal Engineering, Sydney, N.S.W., pp.176—177.

Johnson, D., 1919. Shore Processes and Shoreline Development. Hafner, New York, N.Y., 584 pp.

Jonsson, I.G., 1966. Wave boundary layers and friction factors. Proc. 10th Conf. Coastal Engineering, ASCE, 1: 127—148.

Langford-Smith, T. and Thom, B.G., 1969. New South Wales coastal morphology. J. Geol. Soc. Aust., 16: 572—580.

Madsen, O.S. and Grant, W.D., 1976. Sediment transport in the coastal environment. Ralph M. Parsons Laboratory Rept. No. 209, 105 pp.

Meade, R.H., 1969. Landward transport of bottom sediments in estuaries of the Atlantic Coastal Plain. J. Sediment Petrol., 34: 144—221.

Moody, D.W., 1964. Coastal morphology and processes in relation to the development of submarine sand ridges off Bethany Beach, Delaware. Thesis, Johns Hopkins Univ., 167 pp. (unpublished).

Mooers, C.N.K., 1973. A technique for the cross spectrum analysis of pairs of complex-valued time series, with emphasis on properties of polarized components and rotational invariants. Deep-Sea Res., 20: 1129—1141.

Niedoroda, A.W., 1980. Shoreface-surf zone sediment exchange processes and shoreface dynamics. NOAA Techn. Memo. OMPA-1, 89 pp.

Niedoroda, A.W. and Swift, D.J.P., 1981. Maintenance of the shoreface by wave orbital currents and mean flow: observations from the Long Island coast. Geophys. Res. Lett., 8: 337—340.

Niedoroda, A.W., Mean-Ma, C., Mangarella, P.A., Cross, R., Huntsman, S. and Treadwell, D., 1982. Measured and computed coastal ocean bedload transport. Proc. 18th Int. Conf. Coastal Engineering. Am. Soc. Civ. Eng., 2: 1353—1368.

Scott, J.T., Hopkins, T.S., Pillsbury, R.D. and Divis, E.G., 1978. Velocity and temperature off Shinnecock Long Island in October—November 1976. Brookhaven Natl. Lab. Rep. No. 50895, 35 pp.

Swift, D.J.P., 1976. Coastal sedimentation. In: D.J. Stanley and D.J.P. Swift (Editors), Marine Sediment Transport and Environmental Management. Wiley, New York, N.Y., pp.255—310.

Swift, D.J.P., Niedoroda, A.W., Vincent, C.E. and Hopkins, T.S., 1985. Barrier island evolution, middle Atlantic shelf, U.S.A. Part 1: Shoreface dynamics. In: G.F. Oertel and S.P. Leatherman (Editors), Barrier Islands. Mar. Geol., 63 (in press).

Tucker, W.D., 1974a. Coastal shelf oceanography program. Progress Rept. No. 1, Brookhaven Natl. Lab., 49 pp.

Tucker, W.D., 1974b. Coastal shelf oceanography program. Progress Rept. No. 2, Brookhaven Natl. Lab., 59 pp.

Tucker, W.D., 1975. Coastal shelf oceanography program. Progress Rept. No. 3, Brookhaven Natl. Lab., 145 pp.

Vincent, C.E., Young, R.A. and Swift, D.J.P., 1983. Sediment transport in the Long Island shoreface, North American Atlantic Shelf: Role of waves and currents in shoreface maintenance. Continental Shelf Res., 2: 163—181.

Wells, D.R., 1967. Beach equilibrium and second order wave theory. J. Geophys. Res., 72: 497—509.

Winant, C.D. and Beardsley, R.C., 1979. A comparison of some shallow wind-driven currents. J. Phys. Oceanogr., 9: 218—220.

Wright, L.D. and Coleman, J.M., 1972. River delta morphology: wave climate and the role of the subaqueous profile. Science, 176: 282—284.

Young, R.A., Merrill, J.T., Clark, T.L. and Proni, J.R., 1982. Acoustic profiling of suspended sediments in the marine bottom boundary layer. Geophys. Res. Lett., 9: 175—178.

CONTROL OF BARRIER ISLAND SHAPE BY INLET SEDIMENT BYPASSING: EAST FRISIAN ISLANDS, WEST GERMANY

DUNCAN M. FITZGERALD[1], SHEA PENLAND[2] and DAG NUMMEDAL[3]

[1]*Department of Geology, Boston University, Boston, MA 02215 (U.S.A.)*
[2]*Louisiana Geological Survey, Louisiana State University, Baton Rouge, LA 70803 (U.S.A.)*
[3]*Department of Geology, Louisiana State University, Baton Rouge, LA 70803 (U.S.A.)*

(Received January 17, 1983; revised and accepted September 11, 1983)

ABSTRACT

FitzGerald, D.M., Penland, S. and Nummedal, D., 1984. Control of barrier island shape by inlet sediment bypassing: East Frisian Islands, West Germany. In: B. Greenwood and R.A. Davis, Jr. (Editors), Hydrodynamics and Sedimentation in Wave-Dominated Coastal Environments. Mar. Geol., 60: 355—376.

A study of the East Frisian Islands has shown that the plan form of these islands can be explained by processes of inlet sediment bypassing. This island chain is located on a high wave energy, high tide range shoreline where the average deep-water significant wave height exceeds 1.0 m and the spring tidal range varies from 2.7 m at Juist to 2.9 m at Wangerooge. An abundant sediment supply and a strong eastward component of wave power (4.4×10^3 W m^{-1}) have caused a persistent eastward growth of the barrier islands. The eastward extension of the barriers has been accommodated more by inlet narrowing, than by inlet migration.

It is estimated from morphological evidence that a minimum of 2.7×10^5 m^3 of sand is delivered to the inlets each year via the easterly longshore transport system. Much of this sand ultimately bypasses the inlets in the form of large, migrating swash bars. The location where the swash bars attach to the beach is controlled by the amount of overlap of the ebb-tidal delta along the downdrift inlet shoreline. The configuration of the ebb-tidal delta, in turn, is a function of inlet size and position of the main ebb channel. The swash bar welding process has caused preferential beach nourishment and historical shoreline progradation. Along the East Frisian Islands this process has produced barrier islands with humpbacked, bulbous updrift and bulbous downdrift shapes. The model of barrier island development presented in this paper not only explains well the configuration of the German barriers but also the morphology of barriers along many other mixed energy coasts.

INTRODUCTION

The morphology of barrier islands is influenced by a number of factors including mode of formation, sediment supply, sea-level rise, and wave and tidal processes. When the supply of sediment is sufficient to maintain the barrier system the primary control of its general morphology is the hydrographic regime of the region (Hayes, 1975). Hayes, using the tidal classification of Davies (1964), showed that moderate wave energy barrier island coasts (wave height = 60—150 cm) can be separated into two basic types: microtidal and mesotidal coasts. In his microtidal model ($TR < 2.0$ m), barrier islands are long and continuous and have numerous washovers. Tidal inlets occur infrequently and contain well-developed flood-tidal deltas but poorly developed ebb-tidal deltas. Meso-tidal coasts ($2.0 < TR < 4.0$ m) have short, stubby barrier islands and numerous tidal inlets. Tidal deltas are well formed with ebb deltas more prominently developed. This classification was later redefined by Hayes (1979) and Nummedal and Fischer (1978) and based on a region's tidal range and average wave height. Hayes' mesotidal barrier island coast is now more appropriately termed a mixed energy (tide-dominated) shoreline in which tidal and wave processes are equally responsible for barrier island shape.

It has been observed by Hayes and Kana (1976) that the shape of barrier islands along mixed energy shorelines is similar to that of a drumstick (Fig.1).

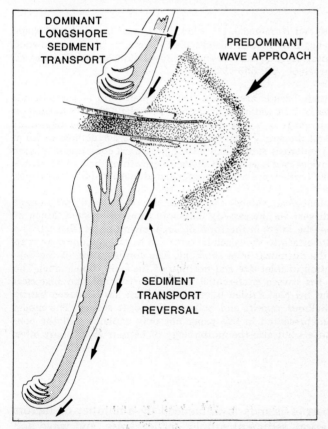

Fig.1. Drumstick barrier island model after Hayes and Kana (1976). The accretionary updrift portion of the barrier is formed from the longshore transport of sand toward the inlet and from the landward migration of swash bars from the ebb-tidal delta. The downdrift end of the barrier develops through spit accretion.

They have identified drumstick barrier islands along the coasts of Alaska, South Carolina, Georgia and The Netherlands. Stephen (1981) has also reported drumstick barriers along the coast of southwest Florida. Despite these examples an examination of any mixed energy coast reveals many other barrier island forms (Fig.2). One such barrier island chain which exhibits an assortment of shapes is the East Frisian Islands off the North Sea coast of West Germany. This coastline is strongly affected by a dominant direction of wave energy flux, abundant sediment supply and tidal inlet processes.

It is the intent of this paper to demonstrate that the variability in shape of the East Frisian barriers is primarily related to processes of inlet sediment bypassing. Sand ultimately bypasses the inlets along this coast in the form of large landward migrating swash bars. Depending upon where these bars attach to the downdrift inlet shoreline the barriers can be drumstick-shaped, humpbacked or even downdrift bulbous-shaped.

PHYSICAL SETTING

The East Frisian Islands consist of seven barrier islands located in the southeast North Sea between the Ems River to the west and the Jade Bay to the east (Fig.3). The barrier system is 90 km long and is separated from the mainland by a 4—12 km wide tidal flat which has been incised by a network of tidal channels.

The wind regime along this section of the North Sea is seasonal. A recorder on the island of Norderney showed that over a 19-yr period (1947—1966) prevailing winds during the fall and winter (September—March) generally blow from the southwest while during the rest of the year (April—August) they blow from the northwest (Luck, 1976a). Dominant winds come from the southwest with an average velocity of 10 m s^{-1}.

Although there exists no reliable, long-term wave records for the nearshore region, the Summary of Synoptic Meteorologic Observations (SSMO data — U.S. Naval Weather Service Command, 1974) indicate that for the Bremerhaven data square the resultant wave power is directed to the east-southeast (azimuth = 101°; Nummedal and Fischer, 1978). This resultant vector yields a net eastward longshore power component of 4.4×10^3 W m^{-1} (Fig.4).

Short-term wave gauge data off the island of Norderney indicate that at the 10 m water depth the average significant wave height ($H_{1/3}$) exceeds 1.6 m (Niemeyer, 1978). Another wave gauge in 8 m of water off the coast of Sylt in the Northern Frisian Islands show that $H_{1/3}$ exceeds 1 m daily and 4.8 m on an annual basis (Dette, 1977). For comparison, the average significant wave height at Atlantic City, New Jersey, is 0.8 m (water depth = 5.2 m) and at Savannah, Georgia, $H_{1/3}$ = 0.9 m (water depth = 15.8 m) (Thompson, 1977).

The mean tidal range along the East Frisian Islands increases in an easterly direction from a low of 2.2 m at Borkum to a high at Wangerooge of 2.6 m

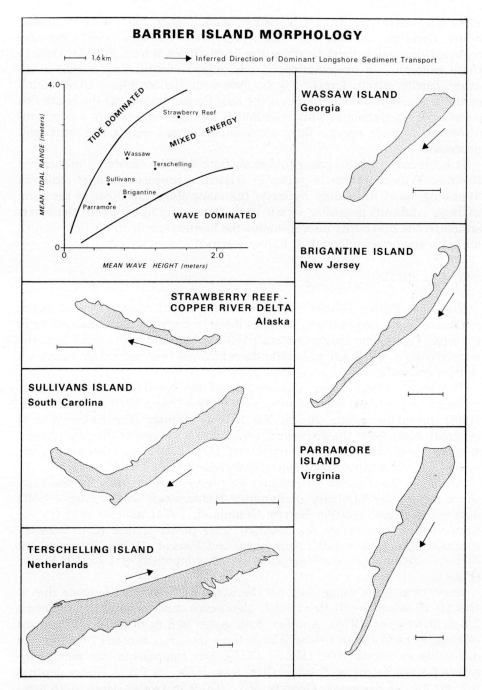

Fig.2. Barrier island morphology along various mixed energy coasts. The coastal classification is after Hayes (1979) and Nummedal and Fischer (1978).

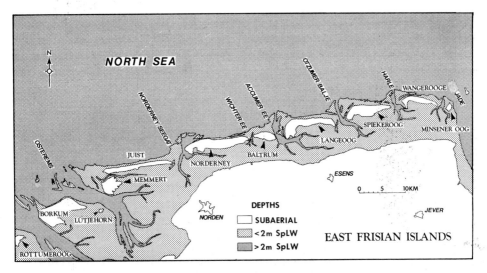

Fig.3. Location map of East Frisian Islands along the West German North Sea coast.

(NOS, 1983). During spring tide conditions these values increase to 2.5 and 2.9 m, respectively. Tides in the North Sea are semidiurnal.

SHORELINE PROCESSES

Historical changes

Morphological changes of the East Frisian Islands, tidal inlets and back-barrier environment have been determined from sequential maps produced by Homeier and Luck (1969) for the years 1660, 1750, 1860 and 1960. A summary of these data is given in Tables I—III and depicted in Figs. 5 and 6. During the past 300 yrs there has been an 80% increase in the areal extent of the barrier islands (Table II). A large proportion of this increase, however, was due to poldering behind the barriers, a process whereby marsh lands and tidal flats are diked from the sea. Still, when the poldered region is subtracted an 18.03 km^2 or a 35% increase in total barrier island area is calculated. These data and reports of substantial dune growth during the past century (Luck, 1975) indicate that there is a positive sand budget for these islands despite rising sea level in the North Sea (Streif and Koster, 1978). The supply of sand for these barriers is believed to come from a variety of sources including: (1) the Rhine and Maas Rivers and erosion of old lobes of the Rhine delta along the Dutch coast (Van Straaten, 1965); (2) sediment discharged from the Ems River (Luck, 1976a); and (3) from offshore glacial deposits (Luck, 1976a). The historical information demonstrates that except for the erosion due to inlet migration, the overall morphology of the East Frisian barriers is a product of accretionary processes.

TABLE I

Historical changes in barrier island length and tidal inlet width (after Luck, 1975)

Barrier islands	Length (m)				Diff. between 1650—1960
	1650	1750	1860	1960	
Juist	10,470	13,010	15,810	14,970	+4500
Norderney	8070	9440	13,470	13,870	+5800
Baltrum	8170	7560	5460	5050	—3120
Langeoog	9600	10,280	11,230	10,920	+1320
Spiekeroog	5230	5180	5910	9810	+4580
Wangerooge	7300	7400	7950	8320	+1020
(total length)	48,840	52,870	59,830	62,940	14,100
Tidal inlets	Width (m)				Diff. between 1650—1960
	1650	1750	1860	1960	
Norderneyer Seegat	6800	4730	2600	2770	—4030
Wichter Ee	2170	2030	710	850	—1320
Accumer Ee	3150	2480	1360	1710	—1440
Otzumer Balje	2440	2400	2440	2410	—30
Harle	5800	5670	4810	2000	—3800
(total width)	20,360	17,310	11,920	9740	—10,620
Total barrier island length and tidal inlet width	69,200	70,180	71,750	72,680	+3480

TABLE II

Areas of the barrier islands and poldered regions (in km²)

Island	1650	1750	1860	1960	Difference between 1650—1960	Poldered region	Difference between 1650—1960 minus poldered region
Juist	11.15*	12.35	14.38	12.60	+1.35	2.6	—1.25
Norderney	11.38	13.25	20.45	24.68	+13.30	5.6	+7.7
Baltrum	6.5	6.05	6.50	6.23	—0.27	2.1	—2.37
Langeoog	9.18	11.30	14.85	20.40	+11.22	7.3	+3.9
Spiekeroog	6.43	7.38	10.10	21.30	+14.87	1.7	+13.17
Wangerooge	7.40	7.58	7.55	8.40	+1.00	4.1	—3.1
Total	52.14	57.91	73.83	93.61	+41.47	23.4	+18.07

*Areas of the barrier islands from Luck (1975).

TABLE III

Drainage areas of the tidal inlets (km²)

	1650	1750	1860	1960	Difference between 1650—1960
Norderneyer Seegat	125	103	98	97	−28
Wichter Ee	60	51	40	28	−32
Accumer Ee	101	100	84	88	−13
Otzumer Balje	57	51	77	73	+16
Harle	154	128	79	62	−92
Totals	497	433	372	348	−149

Longshore sediment transport

The strong easterly wave energy flux that is produced by dominant and prevailing winds out of the westerly quadrant results in a strong eastward movement of sand along the Frisian Island coast (Fig.4). A longshore sediment transport rate of 130,000 m³ yr⁻¹ can be inferred from the average volume of sand, in the form of large bars, that bypassed Norderney Seegat over a 31-yr period of time (Homeier and Kramer, 1957). This is a minimum rate because it does not account for the volume of sand that moved past the inlet without significantly changing the bar forms. A volume of 270,000 m³ yr⁻¹ is calculated from the increase in size of Norderney from

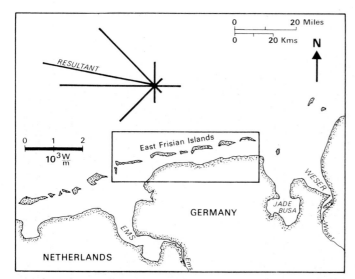

Fig.4. Distribution of deep-water wave power for the Bremerhaven data square as determined from the Summary of Synoptic Meteorologic Observations. The west-northwesterly wave power vector resultant causes a strong easterly longshore transport direction along the East Frisian Island shore.

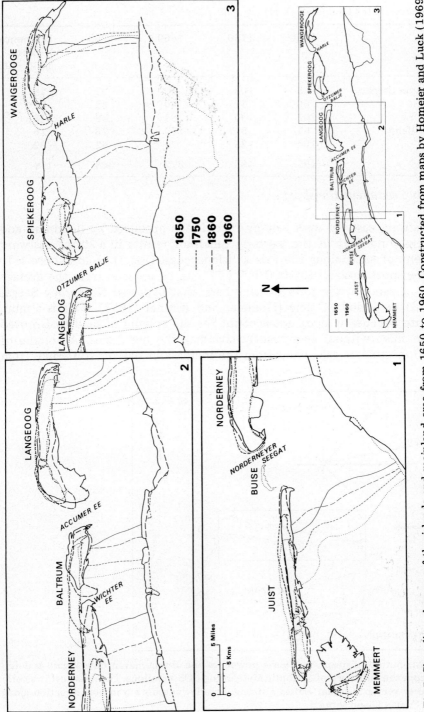

Fig.5. Shoreline changes of the islands and mainland area from 1650 to 1960. Constructed from maps by Homeier and Luck (1969). The lines behind the barrier indicate changes in the position of the drainage divides. Generally, the barriers have prograded eastward and the drainage divides have migrated eastward as well.

1650 to 1960 (Table II). However, during the same period of time, sand must have been bypassing Wichter Ee, so even this value is most likely a minimum estimate.

Over the past 300 yrs the easterly transport of sand has caused a lengthening of the barriers at the expense of inlet width. Although the western ends of the barriers have eroded through inlet migration, this loss in shoreline was more than compensated by the islands' eastward growth. As shown in Table I and Fig.5 total barrier island length increased from 48,840 m in 1650 to 62,940 m by 1960. During the same time, total inlet width decreased from 20,360 to 9740 m. The 3480 m net increase in total length of the inlet-barrier system was due to the westward growth of Juist and the eastward growth of Wangerooge (Fig.5).

The narrowing of the tidal inlets since 1650 can be explained by a decrease in tidal prism which was caused by a 30% decrease in backbarrier tidal flats and channels (Table III). The filling of the backbarrier region has been brought about by: (1) poldering behind the barriers and along the mainland; and (2) landward spit extension at the inlets. The width of the barriers precludes washovers from being deposited in the backbarrier except perhaps along the very eastern ends of the island, where the barriers are young and low in profile.

The eastern ends of the East Frisian barriers have been developed through spit accretion at a relatively rapid rate (ave. = 30 m yr^{-1}) due to an abundant sediment supply and a high longshore sediment transport rate. However, it should be noted that spit accretion has been accommodated more by inlet narrowing than by inlet migration.

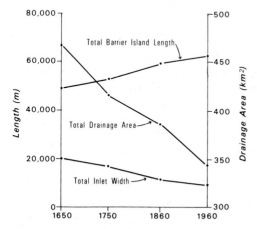

Fig.6. Frisian Island morphological changes from 1650 to 1960 (barrier island and inlet width data from Luck, 1975). The 29% increase in barrier island length has been at the expense of inlet width (52% decrease) and has resulted from a 30% decrease in drainage area.

TIDAL INLET PROCESSES

Inlet sediment bypassing

The pattern of sand accretion along the East Frisian barriers is controlled directly by the location where swash bars attach to the shoreline during the inlet sediment bypassing process. Mechanisms of inlet sediment bypassing were first recognized and described by Bruun and Gerritsen (1959). They documented that sand bypasses inlets by wave action along the terminal lobe or by tidal currents coupled with channel and bar migrations. Detailed studies of transport patterns at Norderneyer Seegat by Luck (1976b) and Nummedal and Penland (1981) and at Harle by Hanisch (1981) have shown that sand bypassing at the East Frisian inlets occurs by a combination of the two processes. Their data consisted of historical morphological information, grain-size analysis, bedform measurements, sand tracer studies and time-series current velocity readings. A general model of inlet sediment bypassing for the East Frisian coast based upon these studies is shown in Fig.7 and summarized below.

Sand moves eastward along the beach toward the inlet through wave induced longshore sediment transport. Once in the vicinity of the inlet, sand bypassing follows the following complex pathway to the downdrift inlet shoreline. Sand entering the inlet is transported by wave action and by flood-tidal currents across the swash platform and through marginal flood channels to the main ebb channel. Some of this sand is then transported to the backbarrier sand flats where it may remain for a period of time (Nummedal and Penland, 1981) or it may be reworked back into the sand transfer system. The main ebb channel at these inlets is dominated by ebb-tidal currents (Kramer, 1961) and thus most of the sand dumped into the inlet is moved in a net seaward direction to the distal northern portion of the ebb delta. At this location, sand is added to the reefbow (Luck, 1974) as a series of swash bars that outline the terminal lobe portion of the ebb-tidal delta. The fate of these bars is tied to the second transport system and will be discussed next.

The western, updrift side of the inlets' ebb-tidal delta is composed of a series of northwest—southeast trending, elongated bars that are separated from one another by shallow (<3 m deep) channels. Some of these channels are dominated by flood-tidal currents. The segregation of flow in these flood and ebb channels (Nummedal and Penland, 1981; Hanisch, 1981) causes the northeasterly, zig-zag transport of sand in a seaward direction. The strong easterly component of wave energy also augments the eastward movement of sediment. Once the sand reaches the outer portion of the ebb delta it becomes incorporated in swash bars that make up the reefbow. Some sand is lost along this pathway when it is dumped into the main ebb channel where, as described earlier, the dominant ebb-tidal currents transport it to the outer portion of the ebb delta.

The inlet sediment bypassing process is completed as bars that comprise

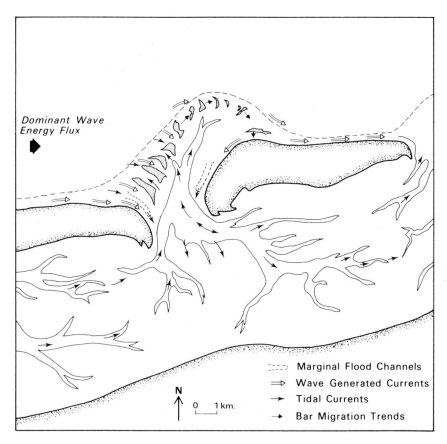

Fig.7. Model of inlet sediment bypassing for the East Frisian Island coast based on historical and morphological data, bedform orientations, sand tracer studies, box cores and tidal current measurements from the investigations of Nummedal and Penland (1981) and Hanisch (1981). Note that one of the end products of inlet sand bypassing is the attachment of large swash bars to the downdrift beach.

the reefbow migrate in an arc along the periphery of the ebb delta to become attached to the landward beach. Maps constructed by Homeier and Kramer (1957) illustrate the movement of the center of gravity of bars at Norderneyer Seegat over a 31-yr period of time (1926—1957) (Fig.8A) and the shore-parallel elongation of the bars as they migrate onshore (Fig.8B). Figure 8A and C demonstrate that the bars have a consistent west—east migration and that their movement has been as great as 400 m yr^{-1}. A photographic time sequence of Wichter Ee and Accumer Ee illustrates the migration of large bars toward the shoreline of Baltrum and Langeoog, respectively (Fig.9). The morphology of a large swash bar welding to the western end of Spiekeroog is shown in Fig.10. From Figs.8—10 it can be seen that as they weld to the beach, the bars are 1—3 km long. It should be noted that an unknown amount of sand is also continuously moved onshore independent of the bar forms.

The Frisian tidal inlets exhibit many of the same mechanisms and pathways of sand bypassing as those of the U.S. East Coast. Mixed energy tidal inlets, including Price Inlet, South Carolina (FitzGerald et al., 1976), North Inlet, South Carolina (Finley, 1976) and Chatham Harbor Inlet, Massachusetts (Hine, 1975), contain ebb- and flood-dominant channels that are similar in hydraulics and morphology to those of the western portion of Frisian ebb-tidal deltas. These are the spillover channels noted by Oertel (1972) and FitzGerald et al. (1978) and the marginal flood channels described by Oertel (1972) and Hayes et al. (1973).

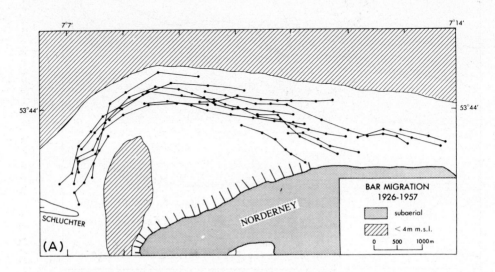

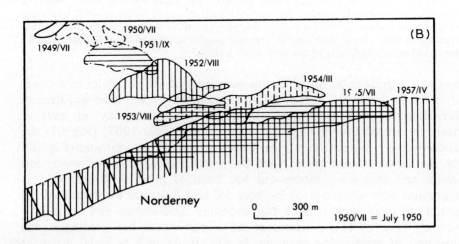

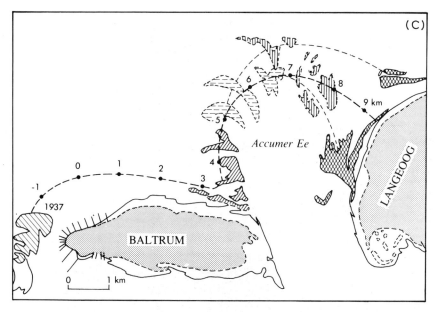

Fig. 8. Bar migrational trends at Norderneyer Seegat (A and B). A. Map of bar movement along the periphery of the ebb-tidal delta between 1926 and 1957. Each dot represents the bar's center of mass on successive years. Note the bar's movement is from west to east and that the outer exposed bars move at a faster rate than the inner bars (from Nummedal and Penland, 1981; after Homeier and Kramer, 1957). B. Morphological changes of a large swash bar as it attached to the beach at Norderney between 1949 and 1957 (from Nummedal and Penland, 1981; after Homeier and Kramer, 1957). C. Bar migrational trends at Accumer Ee. Note that the swash bars attach to the updrift end of Langeoog. The distance between numbers is one kilometer (from Homeier and Kramer, 1957).

Location of bar welding

In the inlet sediment bypassing process, the location where swash bars attach to the barrier island shoreline is controlled by the size and morphology of the ebb-tidal delta. Generally, along the East Frisian Islands the distance from the inlet that swash bars weld to the downdrift shoreline increases as: (1) inlet size increases; and (2) the downdrift skewness of the ebb-tidal delta increases. The asymmetric configuration of the ebb deltas is a product of the strong easterly longshore sediment transport direction, the residual easterly tidal currents (Nummedal and Penland, 1981) and the position of the main ebb channel with respect to the inlet throat.

Using the reefbow as a relative measure of ebb-delta size, it is found that the width of the deltas of the major inlets (Norderneyer Seegat, Accumer Ee, Otzumer Balje and Harle) varies from a high of 8.3 km at Norderneyer Seegat to a low of 5.1 km at Accumer Ee (Table IV). The length of shoreline that the ebb-tidal deltas overlap is also greatest at Norderneyer Seegat (5 km) and least at Accumer Ee (1.9 km). However, this trend is not entirely a function of size. As Table IV and Fig.11 illustrate, the percent

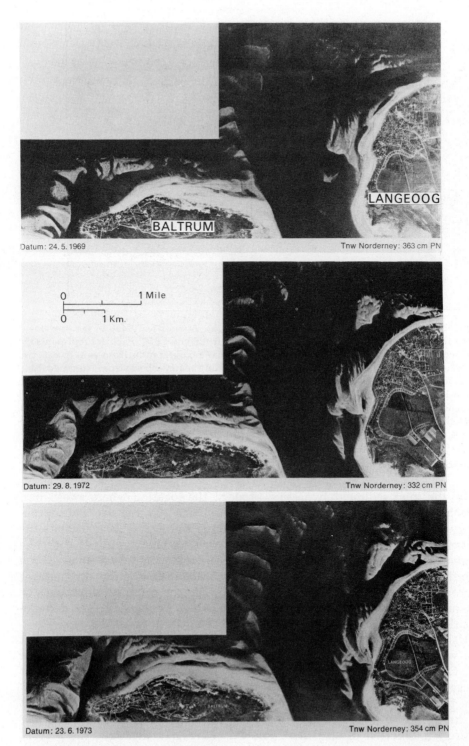

Fig.9. Sequential photographs of the bar welding process along the islands of Baltrum and Langeoog during the period between May 1969 and June 1973 (from Luck, 1974).

Fig.10. Oblique aerial photograph of the western end of Spiekeroog illustrating the morphology of a large, landward migrating swash bar, approximately 1.5 km in length.

TABLE IV

Ebb-tidal delta configurations

Ebb-tidal delta location	Delta width (km)	Downdrift barrier island overlap (km)	% overlap
Norderneyer Seegat	8.3	5.0	61
Accumer Ee	5.1	1.9	37
Otzumer Balje	6.0	3.4	57
Harle*	5.7	3.2	56

*Harle has been stabilized on its east side by 1.4 km length jetty that extends directly southwest into the inlet.

overlap of the deltas is also controlled, in part, by the position of the main ebb channel with respect to the adjacent islands. At Norderneyer Seegat, the main ebb channel abuts the downdrift island of Norderney which results in the ebb-tidal delta being displaced to the east, hence, the large shoreline overlap. The opposite inlet configuration occurs at Accumer Ee. Here the main ebb channel flows against the updrift island of Baltrum resulting in less eastward offset of the delta and thus a smaller length of shoreline overlap. The remaining two inlets, Otzumer Balje and Harle, have main ebb channel positions in the middle of the inlet throat. The percentage of ebb

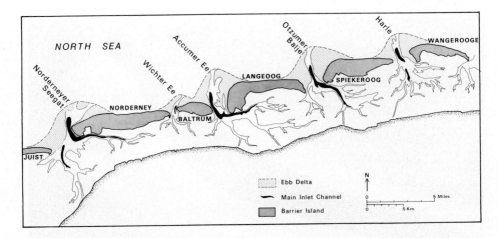

Fig.11. A map illustrating the configurations of the ebb-tidal deltas. The strong easterly component of wave energy flux causes a preferential overlap of the ebb-tidal delta along the downdrift inlet shoreline. The amount of overlap is controlled by inlet size and position of the main channel at the inlet throat.

delta-shoreline overlap of these two inlets is intermediate between that of Norderneyer Seegat and Accumer Ee (Table IV).

BARRIER ISLAND MORPHOLOGY

The morphological variability of the East Frisian Islands is depicted in Fig.12. The island of Juist, the westernmost barrier in the group, is long and straight, and is narrower than the other islands. There has been little land reclamation along the back side of this island compared to the rest. The configurations of Norderney, Spiekeroog and Wangerooge are similar to one another in that their seaward shorelines are humpbacked. The remaining two islands, Baltrum and Langeoog, are bulbous at their downdrift and updrift ends, respectively. With the exception of Baltrum, the narrowest portion of the barriers coincides with their easterly downdrift ends. The landward shorelines of the islands are largely a product of poldering.

Factors affecting barrier island shape

The development of the Frisian Island morphological diversity is explained by the factors summarized in Fig.13. As previously discussed, the location where swash bars attach to the beach coincides with the bulbous portion of the barrier and is controlled by the configuration of the ebb-tidal delta. The symmetry of the ebb-tidal delta has been shown to be a function of the position of the main ebb channel at the inlet throat and the size of the tidal inlet. The first factor is affected by: (1) the easterly longshore sediment transport direction; and (2) the distribution, orientation and magnitude of backbarrier tidal channels. Inlet size is primarily controlled by

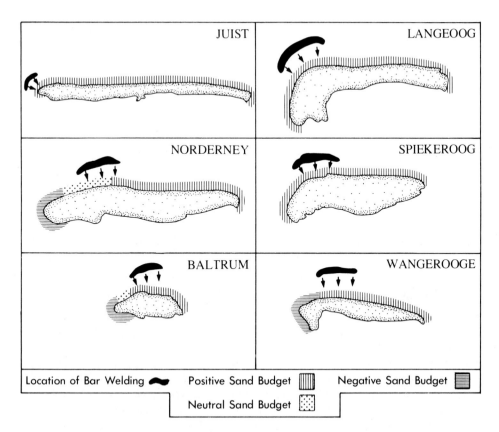

Fig.12. Barrier morphology of the East Frisian Islands. The bulbous portion of the barrier coincides with the position where swash bars attach to the beach. The location of bar welding also correlates well with the erosional-depositional shoreline trends reported by Luck (1975). Generally, the shoreline downdrift of the bar attachment site is stable to depositional while updrift of the bar attachment site is erosional. An exception to the pattern occurs at Langeoog where swash bars migrate into the inlet providing nourishment to the entire western end of the island.

drainage area, as tidal range does not change significantly along this section of coast. The more easterly the location of the main ebb channel at the inlet throat and the larger the size of the tidal inlet, the greater is the overlap of the ebb-tidal delta along the downdrift island shoreline. A highly skewed ebb-tidal delta configuration results in the attachment of swash bars to the shoreline far from the inlet mouth. This condition produces humpbacked and downdrift bulbous barrier islands. The opposite ebb-tidal delta arrangement produces straight barrier islands and drumstick barriers.

Morphology of individual Frisian Islands

The long straight nature of Juist is explained by the fact that bar welding occurs at the westernmost end of the island and not to the northward facing,

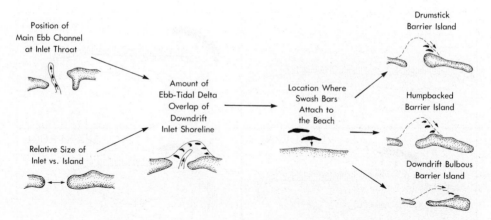

Fig.13. Model for the morphological development of barrier islands. Barriers may have a variety of shapes depending upon the position along the downdrift inlet shoreline where swash bars migrate onshore.

seaward shore (Fig.12). The northwest orientation of the channel between Juist and Memmert controls this sedimentation pattern which has resulted in a 3.0 km westward growth of the island since 1650 (Fig.5).

At Norderneyer Seegat the easterly position of the main ebb channel, abutting the island of Norderney, results in a highly asymmetric configuration of the ebb-tidal delta (Fig.11). Due to this morphology the reefbow meets the downdrift shoreline 5.0 km east of the inlet mouth (Table IV). This region coincides with the humpbacked portion of Norderney and some of the highest dune fields on the island (Fig.12).

Quite a different pattern of sedimentation is found at Langeoog (Fig.12). As discussed earlier, the small size of Accumer Ee, compared to that of Norderneyer Seegat, combined with the westerly position of the main ebb channel at the inlet throat (Fig.11) have produced an ebb-tidal delta that fronts very little of the downdrift shoreline (Table IV). Due to this configuration the onshore migration of swash bars from the reefbow occurs at the western end of Langeoog along the inner inlet shoreline. This pathway of sand movement has accounted for 2 km of shoreline progradation in the southwest portion of the island (since 1860) (Fig.5). The stable to accretionary trend of the inlet shoreline is evidenced by the lack of any groins in this region.

Landward-migrating swash bars at Accumer Ee not only attach to the western end of Langeoog but also migrate toward the inlet throat. This process acts as a feedback mechanism which helps to keep the main ebb channel in a westerly position against Baltrum, thereby maintaining the location of the eastern reefbow close to the inlet mouth. Similar movement of swash bars toward the inlet throat has been recorded at central South Carolina inlets by FitzGerald et al. (1978) and has been attributed to the influence of flood-tidal currents on the wave-induced, bar migrational process.

The mid-throat position of the main-ebb channel at Otzumer Balje and Harle results in an ebb-tidal delta configuration that is intermediate between that of Norderneyer Seegat and Accumer Ee (Fig.11). The resulting humpbacked barrier island morphology of Spiekeroog and Wangerooge is predictably transitional between that of Norderney and Langeoog (Fig.12). The bulbous downdrift shape of Baltrum is caused by the easterly location of the main ebb channel at the inlet throat and the short length of the island (Fig.11). Inlet sediment bypassing at Wichter Ee results in the welding of swash bar to the eastern end of Baltrum (Fig.12).

It should be recognized that the barrier island model explained above (Fig.13) also works well for the East Frisian Islands during their constructional history. For example, note in Fig.5 that as Harle migrated in an easterly direction from 1650 to 1960 so too did the humpbacked portion of Wangerooge. Similar trends are also apparent at Langeoog and Spiekeroog.

DISCUSSION

Hayes and Kana (1976) have put forth the drumstick model (Fig.1) to explain the morphology of barrier islands along mixed energy coasts. In their model the bulbous updrift end of the barrier is attributed to a sediment transport reversal caused by waves refracting around the ebb-tidal delta. They suggest that this reduces the rate at which sand bypasses the inlet resulting in a broad zone of accumulation. The downdrift end of the barrier forms through spit accretion.

In clarification and amplification of Hayes and Kana's (1976) model this study has demonstrated two things. First, it is important to note that deposition on the downdrift inlet shoreline occurs primarily through the attachment of swash bars from the ebb-tidal delta, regardless of whether or not there exists a transport reversal. Also, the zone where bar welding takes place is dependent on inlet size and ebb-tidal delta configuration and can occur some distance from the inlet mouth. The barrier island model that has been developed from this study (Fig.13) can account for barriers that are drumstick-shaped, humpbacked and that have other forms. The East Frisian Islands exemplify this concept well. The bulbous portion of these barriers is a consequence of the location where swash bars move onshore and build up the beach. In accordance with the Hayes and Kana model, the historical data from this study has documented that the downdrift end of these barriers is a product of spit accretion. However, it should be noted that spit growth in the East Frisian Island is accommodated more through inlet narrowing than inlet migration.

The use of our model to explain barrier island morphology along other mixed energy coastlines requires that wave energy be sufficient to cause the formation and landward migration of swash bars on the ebb-tidal delta. In a recent paper (FitzGerald, 1982) it was documented that inlet sediment bypassing occurs through landward bar migrations along the mixed energy coasts of New England, New Jersey, Virginia, South Carolina and the Copper

River Delta region of Alaska. Thus, most of these barriers would be expected to conform to our model.

For example, the shape of Sullivans Island, South Carolina (Fig.2), is a product of an asymmetric ebb-tidal delta configuration and a southerly downdrift location of the main ebb channel. Both of these factors contribute to swash bar attachment some distance from the inlet mouth (1 km).

Although the Georgia coast has barrier islands of similar size compared to the East Frisian Islands, swash bar development does not appear to be an active process at the sounds along this coast. This is due to the large inlet prisms ($10^9-10^{11} m^3$) (Jarrett, 1976), extensive ebb-tidal deltas, and small wave energy ($H_{1/3}$ = 90 cm) (Thompson, 1977) of the Georgia coast. Oertel (1977) has shown that most of the transfer of sand from the delta to the beach occurs very close to the inlet (200 m).

CONCLUSIONS

(1) An abundant sediment supply has led to a 35% increase in size of the East Frisian Islands between 1650 and 1960. Some of this growth has been contemporaneous with a narrowing of the tidal inlets. Poldering of the back-barrier area coupled with natural sedimentation processes has reduced the tidal prism resulting in smaller equilibrium inlet cross-sectional areas.

(2) The strong easterly component of wave energy flux delivers a minimum of 270,000 m^3 of sand to the inlets per year. This sediment bypasses the inlet through tidal and wave-generated currents and is eventually added to the downdrift inlet shoreline in the form of large landward-migrating swash bars (length = 1.0—1.5 km). Swash bar migration accounts for no more than half of the volume of sand that is bypassed.

(3) Swash bars cause a preferential progradation of the beach at the site of attachment. This process has produced a variety of barrier shapes along the East Frisian Island chain. The morphology of individual barriers is controlled by the location where the swash bars migrate onshore which, in turn, is dependent on the amount of overlap of the ebb-tidal delta along the downdrift inlet shoreline. The configuration of the ebb-tidal delta is a function of inlet size and position of the main ebb channel at the inlet throat.

(4) The barrier island model presented in this paper differs from the model proposed by Hayes and Kana (1976) in that barrier shape is dependent on the position along the shoreline where swash bars weld to the beach. Our model can account for not only drumstick-shaped barrier islands but barriers that are humpbacked and other forms as well.

ACKNOWLEDGEMENTS

This investigation was supported by the Office of Naval Research, Coastal Processes Program, through contract no. N00014-78-C-0612 (Dag Nummedal, Principal Investigator).

The authors would like to thank Dr. Gunter Luck, the director of the

Forschungsstelle Norderney, State of Lower Saxony, Germany, for the logistical support that he provided during the field study. Gunter Luck, Hanz Niemeyer and other research personnel at Forschungsstelle Norderney are gratefully acknowledged for their scientific input and advice during our investigation.

REFERENCES

Bruun, P. and Gerritsen, F., 1959. Natural by-passing of sand at coastal inlets. J. Waterways Harbors Div., Am. Assoc. Civ. Eng., 85: 75—107.
Davies, J.L., 1964. A morphogenetic approach to world shorelines. Z. Geomorphol., 8: 27—42.
Dette, H.H., 1977. Ein Vorschlag zur Analyse eines Wellenklimas. Die Küste, 31: 166—180.
Finley, R.J., 1976. Hydraulics and dynamics of North Inlet, South Carolina, 1974—1975. GITI Report, U.S. Army Coastal Eng. Research Center and Waterway Exp. Station, Ft. Belvoir, V., 188 pp.
FitzGerald, D.M., 1982. Sediment bypassing of mixed energy tidal inlets. Am. Soc. Civ. Eng., Proc. 18th Conf. Coastal Engineering, pp.1848—1861.
FitzGerald, D.M., Nummedal, D. and Kana, T.W., 1976. Sand circulation patterns at Price Inlet, South Carolina. Am. Soc. Civ. Eng., Proc. 15th Conf. Coastal Engineering, pp.1868—1880.
FitzGerald, D.M., Hubbard, D.K. and Nummedal, D., 1978. Shoreline changes associated with tidal inlets along the South Carolina Coast. Coastal Ocean Div., Am. Soc. Civ. Eng., Coastal Zone 78, San Francisco, Calif., 1973—1994.
Hanisch, J., 1981. Sand transport in the tidal inlet between Wangerooge and Spiekeroog (W. Germany). Spec. Publ. Int. Assoc. Sedimentol., 5: 175—185.
Hayes, M.O., 1975. Morphology of sand accumulation in estuaries: An introduction to the Symposium. In: L.E. Cronin (Editor), Estuarine Research, Vol. 2. Academic Press, New York, N.Y., pp.3—22.
Hayes, M.O., 1979. Barrier island morphology as a function of tidal and wave regime. In: S.P. Leatherman (Editor), Barrier Islands: From the Gulf of St. Lawrence to the Gulf of Mexico. Academic Press, New York, N.Y., pp.1—28.
Hayes, M.O. and Kana, T.W., 1976. Terrigenous clastic depositional environments. Tech. Report 11, CRD, Dept. of Geology, University of South Carolina, Columbia, S.C., 364 pp.
Hayes, M.O., Owens, E.H., Hubbard, D.K. and Abele, R.W., 1973. The investigation of forms and processes in the coastal zone. In: D.R. Coates (Editor), Coastal Geomorphology. State Univ. of New York, Binghamton, New York, pp.11—41.
Hine, A.C., 1975. Bedform distribution and migration patterns on tidal deltas in the Chatham Harbor Estuary, Cape Cod, Massachusetts. In: L.E. Cronin (Editor), Estuarine Research, Vol. 2. Academic Press, New York, N.Y., pp.235—252.
Homeier, H. and Kramer, J., 1957. Verlagerung der Platen im Riffbogen vor Norderney und ihre Anlandung an den Strand. Jahresber. 1956, Forschungsstelle Norderney, 8: 37—60.
Homeier, H. and Luck, G., 1969. Das Historische Kartenwerk 1:50,000 der Niedersächsischen Wasserwirtschaftsverwaltung, Göttingen.
Jarrett, J.T., 1976. Tidal prism—inlet area relationships, General Investigations of Tidal Inlets. Rept. no. 3., Coastal Eng. Research Center, Ft. Belvoir, Va., 32 pp.
Kramer, J., 1961. Strommessungen im Hafengebiet von Norderney: Jahresber. Forschungsstelle Norderney, 13: 67—94.
Luck, G., 1974. Untersuchungen zur Sedimentbewegung mit Hilfe einer Unterwasserfernanlage. Jahresber. 1973, Forschungsstelle Norderney, 25: 55—78.

Luck, G., 1975. Der Einfluss der Schutzwerke der ostfriesischen Inseln auf die morphologischen Vorgänge im Bereich der Seegaten und ihre Einzugsgebeite. Mitt. Leichtweiss Inst., Braunschweig, 47: 1—22.
Luck, G., 1976a. Inlet changes of the East Frisian Islands. Am. Soc. Civ. Eng., Proc. 15th Conf. Coastal Engineering, New York, II: 1938—1957.
Luck, G., 1976b. Protection of the littoral and seabed against erosion. Fallstudie Norderney, Jahresber. 1975. Forschungsstelle Norderney, 27: 9—78.
National Ocean Survey, NOAA, 1983. Tide tables for Europe and West Coast of Africa including the Mediterranean. 178 pp.
Niemeyer, H.D., 1978. Wave climate study in the region of the East Frisian Islands and coast. Am. Soc. Civ. Eng., Proc. 16th Conf. Coastal Engineering, New York, 1: 134—151.
Nummedal, D. and Fischer, I., 1978. Process-response models for depositional shorelines: The German and Georgia Bights. Am. Soc. Civ. Eng., Proc. 16th Conf. Coastal Engineering, New York, II: 1212—1231.
Nummedal, D. and Penland, S., 1981. Sediment dispersal in Norderneyer Seegat, West Germany. Spec. Publ. Int. Assoc. Sedimentol., 5: 187—210.
Oertel, G.F., 1972. Sediment transport of estuary entrance shoals and the formation of swash platforms. J. Sediment. Petrol., 42: 857—863.
Oertel, G.F., 1977. Geomorphic cycles in ebb deltas and related patterns of shore erosion and accretion. J. Sediment. Petrol., 47: 1121—1131.
Stephen, M.F., 1981. Effects of seawall construction on beach and inlet morphology and dynamics at Caxambas Pass, Florida. Ph.D. Diss., University of South Carolina, Columbia, S.C., 196 pp.
Streif, H. and Koster, R., 1978. The geology of the German North Sea Coast. Die Küste, 32: 30—50.
Thompson, E.F., 1977. Wave climate at selected locations along U.S. coasts. U.S. Army, Coastal Engineering Research Center, Tech. Rept. No. 77-1, 364 pp.
U.S. Naval Weather Command, 1974. Summary of Synoptic Meteorological Observations, Western European Coastal Marine Studies. Nat. Climatic Center, NOAA, Ashville, N.C.
Van Straaten, L.M.J.U., 1965. Coastal barrier deposits in south and north Holland. Meded. Rijks Geol. Dienst, 17: 41—75.

COARSE CLASTIC BARRIER BEACHES: A DISCUSSION OF THE DISTINCTIVE DYNAMIC AND MORPHOSEDIMENTARY CHARACTERISTICS

R.W.G. CARTER and J.D. ORFORD

School of Biological and Environmental Studies, The New University of Ulster, Coleraine, Co. Londonderry BT 52 1SA (Northern Ireland)
Department of Geography, The Queen's University of Belfast, Belfast BT7 1NN (Northern Ireland)

(Received March 1, 1983; revised and accepted August 12, 1983)

ABSTRACT

Carter, R.W.G. and Orford, J.D., 1984. Coarse clastic barrier beaches: A discussion of the distinctive dynamic and morphosedimentary characteristics. In: B. Greenwood and R.A. Davis, Jr. (Editors), Hydrodynamics and Sedimentation in Wave-Dominated Coastal Environments. Mar. Geol., 60: 377—389.

Coarse clastic barriers are common on mid- and high-latitude coasts. They possess a morphosedimentary and dynamic distinctiveness which sets them apart from sandy fine-clastic barrier forms. The reflective nature of the seaward barrier favors the development of zero mode, sub-harmonic edge waves particularly during long period swells (10—20 s), manifest in the formation of high level cusps. In some circumstances the pattern of recent overwashing of the barrier is related to cusps. Lack of distinct tidal passes, due partly to the high seepage potential of coarse barriers, means that very little sediment is transported seaward. Thus the barriers roll steadily onshore, and sections show a variety of washover facies, related to the volume of overwash surges.

INTRODUCTION

Over the last decade there has been an upsurge of interest in barrier coastlines, both from viewpoints of shoreline stability (Leatherman, 1979a; Kaufman and Pilkey, 1979) and as analogues for examining the reservoir potential of ancient sedimentary bodies (McCubbin, 1982).

However, as Zenkovich's (1967) classification shows, there is a number of discrete barrier types, which Zenkovich divided into "free" and "fixed" forms. An unfortunate trend has been to call all barriers "barrier islands" (Glaesner, 1978; Hayes, 1979) or to assume that the barrier island sub-type can be termed simply "barrier" (e.g. Godfrey et al., 1979; Leatherman 1979b). There appear to have been no attempts to classify barriers on the basis of grain size.

This paper concentrates on one type of barrier — the fixed, fringing gravel barrier — with the aim of establishing its distinctiveness in terms of both dynamics and morphosedimentary structure. These barriers are common on wave-dominated, mid- to high-latitude coasts, particularly where glacigenic deposits are being reworked (Armon, 1974; Randall, 1977; Carter and Orford, 1980; Wang and Piper, 1982). Similar structures have not been recorded on low wave energy coasts, except in high latitudes as a result of ice-push processes (Barnes, 1982). Conglomeratic barrier beaches also have a presence in the geological column, albeit a minor one (Ali, 1976; Leckie and Walker, 1982; Wright and Walker, 1981). The paper is largely based on the authors' research experience in southeast Ireland (Carter and Orford, 1980, 1981; Orford and Carter, 1982a, b, 1985) but reference is made to similar barrier structures elsewhere in Britain and Ireland (Fig.1).

Almost all these barrier beaches, around the coasts marginal to the southwestern approaches to the British Isles, are exposed to waves generated within the North Atlantic. The barrier coasts are rarely influenced by storms of exceptional severity, i.e. hurricanes. A more common picture is of repeated lower-magnitude storms associated with the frequent southwest to northeast passage of cyclonic depressions.

Incident waves are predominantly from the southwest or west. Deepwater storm waves may exceed 10 m in height and poststorm decay swell often reaches 12—14 s periods (Hogben and Lumb, 1967) although longer periods have been recorded (>20 s) affecting the gravel barrier at Chesil Beach in Dorset (Golding, 1981). Table I shows the seasonal wave climate

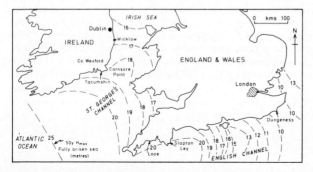

Fig.1. Map of locations mentioned in the text. Fifty year H_{max} for fully arisen sea are also shown to provide a picture of the extreme wave climate.

TABLE I

Seasonal wave conditions at Carnsore Point

	$\bar{H}_s$	$\bar{H}_{max}$	$H_{s(95)}$	$H_{max(95)}$	T_z	$T_{z(95)}$	Spectrum width
Autumn	1.42	2.85	3.55	6.74	4.9	7.7	0.65
Winter	1.72	3.43	3.61	6.97	5.6	8.4	0.69
Spring	1.04	2.08	2.55	5.0	5.2	9.4	0.68
Summer	0.91	1.82	2.00	4.02	4.8	7.5	0.71

Unpublished data from the Electricity Services Board, Dublin. $\bar{H}_s$: mean significant wave height (m), $\bar{H}_{max}$: mean maximum wave height (m), (95): 95th percentile value, T_z: zero-crossing wave period (s).

for southeastern Ireland. Spring tidal ranges on the Irish coasts vary from 1 to 4 m. Graff (1981) has documented the return periods and heights of annual sea-level maxima for Fishguard (on the east side of St. George's Channel, 100 km east of Carnsore — the nearest data source), which shows that the 5—20 yr return period of total surge levels is in the range of 1.8—2.0 m above MHWS. These values should be treated as crude estimates, as coastal configuration is radically different for two sites. Storm surges on more open coasts of the Celtic Sea rarely exceed 0.5 m (Pugh and Vassie, 1978; Flather, 1981).

EVOLUTIONARY FRAMEWORK

Compared to the detailed work on fine clastic barriers (e.g. Kraft, 1971; Moslow and Heron, 1981; Rampino and Sanders, 1981), the Holocene evolution of gravel barriers has not been extensively studied. However, a broad framework may be suggested.

It would appear that most fringing gravel barriers have moved landward under rising post-glacial sea levels. Transgressive sea levels lead to a rolling over of the barrier (Carr and Blackley, 1973; Carter and Orford, 1980) leaving little evidence of earlier shoreline positions.

The magnitude of sediment supply is crucial to the development of gravel beaches as there is limited sediment recycling once material is incorporated in the barrier. Where sediment supply has failed, or is failing, as is the case throughout the British Isles, barriers tend to develop as massive solitary forms, rather than as multi-ridge complexes. The relative proportions of gravel sizes to finer sizes will determine the transportability of the gravel mass. A small gravel component may become either trapped within the larger gravel or boulder component, or buried under sand. Alternatively, the coarse fraction may overpass the fine (Fahnestock and Houshild, 1962; Everts, 1973); and under the asymmetrical wave stress move preferentially onshore.

Through time, supply rate of gravel may change due to the diminution of sources and the maturing of the littoral drift system, the declining rate of sea-level change, or the protuberance of headlands which act as natural groins and disrupt the drift pattern. Examples of these changes have been described around the British Isles; in southeast England, Edison (1983) has argued that the longshore supply of gravel to Dungeness declined asymptotically with a reduction in the rate of sea-level rise (after 5500 B.P.). Carr and Blackley (1974) have described a similar situation for Chesil Beach in Dorset. At Slapton Ley, Devon (Hails, 1975; Morey, 1983), and Carnsore Point, southeast Ireland (Carter and Orford, 1980), the gradual emergence of adjacent headlands has acted to cut off the longshore supply. In the Irish example long-term sea level has been relatively static over the last 5000 yrs (Carter, 1983) and the relative maturity of the drift system, i.e. the establishment of sediment cells (Lowry and Carter, 1982), has also contributed to the diminution of lateral transport.

BARRIER FORM, STRUCTURE AND PROCESS

Wave climate modifications

Gravel-dominated barriers tend to fall within the reflective domain of the morphological beach classifications of Short (1979) and Wright et al. (1979, 1982). This domain is distinguished by low values (<3) of both the surf-scaling parameter, ϵ (Huntley et al., 1977), and the beach stability index ($\Omega < 1$) of Dean (1973). Short (1979) and Wright et al. (1979) outline a number of recognition criteria for reflective beaches, including presence of a steep beach face slope, a break point step, a single swash berm and high level cusps and absence of large-scale nearshore bedforms; points all common to the southeastern Irish barriers. The transitional forms associated with finer clastic beaches and barriers (Wright et al., 1982) do not occur on gravel based barriers.

Reflective topography can engender sub-harmonic edge waves, usually of zero mode (Guza and Davis, 1974). Under long-period swells these waves are easily excited and increase in amplitude. Such long-period swells (>10 s) are common in the North Atlantic (Hogben and Lumb, 1967) forming part of the storm decay spectra. Manifestation of swell activity on the beach appears in the accentuation of run-up, which may lead to berm overtopping and breaching (Short, 1979) or the formation of high-level erosional cusps, with spacings of around half the edge wave length (Darbyshire, 1977; Wright, 1981). The development and persistence of edge-wave controlled topography may assist in both directing subsequent overwash and controlling washover spacing (Orford and Carter, 1984).

For a variety of reasons, gravel-dominated barriers remain relatively stable in the face of wave attack. Of paramount importance is the inability of particles within a gravel mass to become entrained, except under high-energy events. Not only do the individual clasts resist transport, but they give rise to hydrodynamic rough surfaces capable of significantly affecting the inshore wave spectra (Wright et al., 1982).

Inevitably any beach face structures that do develop have a high survival potential into, or even through, normally destructive process phases. Orford and Carter (1985) show that a fair-weather supratidal marine terrace at Carnsore Point [a result of long-term (>1 yr) development], not only survived through a major storm, but also appeared to control both the mode of wave breaking and reformation, to produce a spatially distinctive reflective/dissipative regime (Fig.2) across the upper beach. The terrace acts as a step over which either waves are broken and reformed (similar to the process acting on coral reefs; Suhayada and Roberts, 1977), or swash excursions are amplified (Short, 1979). Ultimately, high-level swash excursions during severe storms lead to barrier crest sedimentation of marine gravels.

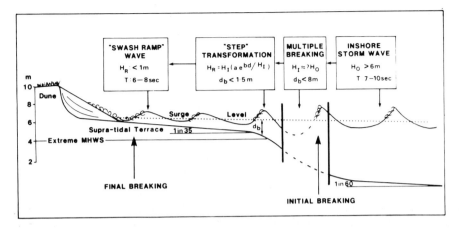

Fig.2. Nearshore wave conditions across a ramp profile at Carnsore Point, Southeast Ireland. Inshore storm waves, that may have already broken offshore, are partially broken again by the outer "step" of the terrace. While a certain amount of wave energy is reflected at this point, low reformed waves also pass across the terrace constructing a swash ramp (Orford and Carter, 1982a). Ultimately this may lead to dune overtopping, up to 6 m above the extreme level of the mean high water Spring tides.

Sedimentation control

Most coarse clastic barriers comprise varying mixtures of gravel and sand, alongshore, crossbeach and at-depth. Where the gravel population dominates, sand becomes a subsidiary, usually interstitial component. On the beach face, sand may become packed into the gravel component, possibly via the vibratory swash mechanism (kinetic sieving) of Middleton (1970). We favor this process to produce the substantial upward coarsening units often encountered in gravel beach sections (especially on the upper beach face), where the ratio of clast size to matrix size can be in excess of thirty times, rather than the dispersive stress mechanism discussed by Sallenger (1979). This latter process may assume more importance under swash/backwash flows on the lower beach face. Continued introduction of fine material into the coarse matrix will lead to reductions in hydrostatic porosity and permeability, so that water storage capacity and throughflow discharge (both from swash inputs and sea—lagoon exchange) will fall. When fine particle infilling reaches matrix capacity, excess sand will provide a source both for the formation of swash-based bedforms, and when dry, eolian ones. Small barrier-top dunes develop at this stage. Eventually the amount of sand, relative to gravel, will increase to the point where it produces a low-angle beach face sufficient for the transition from a stable reflective beach face to one more liable to allow the intermediate reflective/dissipative stages (Wright et al., 1982). Should a small proportion of gravel remain, it may well become more mobile, particularly in the upper flow regime, large clasts moving across (overpassing) the fine-dominated bed. Carr (1971) notes that coarse particles may "outrun" fine particles downdrift, as they are subject to constant "rejection" (Moss, 1962). In circumstances where sediment supply is low, this may lead to coarsening-downdrift beach grading (Zenkovich, 1967).

Structural control

Orford and Carter (1982a) have shown how the barriers in southeast Ireland build-up during storms. This is particularly important at the beach crest, as vertical accumulations reduce the chances of swash breaching and overwashing. Such modifications only arise, however, at a late stage in gravel barrier genesis. Examination of barrier cross-sections and shallow borehole records (Carter and Orford, 1982b) reveals a triparate structure. Figure 3 shows a lenticular lower barrier unit followed by washover gravels and overlain by overtop deposits. This division is related to gradually decreasing volumes of overwash (Orford and Carter, 1982a) as the barrier agrades vertically. The lower unit comprises a ridge-like structure of thin, repetitive coarsening-upward lamina, with a persistent cross-barrier lenticularity. The larger clasts are often imbricated seaward. Occasional massive boulders (-10ϕ) occur in these laminations. The bottom unit is indicative of large-volume swash bores traversing a lower barrier. As the height of the barrier increased relative to sea level, so these lower laminae were replaced by more spatially discrete washover fans of cross-barrier wedge form. The focus of sedimentation is at the ridge crest, rather than on the back of the barrier. Finally the barrier is raised to the point where washover is precluded and only minor overtopping dominates.

The presence and distribution, or absence of certain morphological elements — barrier top dunes, high level cusps and/or supratidal terraces — appear crucial in controlling the pattern of barrier crest and back barrier sedimentation. On the Tacumshin section of the Irish barriers, the spacing of contemporary washover fans appears related to the incidence of high level beach morphology of a periodic nature (Orford and Carter, 1984). Such structural determinism is especially important on gravel-dominated barriers where inheritance from antecedent stages is the norm.

Barrier breaching

Gravel-dominated barriers often enclose brackish or freshwater lagoons. Tidal passes or inlets are generally absent. As a consequence, sediment transport on barriers is mainly onshore, with washover processes dominating. Hayes (1979) considers that lack of tidal inlets is diagnostic of microtidal conditions, yet this, as a rule can not obviously be extended to embrace coarse barriers which appear on both meso- and macrotidal coasts. Absence of tidal inlets is important, because without them the barriers are more capable of moving steadily onshore.

Without tidal inlets, seaward discharge of back-barrier lagoons is effected either by seepage through the barrier or by surface channel outlet. In the case of cross-barrier seepage, discharge occurs along the entire saturated

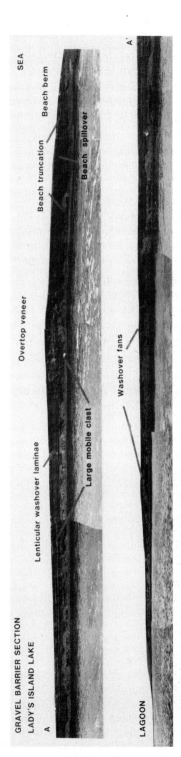

Fig.3. A typical transgressive barrier cross-section from Lady's Island Lake southeast Ireland photographed in March 1980. The section is 260 m long and 9 m hight at the barrier crest. On the seaward (right-hand) side the landward-dipping washover units have been truncated by beach erosion, and some of the material has been reworked into a small marine terrace formed under storm conditions. The washover deposits comprise lenticular coarsening-upward imbricate units many of which may be traced across the entire section. Apart from slight distal fining the units are not graded across the section. In places very large clasts have been transported within the overwash flow, presumably rolling across the mobile substrate. The upper units indicate a decreasing value of overwash, units pinching out regressively up-barrier. In places small lobate fans have formed.

barrier length that is in contact with the lagoon. No distinctive or large-scale morphological seepage forms develop, although the lower beach face may be extensively rilled at low tide. Seepage discharge potential depends on both textural and structural characteristics of the barrier and may be approximated by the Darcy formula, $Q = AK(h/l)$; where Q is discharge, A is the longshore cross-sectional area of the barrier, K is a proportionality coefficient related to the textural attributes of the barrier sediment and hydraulic properties of the fluid, h is the hydrostatic head and l is the barrier width. Interplay between semidiurnal tidal levels and seasonal lagoon levels determines the instantaneous hydraulic gradient and thus the ratio between landward/seaward discharge.

Once the threshold seepage capacity of the barrier is exceeded due either to rising water level or to falling permeability, a surface channel will form. This may be of two types, spillover or incised. A spillover channel simply overtops the barrier and expands out over the seaward face with minimal incision. The seepage flow net remains undisturbed and the channel "closes" as soon as the lagoon drawdown is sufficient. The shift to a more major incised channel would appear to be a function of discharge; in small mixed sediment embayments in Oregon, Clifton et al. (1973) suggest that this transition occurs around 0.01—1 cumecs. A value within the same order of magnitude appears to occur in southeast Ireland. In a survey of twenty barrier dammed streams, with mean annual discharges ranging from 0.08 to 11.05 cumecs, Carter et al. (1984) were able to show that the seepage to channel transition generally occured around 1—2 cumecs. Figure 4 suggests a relationship between modal surface sediment size (d) and discharge as $d \approx Q^{2.5}$.

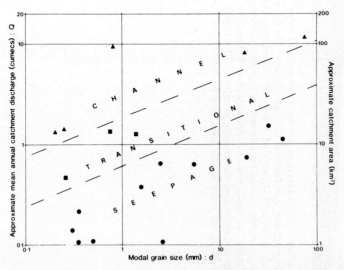

Fig.4. Discharge versus grain size relationship for southeast Ireland barriers (between Wicklow and Tacumshin). Permanent seepage outlets (solid circles), give way, with increasing discharge and/or decreasing grain size to ephemeral channels (solid squares) to permanent channels (solid triangles).

Incised outlet (= dominant freshwater flow) channels may be deflected by longshore drift and cause reworking of the barrier during migration. Closure of incised channels tend to take place by infilling following a decline in seepage discharge associated with drawdown of the lagoon. Only if an outlet becomes a tidal channel could it be maintained on a permanent basis.

Lagoon discharge by seepage removes the necessity for cross-barrier channels and thus diminishes the role of tidalpass sediment recycling. Only a very small proportion of the gravel-dominated barrier mass is stored in inlet related morphology.

Seepage also plays an important part in the landward transfer of sediment and in destabilising the barrier crest. When sea level is elevated relative to lagoon level, significant volumes of water may pass through the barrier emerging as surface flows on the landward side (Arkell, 1955). At Tacumshin, seepage flow has produced extensive back barrier channelling and formed seepage fans along the barrier lagoon margin (Fig.5) Landward discharges of 14 cumecs km^{-1} h^{-1} are feasible during storms (Carter et al., 1984). Gradual excavation of seepage hollows acts to undermine the barrier crest and may presage incipient breach points. The importance of seepage has been recorded on other fringing barriers in the British Isles (Arkell, 1955; Robinson, 1981; Edison, 1983).

Fig.5. Seepage and washover channels on the barrier at Tacumshin, southeast Ireland. The seepage channel, to the left, starts in an erosional hollow about 2 m below the beach crest and expands into a small fan. The washover channel, to the right, is a larger feature originating at a breach in the barrier crest.

DISCUSSION

Gravel-dominated barriers form part of a barrier continuum based on grain size, that has largely gone unnoticed and uninvestigated. Instead efforts have concentrated on wave climate and tidal range variations to explain barrier variation (Hayes, 1979). The barriers of the southwest approaches to the British Isles have developed distinctive morphological facies assemblages and planforms by virtue of the presence of coarse elements in the sediment mass. The importance of seepage in effecting terrestrial drainage has meant a lack of tidal passes, thus removing from the barriers a mechanism by which material may be returned seaward. Although British and Irish washover features are small in area when compared to many on the U.S. coast (e.g. Andrews, 1970; Schwartz, 1975), they are left to dominate the gross transport across the barriers. The inability to achieve onshore/offshore equilibrium means that barriers move steadily onshore. There is no compensatory mechanism to balance the landward-directed effects of overwash processes, while the various factors which lead to constant destabilisation of the barrier crest (Carter and Orford, 1980) simply accentuated any propensity to onshore transport. Lack of tidal inlet-associated ebb deltas not only eliminates an important repository for barrier sediments en route to maintain the shoreface, but also rids the nearshore of significant refraction perturbations.

Gravel-dominated barrier processes are strongly influenced by two factors. One is the appearance of long-period swell waves in the nearshore climate and the other is the inheritance of morphosedimentary structures into the barrier through storm events. The reflective nature of gravel-dominated barrier beaches leads to amplification of the sub-harmonic component of long-period swell thus providing one mechanism for the extending swash run-up up to, over and beyond the crest.

Gravel-beach structures have a high survival potential owing to their high mass entrainment thresholds, aided and abetted by the fact that as strongly reflective beaches, profile variability is low and only a small volume of material moves. Thus fairweather structures (berms, terraces, swash bars) survive well into storms, and act to modify the shoreline wave processes. The presence of extensive berm terraces may, during storm surges, create conditions whereby constructional spilling waves move sediment onto the barrier crest. The comparative lack of exceptionally intense storms is important inasmuch as there is rarely, if ever, a complete reorganisation of the barrier morphosedimentary environments.

The persistent, if slow, landward migration of gravel-dominated barriers probably continues even under stable sea levels, unlike the retreat of some barrier islands discussed by Pilkey (1981). There are enough processes capable of lowering the crest and causing overwashing of gravel-dominated barriers without invoking the need for change in sea level.

CONCLUSIONS

The following conclusions may be noted:

(1) Gravel-dominated barriers constitute a distinctive morphosedimentary environment.

(2) Gravel-dominated barriers fall within the reflective beach stage as defined by Short (1979) and others. The amplification of wave run-up during episodes of long period swell is one mechanism which causes overtopping and over-washing of the barrier crest.

(3) Structural stability means that many features are inherited from antecedent events. This exercises some control over back beach, barrier crest and back barrier processes and sedimentation.

(4) Seepage through the barrier is important, both in draining back barrier lagoons and streams, and during periods when sea levels are elevated above lagoon levels, so that landward seepage takes place.

(5) The comparative importance of seepage and surface outlet channels leads to a lack of tidal passes, so that very little sediment is moved seaward from the barrier. Morphology and structure may be preserved in the absence of migrating inlets.

(6) Washover processes are the major methods by which the barrier migrates. Migration is landward, and it is argued that migration may continue even in the absence of sea level rise.

ACKNOWLEDGEMENTS

We would like to thank our respective universities and departments for continuing field and technical support.

REFERENCES

Ali, M.T., 1976. The significance of a mid-Cretaceous cobble conglomerate, Beer District, South Devon. Geol. Mag., 13: 151—158.
Amin, M., 1982. On analysis and forecasting of surges on the west coast of Britain. Geophys. J.R. Astron. Soc., 68: 79—94.
Andrews, P.B., 1970. Facies and genesis of a hurricane-washover fan, St. Joseph's Island, Central Texas Coast. Univ. Texas Bur. Econ. Geol., R.I. 67, 167 pp.
Arkell, W.J., 1955. Effects of storms on Chesil Beach in November 1954. Proc. Dorset Nat. Hist. Arch. Soc., 76: 141—145.
Armon, J.W., 1974. Late Quaternary shore lines near Lake Ellesmere Canterbury, New Zealand. N.Z. J. Geol. Geophys., 17: 63—73.
Barnes, P.W., 1982. Marine ice-pushed boulder ridge, Beaufort Sea, Alaska. Arctic, 35: 312—316.
Carr, A.P., 1971. Experiments on the longshore transport and sorting of pebbles. J. Sediment. Petrol., 41: 1084—1104.
Carr, A.P. and Blackley, M.W.L., 1973. Investigations bearing on the age and development of Chesil Beach, Dorset, and the associated area. Trans. Inst. Brit. Geogr., 58: 99—112.
Carr, A.P. and Blackley, M.W.L., 1974. Ideas on the origin and development of Chesil Beach, Dorset. Proc. Dorset Nat. Hist. Arch. Soc., 95: 9—17.
Carter, R.W.G., 1983. Raised coastal landforms as products of modern process variations

and their relevance to eustatic sea level studies: examples from eastern Ireland. Boreas, 12: 167—182.
Carter, R.W.G. and Orford, J.D., 1980. Gravel barrier genesis and management: a contrast. Proc. Coastal Zone 80, Am. Soc. Civ. Eng.: 1304—1320.
Carter, R.W.G. and Orford, J.D., 1981. Overwash processes along a gravel beach in southeast Ireland. Earth Surf. Proc., 6: 413—426.
Carter, R.W.G., Johnson, T.W. and Orford, J.D., in press. Formation and significance of stream outlets on the mixed sand and gravel barrier coasts of southeast Ireland. Z. Geomorphol.
Clifton, H.E., Phillips, R.L. and Hunter, R.E., 1973. Depositional structures and processes in the mouths of small coastal streams, southwestern Oregon. In: D.R. Coates (Editor), Coastal Geomorphology. Publications in Geomorphology, State Univ. of New York, Stoney Brook, N.Y., pp.115—140.
Darbyshire, J., 1977. An investigation of beach cusps in Hell's Mouth Bay. In: M. Angel (Editor), A Voyage of Discovery. Pergamon, Oxford, pp.405—427.
Dean, R.G., 1973. Heuristic models of sand transport in the surf zone. Proc. Conf. Eng. Dynamics in Surf Zones, Sydney, N.S.W., pp.207—214.
Edison, J., 1983. Flandrian barrier beaches off the coast of Sussex and southeast Kent. Q. Newslet., 39: 25—29.
Everts, C.H., 1973. Particle overpassing on flat granular boundaries. J. Water. Harbour Div. Am. Soc. Civ. Eng., 99 (WW4): 425—438.
Fahnestock, R.K. and Houshild, W.L., 1962. Flume studies of the transport of pebbles and cobbles on a sand bed. Geol. Soc. Am. Bull., 73: 1431—1436.
Flather, R.A., 1981. Practical surge prediction using numerical models. In: D.H. Peregrine (Editor), Floods Due to High Winds and Tides. Academic Press, London, pp.21—43.
Glaesner, J.D., 1978. Global distribution of barrier islands in terms of tectonic setting. J. Geol., 86: 283—298.
Godfrey, P.J., Leatherman, S.P. and Zaremba, R., 1979. A geobotanical approach to the classification of barrier beach systems. In: S.P. Leatherman (Editor), Barrier Islands. Academic Press, New York, N.Y., pp.99—126.
Golding, B., 1981. A meteorological input to surge and wave prediction. In: D.H. Peregrine (Editor), Floods Due to High Winds and Tides. Academic Press, London, pp.9—20.
Guza, R.T. and Davis, R.E., 1974. Excitation of edge waves by waves incident on a beach. J. Geophys. Res., 79: 1285—1291.
Graff, J., 1981. An investigation of the frequency distributions of annual sea level maxima at Ports around Great Britain. Estuarine Coastal Shelf Sci., 12: 289—349.
Hails, J.R., 1975. Some aspects of the Quaternary history of Start Bay, Devon. Field Stud., 4: 207—222.
Hayes, M.O., 1979. Barrier island morphology as a function of tidal and wave regime. In: S.P. Leatherman (Editor), Barrier Islands. Academic Press, New York, N.Y., pp.1—28.
Hogben, N. and Lumb, F.E., 1967. Ocean Wave Statistics. H.M.S.O. London, 274 pp.
Huntley, D.A., Guza, R.T. and Bowen, A.J., 1977. A universal form for shoreline run-up spectra? J. Geophys. Res., 82: 2577—2581.
Kaufman, W. and Pilkey, O.H., 1979. The Beaches are Moving. Anchor Press, New York, N.Y., 326 pp.
Kraft, J.C., 1971. Sedimentary facies patterns and geologic history of a Holocene marine transgression. Geol. Soc. Am. Bull., 82: 2131—2158.
Leatherman, S.P., 1979a. Barrier Islands. Academic Press, New York, N.Y., 325 pp.
Leatherman, S.P., 1979b. Barrier dunes: a reassessment. Sediment. Geol., 24: 1—16.
Leckie, D.A. and Walker, R.W., 1982. Storm and tide dominated shorelines in Cretaceous Moosebar—Lower Gates interval — Outcrop equivalents of Deep Basin Gas Trap in Western Canada. Bull. Am. Assoc. Pet. Geol., 66: 138—157.
Lowry, P. and Carter, R.W.G., 1982. Computer simulation and delimitation of littoral drift cells on the south coast of Co. Wexford, Ireland. J. Earth Sci. R. Dublin Soc., 4: 121—132.

McCubbin, D.G., 1982. Barrier-island and strand-plain facies. In: P.A. Scholle and D.A. Spearing (Editors), Sandstone Depositional Environments. Am. Assoc. Pet. Geol. Tulsa, Okla., pp.247—279.

Middleton, G.V., 1970. Experimental studies related to problems of flysch sedimentation. Geol. Assoc. Can., Spec. Publ., 7: 253—272.

Morey, C.R., 1983. Barrier stability. Q. Newslet., 40: 23—27.

Moslow, T.F. and Heron, D., 1981. Holocene deposition and history of a microtidal cuspate foreland cape: Cape Lookout, North Carolina. Mar. Geol., 41: 251—270.

Moss, A.J., 1962. The physical nature of common sandy and pebbly deposits. Part I. Am. J. Sci., 260: 337—373.

Orford, J.D. and Carter, R.W.G., 1982a. Crestal overtop and washover sedimentation on a fringing sandy gravel barrier coast, Carnsore Point, Southeast Ireland. J. Sediment. Petrol., 52: 265—278.

Orford, J.D. and Carter, R.W.G., 1982b. Geomorphological changes on the barrier coasts of south Wexford. Ir. Geogr., 15: 70—84.

Orford, J.D. and Carter, R.W.G., 1984. Models of longshore washover spacing on coarse clastic barriers. Mar. Geol., 56: 207—226.

Orford, J.D. and Carter, R.W.G., 1985. Storm-generated dune armouring on a sand-gravel barrier system, southeastern Ireland. Sediment. Geol., 42 (in press).

Pilkey, O.H., 1981. Geologists, engineers, and a rising sea level. Northeast. Geol., 3: 150—158.

Pugh, D.T. and Vassie, J.M., 1978. Extreme sea levels from tide and surge probability. Proc. 16th Conf. Coastal Engineering, Am. Soc. Civ. Eng., pp.911—930.

Rampino, M.R. and Sanders, J.E., 1981. Evolution of the barrier islands of southern Long Island, New York. Sedimentology, 28: 37—48.

Randall, R.E., 1977. Shingle foreshores. In: R.S.K. Barnes (Editor), The Coast. Wiley, London, pp. 49—61.

Robinson, I.S., 1981. Salinity structure and tidal flushing of the Fleet. In: M. Ladle (Editor), The Fleet and Chesil Beach. Dorset County Council, Dorchester, pp.33—38.

Sallenger, A.H. 1979. Inverse grading and hydraulic equivalence in grain-flow deposits. J. Sediment. Petrol., 49: 553—562.

Schwartz, R.K., 1975. Nature and genesis of some storm washover deposits. U.S. Army Corps of Eng., Tech. Memo 61, 70 pp.

Short, A.D., 1979. Three dimensional beach stage model. J. Geol., 87: 553—571.

Suhayada, J.N. and Roberts, H.H., 1977. Wave action and sediment transport on fringing reefs. Proc. 3rd Int. Coral Reef Symp., 1: 65—70.

Wang, Y. and Piper, D.J.W., 1982. Dynamic geomorphology of the Drumlin coast of southeast Cape Breton Island. Marit. Sediment. Atlant. Geol., 18: 1—28.

Wright, L.D., 1982. Field observations of long period surf zone oscillations in relation to contrasting beach morphologies. Aust. J. Mar. Freshwater Res., 33: 181—201.

Wright, L.D., Chappell, J., Thom, B.G., Bradshaw, M.P. and Cowell, P., 1979. Morphodynamics of reflective and dissipative beach and inshore systems: southeastern Australia. Mar. Geol., 32: 105—140.

Wright, L.D., Short, A.D. and Nielsen, P., 1982. Morphodynamics of high energy beaches and surf zones: A brief synthesis. Coastal Studies Unit, Univ. of Sydney, Rep. 82/5, 64 pp.

Wright, M.E. and Walker, R.G., 1981. Cardium Formation (Upper Cretaceous) and Seebee, Alberta — storm transported sandstones and conglomerates in shallow marine depositional environments below fairweather wave base. Can. J. Earth Sci., 18: 795—809.

Zenkovich, V.P., 1967. Processes of Coastal Development. Oliver and Boyd, Edinburgh, 738 pp.

SHOREFACE TRANSLATION AND THE HOLOCENE STRATIGRAPHIC RECORD: EXAMPLES FROM NOVA SCOTIA, THE MISSISSIPPI DELTA AND EASTERN AUSTRALIA

RON BOYD and SHEA PENLAND

Geology Department, Dalhousie University, Halifax, N.S. B3H 3J5 (Canada)
Louisiana Geological Survey, Box G, University Station, Baton Rouge, LA 70803 (U.S.A.)

(Received April 11, 1983; revised and accepted October 7, 1983)

ABSTRACT

Boyd, R. and Penland, S., 1984. Shoreface translation and the Holocene stratigraphic record: Examples from Nova Scotia, the Mississippi Delta and eastern Australia. In: B. Greenwood and R.A. Davis, Jr. (Editors), Hydrodynamics and Sedimentation in Wave-Dominated Coastal Environments. Mar. Geol., 60: 391—412.

Classic descriptive models of barrier sedimentation have been developed with data from the Atlantic and Gulf coasts of the United States. These models are dominated by low to moderate rates of relative sea level (RSL) rise and wave energy. Barriers respond by landward recycling of sediment through the mechanism of shoreface retreat. Sedimentation processes on the central coast of New South Wales (N.S.W.), Australia, consist of rapid RSL rise in early Holocene times followed by a stillstand since 6500 B.P. Wave energy is relatively high year-round and sand sources for barrier formation are only found on the inner shelf. Barrier sedimentation on the central coast of N.S.W. exhibits a thick, composite sequence composed of a basal marine transgressive sand overlain by regressive beach and dune facies.
The Louisiana coast surrounding the Mississippi delta is underlain by compacting deltaic muds which generate very rapid rates of RSL rise. The Louisiana coast experiences low wave energy punctuated by high-energy tropical and extra-tropical storm events. Barrier sediments accumulate from the erosion of deltaic headlands and undergo a transformation from subaerial barrier island systems to subaqueous shoals located on the inner shelf. Drumlins experience coastal erosion on the Eastern Shore of Nova Scotia and provide a sediment source for compartmented estuary mouth barriers. An ongoing, moderate rise of RSL results from the passage of a glacial forebulge. Wave energy is intermediate between Louisiana and N.S.W. and displays a seasonal pattern dominated by frequent winter storms. Coastal barrier sedimentation is episodic, consisting of a period of beach ridge progradation followed by barrier destruction and re-establishment further landward.
The three contrasting sedimentary sequences found in examples from Louisiana, N.S.W. and Nova Scotia indicate that presently available sedimentation models from locations such as the middle Atlantic or Texas coasts of the United States may only represent well-documented regional case studies. A true generalised coastal sedimentation model is required which can identify the parameters controlling vertical and horizontal translation of the depositional surface and provide relationships between these parameters which quantitatively predict the genesis, distribution and geometry of coastal sedimentary facies.

0025-3227/84/$03.00   © 1984 Elsevier Science Publishers B.V.

INTRODUCTION

A complex set of parameters control shoreface translation and the generation of coastal facies during sea-level regressions and transgressions. Recent reviews indicate that existing coastal facies models are descriptive and represent generalisations based on a few well-documented case studies (Reading, 1978; Reinson, 1979). The next step in understanding coastal sedimentation is to identify the dominant factors which control the vertical and horizontal translation of the depositional surface. Once identified, relationships between these parameters can be established which are capable of quantitatively predicting the geometry and distribution of coastal stratigraphic sequences in modern and ancient environments.

Initial contributions by Sloss (1962) and Allen (1964) identified the concepts of facies generation during sea-level regression and transgression. Sloss (1962) expressed the shape of a body of sedimentary rocks as a function of the quantity (Q) of material supplied to the depositional site, the rate of subsidence (R) at the site, the rate of sediment dispersal (D) from the site and the nature of the materials supplied (M), or

shape = $f(Q, R, D, M)$

Allen (1964) derived geometrical expressions for the thickness of transgressive and regressive layers under conditions of variable sediment supply (Q) and variation in relative sea level depending on the basinward subsidence from a clastic wedge hinge line. Transgressive units were found to be commonly thinner than regressive units and their thickness decreased with increasing rates of transgression. Swift et al. (1972) grouped Sloss's variables in quasi-quantitative fashion

$$\left(\frac{Q}{E}\right)_G - R = K$$

and replaced D with the concept of energy input (E), indicating that, for the coastline position to remain constant (K), the ratio of sediment input to the energy available for its dispersal must be balanced by an equivalent change in relative sea level. Curray (1964) was able to plot a conceptual graph of erosion and deposition versus change in relative sea level and hence classify eight types of transgression and regression. Examples of variation in Q and R were given but no attempt to quantify the variables was attempted. From these earlier investigations, the two most significant parameters to emerge in controlling coastal sedimentation were the rate of sediment input and the relative sea level. Physical processes responsible for sediment dispersal had not received widespread attention. Swift (1975) restated the problem in the form of the sediment continuity equation

$$\frac{\partial c}{\partial t} + VC = \frac{\rho}{d} \frac{\partial h}{\partial t}$$

where the time rate of change of the coastal sediment—water interface $\partial h/\partial t$ is proportional to the time rate of change of sediment concentration in a unit volume $\partial c/\partial t$ plus the net difference of sediment advected into and out of the unit volume. Here C represents sediment concentration, V a velocity vector, ρ sediment density, and d water depth. Swift (1975) also identified the processes responsible for the sediment flux as: (1) wave driven longshore currents; (2) a coast-parallel wind driven flow; and (3) wind and wave setup driving shore normally oriented upwelling and downwelling events. A more recent coastal sedimentation model (Belknap and Kraft, 1981) examined the preservation potential of transgressive coastal lithosomes and extended the controlling factors to include pre-existing topography, erosion resistance, and tidal range. Belknap and Kraft (1981) concluded that the depth of shoreface erosion was related to the rate of sea-level rise with faster rates of sea-level rise capable of greater preservation — the opposite conclusion to that reached by Allen (1964).

Many studies have identified the characteristics of the shore-parallel and shore-normal processes defined in Swift (1975). Wave-derived longshore transport is widely accepted as controlling shore-parallel sedimentation in the surf zone in response to oblique wave approach (see for example Komar, 1976). Further seaward shore-parallel or shore-oblique transport derived from wind forcing is known to occur across the shoreface (e.g. Boyd, 1980a; Niederoda and Swift, 1981) and the continental shelf (e.g. Sternberg and McManus, 1972) but the sedimentary responses to this wind forcing are less well documented. Bowen (1980) has recently summarised the processes controlling shore-normal sediment dispersal on the shoreface in a revision of Cornaglia's (1898, in Munch-Peterson, 1938) original null point hypothesis. Shore-normal transport here is seen as a balance between the force of gravity and an oscillatory wave force combined with a perturbation component.

To summarise, Sloss's (1962) original variables (Q, R, D, M) seem conceptually appropriate for the formation of coastal stratigraphic models. Each of the sediment supply, RSL and sediment dispersal variables is seen to consist, in turn, of a complex subset of further variables (see for example the list in Kraft, 1978, table 1).

Three regional examples from the central coast of New South Wales, Louisiana and Nova Scotia are presented here to provide a wide spectrum of contrasting case studies. For each example the variables controlling sedimentation are identified. These variables range from: (1) a rising RSL followed by stability under a moderate- to high-energy wave climate in N.S.W.; to (2) very rapid continuing rates of RSL rise, and low wave energy punctuated by storm events on the low-gradient Louisiana coast; to (3) drumlin point sources of sediment supplied to topographically controlled estuaries by rising RSL and moderate levels of wave energy in Nova Scotia. The resulting coastal stratigraphic record is also presented for each example to illustrate how contrasting process interactions can produce a spectrum of barrier sedimentary sequences.

THE CENTRAL COAST OF NEW SOUTH WALES, AUSTRALIA

Relative sea level

RSL in southeast Australia rose from −130 m between 17,000 and 24,000 B.P. (Phipps, 1970) to around −20 m at 9000 B.P. (Fig.1). RSL continued to rise rapidly until reaching and maintaining a stillstand within a meter of its present position by 6000 B.P. (Thom and Chappell, 1975). Using an average distance of 14 km between the 130 and 20 m isobaths on this shelf (Boyd, 1980b), the average rate of RSL rise between 17,000 and 9000 B.P. was 1.40 m per century and the average rate of landward shoreline displacement was 1.75 m yr^{-1}. Between 9000 and 6000 B.P. the average rate of RSL rise was 0.67 m per century and the minimum average rate of shoreline displacement was around 0.35 m yr^{-1}. Due to subsequent shifts in the shoreface equilibrium profile and corresponding uncertainty regarding the original location of the −20 m shoreline, rates of shoreline displacement between 9000 and 6000 B.P. are likely to have been faster by a factor of 2−3.

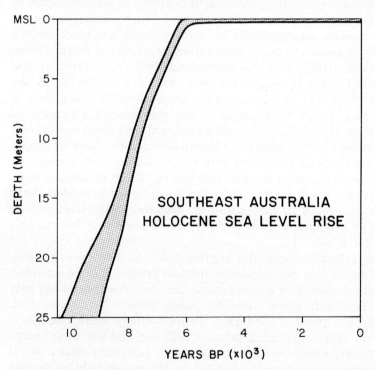

Fig.1. Holocene sea level relative to south east Australia. RSL rise was rapid prior to establishing a stillstand within 1 m of its present position by 6300 B.P. (redrawn from Thom and Chappell, 1975).

Sediment input

Since at least 9000 B.P., there appear to have been no external sediment sources for most central N.S.W. coastal compartments except for onshore transport from the inner continental shelf located directly seaward. This situation arises from the inhibition of longshore sediment transport between deeply embayed coastal compartments by shore-normal submarine bedrock ridges and the trapping of fluvial sand input at the upstream estuary margins of drowned river valleys. Numerous studies along the coastline north and south from Sydney have established the presence of continued bedrock ridges extending to the 60 m isobath from seismic data (MWSDB, 1976), bathymetric profiles and Scuba surveys (Reffell, 1978; Hann, 1979; Boyd, 1980b). Water depths of over 20 m are commonly encountered on outcropping bedrock less than 200 m seaward of all headlands separating major coastal compartments in the Sydney region. Adjacent coastal compartments exhibit wide variations in sediment lithology as demonstrated by mean grainsize statistics and calcium-carbonate content. (Hann, 1979; Boyd, 1980b). Only two central N.S.W. coastal rivers have succeeded in establishing marine Holocene deltas. Of these the Hunter River does not supply sand to the present coastline (Roy, 1975) while a minor sand source from the Shoalhaven River (Wright, 1970; Boyd, 1980b) is contained within the Shoalhaven Bight compartment. Headlands along the central N.S.W. coast are mainly Carboniferous—Triassic massive sandstones and volcanics (Packham, 1969) which contribute an insignificant amount of sediment to the coastal sand budget.

Coastal processes

A high-energy wave climate and shore normal wind-forced bottom flow are the dominant processes influencing sediment transport on the central N.S.W. coast. The highest probability occurrence of wave height-period combinations is for wave periods of 7—9 s and deep-water wave heights of 1—1.5 m from the southeast direction. Wave height exceeds 1 m around 80% of the time and exceeds 4 m for 1% of the time (Lawson and Abernethy, 1975). However, the modal wave, for which the product of wave power × frequency (see Fig. 2) is at a maximum is for waves of 2—2.5 m and periods of 8—10 s (Wright et al., 1980) whereas the average wavepower is 2.7×10^4 W m^{-1}. Because of the steep, narrow continental shelf only 4% of the deep-water incident wave power is dissipated before reaching the surf zone of the central N.S.W. coast in contrast to 29 and 58% for comparable east coast swell environments at Sergipe, Brazil and Cape Hatteras, North Carolina (Wright, 1976). Boyd (1980a) found that for wind velocities in excess of 25 km h^{-1}, wind orientation to the coastline controlled the near-bottom flow field and the resulting sediment transport. Onshore winds dominantly produced offshore bottom flows and offshore winds produced onshore flows at the bed. The sandy barrier shoreface profile in this region (No. 3, Fig.3)

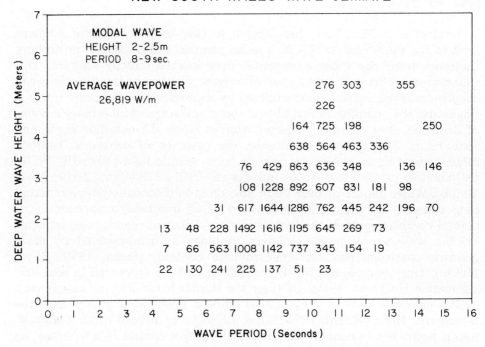

Fig.2. A wave-climate matrix in which individual data values represent the product of wavepower and probability of occurrence expressed in units of W m^{-1} for the full range of wave height and period combinations on the central N.S.W. coast.

is concave up and decreases exponentially between depths of around 3 and 30 m, over a distance of less than 2 km. Gradients average 1:40—1:70 and a corresponding decrease in mean sediment grainsize from 1.4 to 2.0 ϕ parallels the increase in depth from the upper to lower shoreface.

Depositional history

Sedimentation along the central N.S.W. shelf is characterised by the development of sandy bay barriers composed of beach-ridge plains and dune complexes (Thom, 1965; Thom et al., 1978; Roy et al., 1980). Construction of these barriers began around 6000—6500 B.P. and in most cases was over 75% complete by 3000 B.P. (Thom et al., 1981). Barriers are typically 20—30 m thick and occupy coastal compartments ranging from 1 to 30 km long. Barrier widths range from 1 to 3.3 km. Shoreline progradation slowed during formation, from initial values of around 1.2—1.45 m yr^{-1} to later values of 0.5—0.25 m yr^{-1} (Thom et al., 1981).

The pattern of barrier genesis and evolution on this coast appears to have begun with establishment of the present RSL around 6000—6500 B.P. During the following 3000 yrs sediment derived from the adjacent shelf and

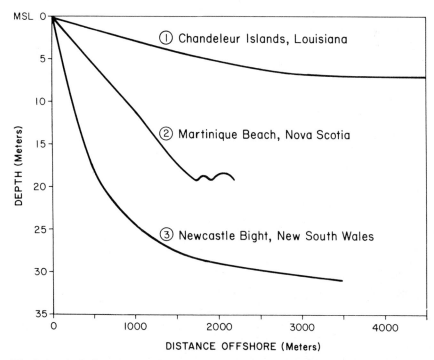

Fig. 3. Typical shoreface bathymetric profiles from. *1*: Chandeleur Islands, Louisiana; *2*: Martinique Beach, Nova Scotia — note the boulder retreat shoal at the seaward termination of the profile; and *3*: Newcastle Bight, N.S.W. The increase in profile gradient from top to bottom is paralleled by an increase in wave power.

shoreface was transported landward and accumulated for a finite period as prograding bay barriers in estuary mouths. Since 3000 B.P. most barriers have remained stable or undergone net erosion. The implication of this evolutionary sequence is that landward rates of shoreline translation exceeded the capabilities of landward sediment transfer in the period from 9000 to 6000 B.P. Following the establishment of stable RSL at 6000 B.P., wind- and wave-dominated sediment dispersal transformed a shallow disequilibrium shoreface profile into a relatively steep, narrow profile which achieved equilibrium with the high-energy wave climate. Excess sediment accumulated as progradational shoreline deposits. Equilibrium was apparently established by 3000 B.P. and major shoreline progradation on the central N.S.W. coast ceased. A conceptual model (Fig. 4) describing this barrier evolution sequence for N.S.W. has been developed by Roy and Thom (1981).

MISSISSIPPI DELTA COASTAL BARRIER SYSTEMS, LOUISIANA U.S.A.

Relative sea level

RSL in the Gulf of Mexico is inferred to have risen from depths of -130 m, 15,000 B.P. (Curray, 1960) to around -9 m by 8000 B.P. A subsequent rate

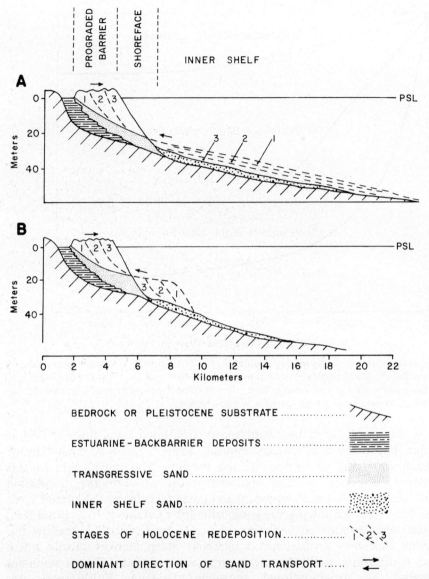

Fig.4. A conceptual model for barrier formation on the N.S.W. coast (from Roy and Thom, 1981). Two alternative mechanisms are identified to produce stages of shoreline progradation by landward sediment transfer from the shoreface.

of RSL rise around 20 cm per century then occurred (Fig.5) until eustatic sea level reached its present position around 3650 B.P. (Coleman and Smith, 1964). Since this time RSL along the Mississippi Delta coastline of Louisiana has continued to rise in response to subsidence of the land surface. Subsidence is primarily caused by the compactional subsidence of deltaic deposits and varies as a function of sediment thickness and age. Present rates range

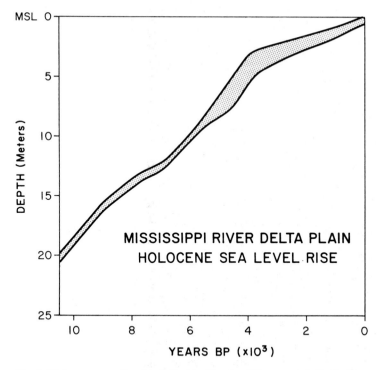

Fig.5. Holocene sea level relative to the Louisiana coast. Eustatic components controlled the rise in RSL until 3600 B.P. since when compactional subsidence has predominated (based on Coleman and Smith, 1964, and Frazier, 1974).

from 7.5 cm per century for old, shallow-water delta complexes (Coleman and Smith, 1964), to 60 cm per century for intermediate age deltas, to 600 m per century for the presently active deep-water Balize Delta (Kolb and Van Lopik, 1958). Rates of shoreline retreat for abandoned delta complexes exhibit a range from less than 1 to over 15 m yr^{-1} with extreme rates of over 50 m yr^{-1} recorded in 1979 (Penland and Boyd, 1981), reflecting the extremely low-gradient characteristic of deltaic coastlines.

Sediment input

Sediment has been supplied to the southeast Louisiana coast by the Mississippi River in a series of spatially and temporally varying deltaic depocentres (Frazier, 1967). The only Holocene sand-sized sediment suitable for coastal barrier formation is supplied from digitate or arcuate distributary mouth bars (Fisk, 1955). Following upstream diversion, delta complexes become abandoned and ongoing subsidence induces sea-level transgression. During transgression the only source of sand for barrier formation is from reworking (see Fig. 6) of distributary mouth bars and pre-existing shoreline sand bodies by shoreface retreat (Swift, 1975). As subsidence and transgression continue, shoreface retreat becomes progressively less effective until

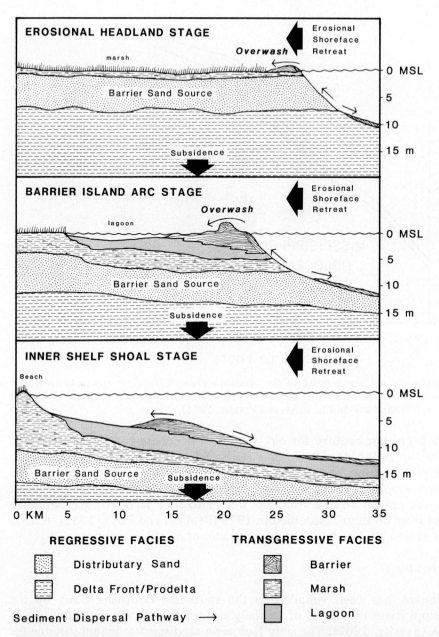

Fig.6. Sediment input to coastal barrier systems of the Mississippi Delta is controlled by the processes of subsidence and erosional shoreface retreat. The original sand source lies in distributary and pre-existing barrier deposits. During transgression this source is reworked to form new barrier deposits by erosional shoreface retreat until ongoing compactional subsidence removes the sand source below the base of the shoreface resulting in eventual barrier submergence.

the base of the shoreface no longer penetrates to the underlying deltaic sand source. Barriers are then decoupled from their only sediment input, and subaerial barrier volume continuously decreases as sand is lost to sediment sinks located in spits, tidal inlets, washovers and the inner shelf (Penland and Boyd, 1982).

Coastal processes

The Louisiana Gulf Coast is a storm-dominated environment and energy levels resulting from wind and wave processes are low except for the winter passage of cold front systems and the summer occurrence of hurricanes and tropical storms. Modal wave conditions in deep water offshore from the Mississippi Delta are characterised by wave heights of 1 m and wave periods of 5—6 s (Fig.7). On average, modal wave conditions occur for 4% of the time (Bretschneider and Gaul, 1956) and average deep-water wave power is only 1.8×10^3 W m^{-1} Wright et al. (1974) indicated that over 99% of the deep-water wave power offshore from the Mississippi Delta is dissipated before reaching the shoreline. Locally generated high-energy storm wave conditions are usually accompanied by strong onshore winds. Murray (1970,

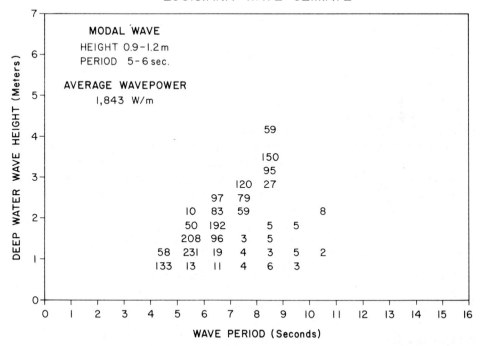

Fig.7. A wave climate matrix for Mississippi Delta barrier coasts (same method as Fig.2). Low probabilities of occurrence result from the variable directions of locally generated waves during hurricanes and winter fronts. Values shown represent waves propagating onshore only.

1972) recorded mainly offshore and longshore-directed bottom currents both during the passage of a moderately energetic front in 20 m depth east of the delta and during Hurricane Camille. Shelf sediment surveys conducted by Krawiec (1966) identified the deposition of a large sand sheet seaward from Grand Isle between the 5 and 12 m isobaths supplied by offshore transport during Hurricane Betsy. In contrast, wave-refraction analyses for the Louisiana coast (Penland and Boyd, 1982) found significant sediment transport in non-storm conditions was restricted to the upper shoreface landward of the 5 m isobath. The base of the shoreface lies at a depth of 5—8 m seaward from most Louisiana barrier islands and exhibits a flattened gradient of 1:3000 (Fig.3). Mean sediment size decreases from 2.6 ϕ at the shoreline to 3.9 ϕ at the base of the shoreface before passing into prodelta or shelf muds. Tides in the region are mixed and predominantly diurnal with a microtidal range of 30 cm. Tidal inlets develop as the delta lobe subsides, exchanging the gradually increasing tidal prism volumes between subsiding intra-deltaic lagoons and the Gulf of Mexico. Storm surges frequently accompany tropical cyclones and winter fronts, which generate storm surge elevations of 0.5 to over 8 m with concurrent intense overwash events (Boyd and Penland, 1981). Dominant wave approach is from the southeast but the local wind wave variability combined with a complex coastal orientation results in uninterrupted longshore transport *within* most barrier systems.

Depositional history

Abandoned delta complexes undergoing transgression on the Louisiana coast are characterised by sandy barrier system genesis and subsequent evolution (Fig.8) through a well-defined three-stage cycle (Penland et al., 1981). Distributary sand bodies reworked by shoreface retreat provide sediment from an erosional headland source for longshore transport into accumulating flanking spits and barrier islands (Stage 1). Subsidence of the delta plain behind the barriers forms an intra-deltaic lagoon separating the retreating transgressive barrier island arc (Stage 2) from the mainland. With continued transgression and loss of the deltaic sand source, barrier migration cannot keep pace with subsidence and lagoon shoreline retreat. The resulting subaqueous inner shelf shoal forms the final phase, Stage 3 of the evolutionary model. The presence of Mississippi Delta complexes of varying Holocene age has produced barrier systems in all stages of evolution on the present Louisiana coast. Typical time scales for the transition from one evolutionary stage to another lie between 600 and 1600 yrs.

EASTERN SHORE OF NOVA SCOTIA, CANADA

Relative sea level

Holocene RSL in Atlantic Canada, has been dominated by advance and ablation of continental ice masses. Large volumes of water are withdrawn

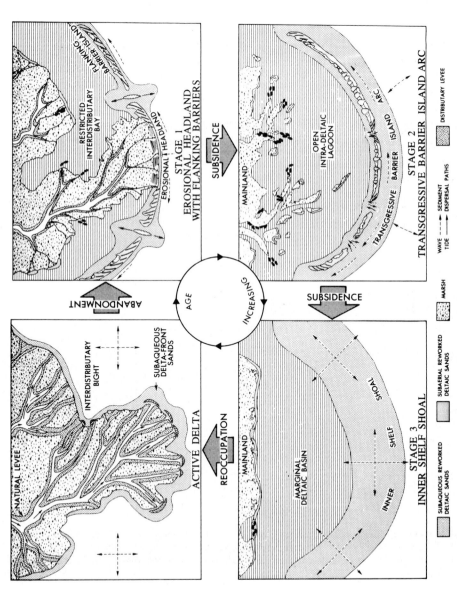

Fig.8. The transgressive depositional history of an abandoned Mississippi River Delta can be explained in three distinct evolutionary stages generated by abandonment, subsidence and marine reworking. An abandoned delta complex is successively transformed from an erosional headland with flanking barriers (Stage 1) to a transgressive barrier island arc (Stage 2) and finally to an inner shelf shoal (Stage 3).

from ocean basins as ice thickness increases on land, causing a fall in RSL. Areas under thick continental ice accumulation also experience an isostatic sinking due to the extra crustal loading. Areas such as Nova Scotia at the ice margin experience less RSL fall due to the development of a glacial forebulge (Quinlan and Beaumont, 1981). Following ice recession the glacial forebulge migrates inward through the marginal ice zone and initially causes a fall in RSL followed by a rise after passage of the forebulge crest. A glacial advance reached its maximum in Nova Scotia between 32,000 and 12,000 yrs B.P. During subsequent forebulge migration RSL on the Eastern Shore fell to a minimum of at least −27 m around 8000—12,000 yrs B.P. and since that time has risen continuously (Fig. 9) at an average rate of 35 cm per century (Scott and Medioli, 1982). Rates of shoreline retreat associated with this transgression have averaged around 1 m yr^{-1} but, as detailed below, temporary reversals and wide fluctuations have been common.

Sediment input

Sea-level transgression on the Eastern Shore of Nova Scotia has confined post-glacial, fluvial sediment supply to freshwater lakes and the upper

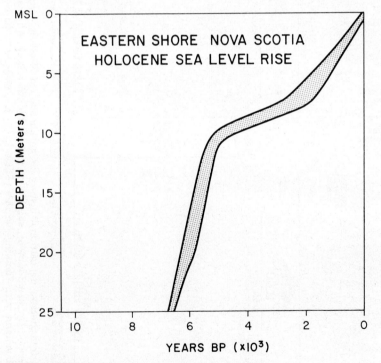

Fig.9. Holocene sea level relative to the Eastern Shore, Nova Scotia. A regression prior to 7000 B.P. was followed by an ongoing transgression resulting in rates of RSL rise of around 35 cm per century. Unpublished and published data courtesy D.B. Scott (Scott and Medioli, 1982).

reaches of glacial estuaries. Longshore sediment transport is also ineffective on this coast because of the highly irregular shoreline orientation and the recessed nature of most estuary mouth barriers. Sediment input to the coastal zone is derived almost exclusively from marine erosion of unconsolidated glacial shoreline deposits.

Throughout the Eastern Shore of Nova Scotia resistant bedrock lithologies of the Cambro-Ordovician Meguma Group are covered by 3—4 m thick accumulations of glacial till (Stea and Fowler, 1979) except where drumlin fields occur. Drumlins consist of glacial till and glaciofluvial sediments composed of a wide range of grain size. Drumlin fields often contain several hundred individuals with average sediment volumes in the order of $7.5 \times 10^6 m^3$. Major barrier systems on the Eastern Shore are restricted to locations where the transgressing shoreline intersects a drumlin field. Cliff retreat rates along exposed drumlins average 1—5 m yr^{-1}, supplying sediment volume at the rates of between 5×10^2 and $7.5 \times 10^4 m^3$ yr^{-1}. Sediment supply from individual drumlins displays a skewed Gaussian distribution with typical time scales of several hundred years for the total contribution of any one drumlin.

Coastal processes

The Eastern Shore of Nova Scotia is located in an east coast swell environment (Davies, 1980) which experiences frequent winter storm conditions. Data collected during the Canadian Marine Environmental Data Service's Wave climate Study for locations on the Nova Scotian Atlantic coast show that the most frequently occurring wave conditions are for 0.6—0.9 m height and 8—9 s period. The product of frequency and wave power is greatest for modal wave conditions of 1.2—1.5 m height and 9—10 s waves whereas the average year-round wave power is 1.68×10^4 W m^{-1} (Fig. 10). Winter storms are frequently accompanied by strong onshore winds but no detailed data on the near-bottom flow field have been collected.

Coastal sand accumulation on the Eastern Shore generally forms highly embayed estuary mouth barriers with little sediment exchanged alongshore between individual compartments. The location of barriers across the mouths of large estuaries and the mesotidal (around 2 m) range results in the development of tidal inlets and extensive flood tide delta systems which often contain more sediment than the barrier itself.

The shoreface seaward of Eastern Shore barriers is frequently terminated by till or bedrock, creating a ponded sediment accumulation (see Fig.3). The base of the shoreface commonly is located at the 15—20 m isobath less than 2000 m seaward of the shoreline, creating a gradient of 1:100 (Fig.3). Sediment size ranges from fine to medium-grained sand (1—2.5 ϕ) at the beach to fine-grained sand and some silt (mean size 2.8—3.5 ϕ) at —20 m on the shelf.

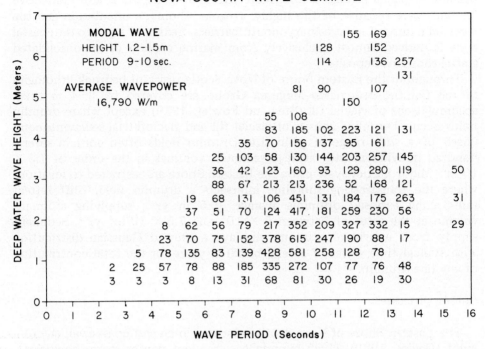

Fig.10. Wave-climate matrix for the Atlantic coast of Nova Scotia. Data values represent the product of wavepower and its probability of occurrence expressed in units of W m^{-1}. Nova Scotian wavepower values are intermediate between those of N.S.W. and the Mississippi Delta.

Depositional history

Nova Scotia coastal sedimentation has been dominated by repeated glaciations (Fig.11, Stage 1). The resulting coastal environment consists of thin lodgement — ablation till and thicker drumlin till overlying a resistant Meguma Group bedrock erosion surface of variable relief.

Sea-level transgression following passage of the glacial forebulge crest resulted in exposure and reworking of till, formation of estuaries in basement depressions (Fig.11, Stage 2) and concentration of sand and gravel-sized glacial sediment in estuary mouth barriers (Fig.11, Stage 3). The subsequent depositional history of the Eastern Shore has been a cyclic repetition of barrier retreat, destruction and re-establishment further landward (Fig.11, Stages 4—6).

This Eastern Shore example shows the input of a limited, spatially variable sediment supply to an ongoing regional transgression. Sediment temporarily accumulates in compartment barriers while supply exceeds RSL rise. Following depletion of the sediment source, barrier systems are destroyed and migrate landward. Seismic surveys of the inner shelf reveal only a thin (1—2 m) sand veneer overlying till at the location of former barriers. This

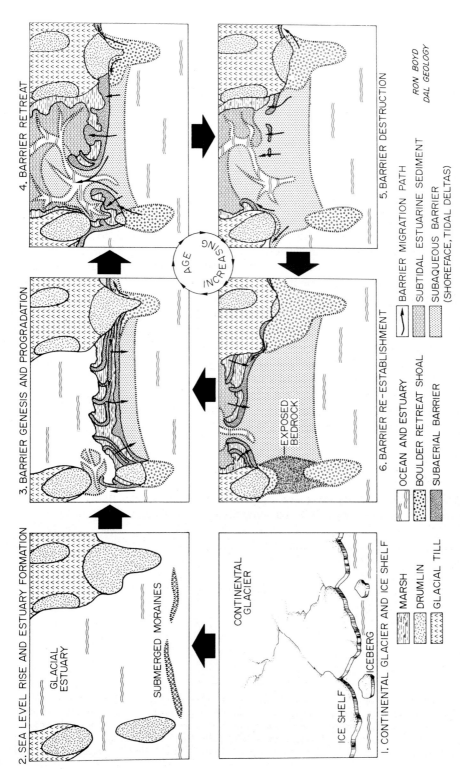

Fig.11. Coastal sedimentation in Nova Scotia can be considered as a 6-stage transgressive sequence of barrier genesis, destruction and re-establishment dominated by localised glacial sediment sources, rising RSL and tidal inlet processes. Initial barrier genesis (Stages 1 and 2) was related to advance and ablation of continental ice sheets. Subsequent evolution consists of a cyclic repetition of Stages 3–6.

indicates conservation of barrier sediment during transgression up the estuary. Landward sediment transfer appears to be accomplished primarily by transport through tidal inlets into flood tide deltas, and transport across the barrier by overwash and aeolian processes. While under the influence of wave action on the upper shoreface, the submerged barrier moves landward as a series of subtidal and intertidal shoals until encountering a new set of headland anchor points which provide the stability for barrier re-establishment.

SUMMARY AND CONCLUSIONS

In qualitative terms, the problem of coastal sedimentation during transgression and regression is determined by the parameters which control the shape of the shoreface profile and the factors which result in profile translation. Vertical profile translation (v) is controlled by RSL. The vertical profile translation may be taken as the vertical translation of an origin fixed at the intersection of the mean sea level and the regional land surface. The vertical position of the origin is determined by RSL since marine processes are incapable of elevating sediments far above high tide level.

The horizontal profile translation may result from net vertical changes within the profile and both net horizontal changes *of* and *within* the profile. Profile shape and its horizontal translation (u) is controlled by regional gradient (θ) and the equilibrium relationship between sediment flux (Q) across the profile, a wave climate parameter such as $H(\omega T)^{-1}$ and the inner shelf oceanography O_{IS} (primarily a result of wind and tidal forcing). This may also be stated as (see Fig.12):

$$v = f(\text{RSL})$$

$$u = f[Q, H(\omega T)^{-1}, O_{IS}, \theta]$$

Wright et al. (1982) have achieved considerable success using the dimensionless parameter $H(\omega T)^{-1}$ to characterise surf-zone morphodynamics.

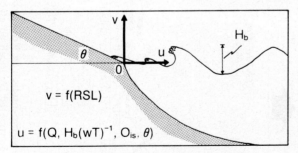

Fig.12. Schematic representation of shoreface profile translation. The vertical component of translation (v) from an origin at the shoreline (O) is seen to be a function of relative sea level (RSL). The horizontal component of translation (u) is a function of sediment flux (Q) a parameter relating wave effects (Hb, local wave height; T, wave period) and sediment response (w, sediment settling velocity), inner shelf oceanography (O_{IS}) and regional coastal gradient θ.

Here H is local wave height, T wave period and ω is sediment settling velocity. It is felt that a parameter like $H(\omega T)^{-1}$ may also be suitable for shoreface profile characterisation since H and T are dominant parameters determining wave mass transport effects (e.g. Longuet-Higgins, 1953) while ω and T strongly influence the location of initial sediment motion and hence the base of the shoreface (see Komar and Miller, 1973; Hallermeier, 1981).

Calculations of the $H(\omega T)^{-1}$ parameter result in low values of 3—6 for central N.S.W., intermediate values of 6—9 for Nova Scotia, and high values of 20—25 for Louisiana. These values of $H(\omega T)^{-1}$ also correspond to the sequence of decreasing shoreface gradient and decreasing depth to the base of the shoreface observed (Fig.3) from N.S.W. to Louisiana. Since the maximum potential barrier thickness and the critical depth of shoreface erosion both depend on the shoreface gradient and the depth at its base, the $H(\omega T)^{-1}$ parameter shows a potential for predicting barrier facies geometry.

Using the three contrasting regional examples presented here a sequence of sedimentary processes and their resulting barrier stratigraphy is also apparent. In Nova Scotia a continuing rise in RSL results in barrier progradation followed by destruction. During the destruction phase estuarine headlands act as effective barriers to longshore transport and sediment is moved landward by aeolian, overwash and tidal inlet processes. Sediment remaining on the shoreface is still within the range of wave-induced transport and most appears to move landward, leaving behind a thin shelf sand veneer. Transgressive sedimentation on the Eastern Shore of Nova Scotia thus illustrates a form of Swift's (1975) shoreface retreat concept.

On the central coast of N.S.W. rapid RSL rise occurred prior to the development of a stillstand after 6500 B.P. Sediment supplied from the adjacent disequilibrium shelf formed regressive barrier deposits. Barrier progradation slowed or ceased as the sediment supply diminished and the resulting barrier sequence is one of a thin basal transgressive sand sheet overlain by regressive beach ridge and dune complexes. Sediment dispersal on the central N.S.W. coast seems to follow a pattern of rapid RSL rise during which barrier sediments are retained on the inner shelf following the model of Belknap and Kraft (1981). However, during stillstand, abundant wave energy is available for landward transport of sediment from the inner shelf to prograding barriers. This overall pattern of barrier sedimentation on the central N.S.W. coast is therefore a form of punctuated, stepwise shoreface retreat.

In Louisiana, extremely rapid rates of RSL rise generate barriers during reworking of underlying deltaic sand sources. As rapid transgression continues over a very low regional gradient, the sandy barrier systems are transformed into subaqueous shoals. Low wave energy in the northern Gulf of Mexico is incapable of sediment transfer at a rate comparable to that of landward shoreline translation. The resulting barrier stratigraphy is a dispersed shelf sand lens overlying lagoonal muds (see, for example, Frazier, 1967). Coastal sedimentation processes in Louisiana therefore incorporate reworked barrier sandbodies into the shelf stratigraphic record.

The presentation here of three contrasting case studies of coastal sedimentation highlights some problems arising from the construction of coastal facies models. Sufficient variability exists within the three studies to suggest they are not fully described by presently available concepts of barrier genesis and evolution. Future coastal sedimentation models require a more quantitative approach to firmly establish the relationships between dominant sedimentary processes within a region and the resulting coastal stratigraphy.

ACKNOWLEDGEMENTS

Funding for the present study was provided through Canadian Natural Sciences and Engineering Research Council Grant A8425, Canadian Department of Energy, Mines and Resources Research Agreement 251 and from the Coastal and Fluvial Geology Program of the Louisiana Geological Survey. Useful data and discussions for the Nova Scotia case study were provided by D.B. Scott and A.J. Bowen.

REFERENCES

Allen, P., 1964. Sedimentological models. J. Sediment. Petrol., 34: 289—293.
Belknap, D.F. and Kraft, J.C., 1981. Preservation potential of transgressive coastal lithosomes on the U.S. Atlantic shelf. Mar. Geol., 42: 429—442.
Bowen, A.J., 1980. Simple models of nearshore sedimentation; beach profiles and longshore bars. In: S.B. McCann (Editor), The Coastline of Canada. Geol. Surv. Can., Pap. 80-10.
Boyd, R., 1980a. Sediment dispersal on the N.S.W. continental shelf. Proc. 17th Int. Conf. Coastal Eng., Am. Soc. Civ. Eng., pp.1364—1381.
Boyd, R., 1980b. Sediment dispersal on the central New South Wales continental shelf. Geology Department, University of Sydney, unpubl. Ph.D. thesis, 319 pp.
Boyd, R. and Penland, S., 1981. Washover of deltaic barriers on the Louisiana coast. Trans. Gulf Coast Assoc. Geol. Socs., 31: 243—248.
Bretscheider, C.L. and Gaul, R.D., 1956. Wave statistics for the Gulf of Mexico off Burrwood, Louisiana. U.S. Army Corps of Engineers, Tech. Mem. 87.
Coleman, J.M. and Smith, W.G., 1964. Late recent rise of sea level. Geol. Soc. Am. Bull., 75: 833—840.
Curray, J.R., 1960. Sediments and history of the Holocene transgression, continental shelf, Northern Gulf of Mexico. In: F.P. Shepard (Editor), Recent Sediments, Northwest Gulf of Mexico. Am. Assoc. Pet. Geol., Tulsa, Okla., pp. 221—266.
Curray, J.R., 1964. Transgressions and regressions. In: R.L. Miller (Editor), Papers in Marine Geology (Shepard Commemorative Volume). McMillan, New York, N.Y., pp.175—203.
Davies, J.L., 1980. Geographical Variation in Coastal Development (2nd ed.). Hafner, New York, N.Y., 204 pp.
Fischer, A.G., 1961. Stratigraphic record of transgressing seas in the light of sedimentation on the Atlantic coast of New Jersey. Bull. Am. Assoc. Pet. Geol., 45: 1656—1666.
Fisk, H.N., 1955. Sand facies of Recent Mississippi delta deposits. Fourth World Petroleum Congress, Rome, Section 1-C: 377—398.
Frazier, D.E., 1967. Recent deltaic deposits of the Mississippi River: their development and chronology. Trans. Gulf Coast Assoc. Geol. Socs., 27: 287—315.
Hallermeier, R.J., 1981. Seaward limit of significant sand transport by waves: An annual zonation for seasonal profiles. Coastal Engineering Research Centre, U.S. Army Corps of Engineers, Technical Aid 81-2.

Hann, J., 1979. Offshore bathymetry and sediment patterns off Sydney beaches. School of Earth Science, Macquarie University, unpubl. B.S. (Hons.) thesis.

Hoyt, J.H., 1967. Barrier island formation. Geol. Soc. Am. Bull., 78: 1125—1136.

Kolb, C.R. and Van Lopik, J.R., 1958. Geology of the Mississippi River Deltaic Plain, Southeast Louisiana. Tech. Rep. 3-483. U.S. Army Corps of Engineers Waterways Experiment Station, Vicksburg, La., 120 pp.

Komar, P.D., 1976. Beach Processes and Sedimentation. Prentice-Hall, Englewood Cliffs, N.J., 429 pp.

Komar, P.D. and Miller M.C., 1973. The threshold of sediment movement under oscillatory water waves. J. Sediment. Petrol., 43: 1101—1110.

Kraft, J.C., 1978. Coastal stratigraphic sequences. In: R.A. Davis (Editor), Coastal Sedimentary Environments. Springer, New York, N.Y., 420 pp.

Krawiec, W., 1966. Recent sediment of the Louisiana inner continental shelf. Rice University, unpubl. Ph.D. thesis, 50 pp.

Lawson, N.V. and Abernethy, C.L., 1975. Long term wave statistics off Botany Bay. Second Australian Conf. on Coastal and Ocean Engineering, The Institute of Engineers, 75/2: 167—176.

Longuet-Higgins, M.S., 1953. Mass transport in water waves. Philos. Trans. R. Soc. London, Ser. A, 245: 535—581.

Munch-Peterson, 1938. Littoral drift formula. U.S. Army Corps Eng., Beach Erosion Board Bull. 4(4) (1950): 1—36.

Murray, S.P., 1970. Bottom currents near the coast during hurricane Camille. J. Geophys. Res., 75(24): 4579—4582.

Murray, S.P., 1972. Observations on wind, tidal and density driven currents in the vicinity of the Mississippi River delta. In: D.J.P. Swift, D.B. Duane and O.H. Pilkey (Editors), Shelf Sediment Transport: Process and Pattern. Dowden, Hutchinson and Ross, Stroudsburg, Pa.

M.W.S.D.B. (Metropolitan Water Sewerage and Drainage Board), 1976. Report on submarine outfall studies. M.S.W.D.B., Sydney, N.S.W., 288 pp.

Niedoroda, A.W. and Swift, D.J.P., 1981. Maintenance of the shoreface by wave orbital currents and mean flow: observations from the Long Island coast. Geophys. Res. Lett., 8(4): 337—340.

Packham, G.H., 1969. The geology of New South Wales. J. Geol. Soc. Aust., 16(1): 654 pp.

Penland, S. and Boyd, R., 1981. Shoreline changes on the Louisiana barrier coast. IEEE, Oceans '81: 209—219.

Penland, S.P. and Boyd, R., 1982. Assessment of geological and human factors responsible for Louisiana coastal barrier erosion. In: D.F. Boesch (Editor), Proceedings of the Conference on Coastal Erosion and Wetland Modification in Louisiana: Causes, Consequences and Options. U.S. Fish and Wildlife Services, Biological Services Program, Washington, D.C., FWS/OBS — 82/59, pp. 14—39.

Penland, S., Boyd, R., Nummedal, D. and Roberts, H.H., 1981. Deltaic barrier development on the Louisiana coast. Trans. Gulf Coast Assoc. Geol. Socs., 31 (Suppl.): 471—476.

Phipps, C.V.G., 1970. Dating of eustatic events from cores taken in the Gulf of Carpentaria, and samples from the New South Wales continental shelf. Aust. J. Sci., 32: 329—330.

Quinlan, G. and Beaumont, C., 1981. A comparison of observed and theoretical postglacial sea levels in Atlantic Canada. Can. J. Earth Sci., 18: 1146—1163.

Rampino, M.R. and Sanders, J.E., 1980. Holocene transgression in south central Long Island. J. Sediment. Petrol., 50: 1063—1079.

Rampino, M.R. and Sanders, J.E., 1981. Evolution of the barrier islands of southern Long Island, New York. Sedimentology, 28: 37—48.

Reading, H.G., 1978. Sedimentary Environments and Facies. Elsevier, New York, N.Y., 557 pp.

Reffell, G., 1978. Descriptive analysis of the subaqueous extensions of subaerial rock platforms. Geography Department, University of Sydney, unpubl. B.A. (Hons.) thesis.

Reinson, G.E., 1979. Facies models 6: barrier island systems. In: R.G. Walker (Editor), Facies Models. Geosci. Can., Reprint Ser. 1, Toronto, Ont., pp.57—75.

Roy, P.S., 1975. Does the Hunter River supply sand to the New South Wales coast today? J. Proc. R. Soc. N.S.W., 110: 17—25.

Roy, P.S. and Thom, B.G., 1981. Late Quaternary marine deposition in New South Wales and southern Queensland — an evolutionary model. J. Geol. Soc. Aust., 28: 471—489.

Roy, P.S., Thom, B.G. and Wright, L.D., 1980. Holocene sequences on an embayed high energy coast: an evolutionary model. Sediment. Geol., 26: 1—19.

Scott, D.B. and Medioli, F.S., 1982. Micropaleontological documentation for early Holocene fall of relative sea level on the Atlantic coast of Nova Scotia. Geology, 10: 278—281.

Sloss, L.L., 1962. Stratigraphic models in exploration. J. Sediment. Petrol., 32: 415—422.

Stea, R.R. and Fowler, J.H., 1979. Minor and trace element variations in Wisconsinan tills, Eastern Shore, Nova Scotia. Nova Scotia Department of Mines and Energy, Pap. 79-4, 30 pp.

Sternberg, R.W. and McManus, D.A., 1972. Implications of sediment dispersal from longterm, bottom current measurements on the continental shelf off Washington. In: D.J.P. Swift, D.B. Duane and O.H. Pilkey (Editors), Shelf Sediment Transport: Process and Pattern. Dowden, Hutchinson and Ross, Stroudsburg, Pa., pp.181—194.

Swift, D.J.P., 1975. Barrier island genesis: evidence from the central Atlantic shelf, eastern U.S.A. Sediment. Geol., 14: 1—43.

Swift, D.J.P. and Moslow, T.F., 1982. Holocene transgression in South Central Long Island, New York — Discussion. J. Sediment. Petrol., 52(3): 1014—1019.

Swift, D.J.P., Kofoed, J.W., Saulsbury, P.J. and Sears, P., 1972. Holocene evolution of the shelf surface, certral and southern Atlantic shelf of North America. In: D.J.P. Swift, D.B. Duane and O.H. Pilkey (Editors), Shelf Sediment Transport: Process and Pattern. Dowden, Hutchinson and Ross, Stroudsburg, Pa., pp.499—574.

Thom, B.G., 1965. Late Quaternary coastal morphology of the Port Stephens — Myall Lakes area, N.S.W. J. Proc. R. Soc. N.S.W., 98: 23—36.

Thom, B.G. and Chappell, J., 1975. Holocene sea levels relative to Australia. Search, 6: 90—93.

Thom, B.G., Polach, H. and Bowman, G., 1978. Holocene age structure of coastal sand barriers in New South Wales, Australia. Report of Department of Geography, Faculty of Military Studies, University of N.S.W., Duntroon, N.S.W., 86 pp.

Thom, B.G., Bowman, G.M. and Roy, P.S., 1981. Late Quaternary evolution of coastal sand barriers, Port Stephens — Myall Lakes area, central New South Wales, Australia. Quat. Res., 15: 345—364.

Wright, L.D., 1970. The influence of sediment availability on patterns of beach ridge development in the vicinity of the Shoalhaven River delta, N.S.W. Aust. Geogr., 11: 336—348.

Wright, L.D., 1976. Nearshore wave power dissipation and the coastal energy regime of the Sydney—Jervis Bay region. New South Wales: a comparison. Aust. J. Mar. Freshwater Res., 27: 633—640.

Wright, L.D., Coleman, J.M. and Erickson, M.W., 1974. Analysis of major river systems and their deltas: morphologic and process comparisons. Louisiana State University, Coastal Studies Institute, Tech. Rep. 156.

Wright, L.D., Coffey, F.C. and Cowell, P.J., 1980. Nearshore Oceanography and Morphodynamics of the Broken Bay — Palm Beach Region, N.S.W.: Implications for Offshore Dredging. Coastal Studies Unit, Department of Geography University of Sydney, Tech. Rep. 80/1, 210 pp.

Wright, L.D., Short, A.D. and Nielsen, P., 1982. Morphodynamics of high energy beaches and surf zones: a brief synthesis. Department of Geography, University of Sydney, Coastal Studies Unit, Tech. Rep. 82/5.

HOLOCENE SEDIMENTATION OF A WAVE-DOMINATED BARRIER-ISLAND SHORELINE: CAPE LOOKOUT, NORTH CAROLINA

S. DUNCAN HERON, Jr., THOMAS F. MOSLOW[1], WILLIAM M. BERELSON[2], JOHN R. HERBERT[3], GEORGE A. STEELE III[4] and KENNETH R. SUSMAN[5]

Department of Geology, Duke University, Durham, NC 27708 (U.S.A.)

(Received February 7, 1983; revised and accepted July 14, 1983)

ABSTRACT

Heron Jr., S.D., Moslow, T.F., Berelson, W.M., Herbert, J.R., Steele III, G.A. and Susman, K.R., 1984. Holocene sedimentation of a wave-dominated barrier-island shoreline: Cape Lookout, North Carolina. In: B. Greenwood and R.A. Davis, Jr. (Editors), Hydrodynamics and Sedimentation in Wave-Dominated Coastal Environments. Mar. Geol., 60: 413—434.

The sedimentary record of 130 km of microtidal (0.9 m tidal range) high wave energy (1.5 m average wave height) barrier island shoreline of the Cape Lookout cuspate foreland has been evaluated through examination of 3136 m of subsurface samples from closely spaced drill holes. Holocene sedimentation and coastal evolution has been a function of five major depositional processes: (1) eustatic sea-level rise and barrier-shoreline transgression; (2) lateral tidal inlet migration and reworking of barrier island deposits; (3) shoreface sedimentation and local barrier progradation; (4) storm washover deposition with infilling of shallow lagoons; and (5) flood-tidal delta sedimentation in back-barrier environments.

Twenty-five radiocarbon dates of subsurface peat and shell material from the Cape Lookout area are the basis for a late Holocene sea-level curve. From 9000 to 4000 B.P. eustatic sea level rose rapidly, resulting in landward migration of both barrier limbs of the cuspate foreland. A decline in the rate of sea-level rise since 4000 B.P. resulted in relative shoreline stabilization and deposition of contrasting coastal sedimentary sequences. The *higher energy, storm-dominated northeast barrier limb* (Core and Portsmouth Banks) has migrated landward producing a transgressive sequence of coarse-grained, horizontally bedded washover sands overlying burrowed to laminated back-barrier and lagoonal silty sands. Locally, ephemeral tidal inlets have reworked the transgressive barrier sequence depositing fining-upward spit platform and channel-fill sequences of cross-bedded, pebble gravel to fine sand and shell. Shoreface sedimentation along a portion of the *lower energy, northwest barrier limb* (Bogue Banks) has resulted in shoreline progradation and deposition of a coarsening-up sequence of burrowed to cross-bedded and laminated, fine-grained shoreface and foreshore sands. In contrast, the adjacent barrier island (Shackleford Banks) consists almost totally of inlet-fill sediments

---

Present addresses:
[1] Louisiana Geological Survey, Louisiana State University, Baton Rouge, LA 70893, U.S.A.
[2] University of Southern California, Los Angeles, CA 90007, U.S.A.
[3] Penzoil Company, Houston, TX 77001, U.S.A.
[4] Marathon Oil Company, Casper, WY 82601, U.S.A.
[5] San Francisco State University, San Francisco, CA 94132, U.S.A.

0025-3227/84/$03.00   © 1984 Elsevier Science Publishers B.V.

deposited by lateral tidal inlet migration. Holocene sediments in the shallow lagoons behind the barriers are 5—8 m thick fining-up sequences of interbedded burrowed, rooted and laminated flood-tidal delta, salt marsh, and washover sands, silts and clays.

While barrier island sequences are generally 10 m in thickness, inlet-fill sequences may be as much as 25 m thick and comprise an average of 35% of the Holocene sedimentary deposits. Tidal inlet-fill, back-barrier (including flood-tidal delta) and shoreface deposits are the most highly preservable facies in the wave-dominated barrier-shoreline setting. In the Cape Lookout cuspate foreland, these three facies account for over 80% of the sedimentary deposits preserved beneath the barriers. Foreshore, spit platform and overwash facies account for the remaining 20%.

INTRODUCTION

The Outer Banks of North Carolina are a classic wave-dominated microtidal barrier island shoreline (Fig.1). For the past two decades numerous investigations have examined the morphology, processes, vegetative patterns and historical development of this shoreline setting (Fisher, 1962, 1967; Pierce, 1969; Knowles et al., 1973; Godfrey and Godfrey, 1976; Mixon and Pilkey, 1976; Cleary and Hosier, 1979). Prior to 1975 there had been only one subsurface study of Holocene sediments on the North Carolina coastline (Pierce and Colquhoun, 1970). Since this time, however, there have been several coordinated investigations of the Holocene stratigraphy along the 130 km of barrier shoreline that forms the Cape Lookout cuspate foreland (Susman, 1975; Moslow, 1977; Herbert, 1978; Berelson, 1979; Steele, 1980). This paper is a synthesis of these studies and provides insight into the

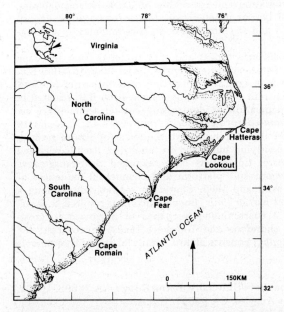

Fig.1. Morphology of the southeastern United States Atlantic coastline showing major cuspate forelands and cape systems. The Cape Lookout cuspate foreland is enclosed within the rectangular box and shown in detail in Fig.3.

record of Holocene sedimentation for a clastic wave-dominated barrier-island shoreline.

Earliest stratigraphic studies of barrier islands were conducted on the Texas Gulf Coast (Fisk, 1959; Bernard et al., 1962), and resulted in the first depositional models of barrier island facies. Even today Galveston Island is considered by some to be the "type" example of barrier island stratigraphy (McCubbin, 1982). Since the development of these early models, however, subsurface studies of U.S. East Coast barriers have shown that there is a wide disparity of stratigraphic motifs and facies relationships within barrier island complexes (Hoyt and Henry, 1967; Kraft, 1971; Kraft et al., 1978; Susman and Heron, 1979; Leatherman, 1979; Moslow and Colquhoun, 1981). From these studies it has been recognized that the Gulf Coast barriers are one of a wide spectrum of stratigraphic types. As with other clastic depositional systems the morphology, depositional environments and internal stratigraphy of barrier islands vary dramatically as a function of waves, tides and sediment input.

This paper is a synthesis of closely spaced drill hole data gathered to examine the Holocene stratigraphy and sedimentation of a chain of barrier islands in a low tidal-range, high wave-energy (wave-dominated) setting. The major objectives of this study are twofold: (1) document the wide variation in preserved vertical sedimentary sequences within the wave-dominated shoreline of the Cape Lookout cuspate foreland; and (2) provide a process-response depositional model for wave-dominated barrier islands in sand-rich shoreline environments.

Several coordinated investigations of barrier island stratigraphy in the study area have provided an extensive data base that includes detailed examination of approximately 3000 m of Holocene and Pleistocene sediments from 230 auger, wash-bore and vibracore holes. Sedimentary structures, lithology, grain size and fossil assemblages were analyzed to determine environments of deposition. Analysis of historic maps and charts, and a late Holocene sea-level curve constructed from 25 carbon-14 dated samples (Fig.2) provided the basis for interpreting the depositional history and shoreline evolution of the study area.

Detailed explanations of drill hole methodology and sediment analyses are provided in Susman and Heron (1979) and Moslow and Heron (1979, 1981).

GEOLOGIC SETTING

The barrier islands of the Cape Lookout area form one of the "cuspate forelands" that are dominant features of coastal North and South Carolina (Fig.1). These more prominent cape systems are associated with large subaqueous shoals that protrude into the Atlantic Ocean at roughly 90° to the shoreline. Cuspate forelands such as Cape Lookout are best developed in clastic shoreline settings where wave energy prevails over tidal or fluvial processes. Classic examples of cuspate-foreland systems on the east coast of

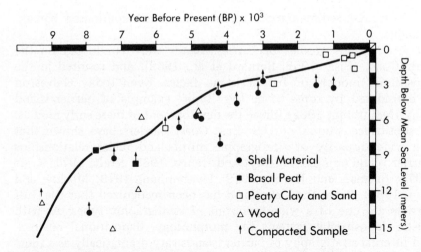

Fig.2. Holocene sea-level curve for the Cape Lookout cuspate foreland as determined from radiocarbon-dated samples of peat, wood and shell material cored beneath Bogue Banks, Shackleford Banks, Back Sound, Core Banks and Portsmouth Island. Peat samples are given preferential consideration as more accurate indicators of past sea-level positions. Note the decreased rate in sea-level rise since approximately 4000 B.P.

the United States (Capes Hatteras, Lookout and Fear) are all found in the wave-dominated Outer Banks of North Carolina (Fig.1). The origin and even-spacing of these "Carolina Capes" (120—140 km) have been related to a number of phenomena including eddy patterns of ocean currents (Dolan and Ferm, 1968; Shepard and Wanless, 1971) and erosional remnants of Pleistocene deltas (Hoyt and Henry, 1971). Recent studies on the North Carolina continental shelf, however, suggest that the shoreline orientation, origin and distribution of the Carolina capes and cuspate forelands are controlled by the structure and pattern of pre-Holocene erosion-resistant strata (Blackwelder et al., 1982). It is therefore quite conceivable that antecedant topography and subtle structural features are an important control on present shoreline configuration and sedimentation even along a passive, trailing-edge continental margin. Pleistocene beach ridges and shoreline orientation suggest that the Cape Lookout area was the site of at least one pre-Holocene cuspate foreland (Fisher, 1967; Blackwelder and Cronin, 1981; Mixon and Pilkey, 1976; and Fig.3).

COASTAL PROCESSES AND MORPHOLOGY

Waves, tides and storms

The Cape Lookout cuspate foreland consists of two barrier island limbs whose contrasting morphologies reflect local variations in wave energy, tidal range and storm response. Mean tidal range in the Cape Lookout area varies from 0.3 to 1.1 m (U.S. Department of Commerce, 1982), and therefore is classified as microtidal (<2 m tidal range; Davies, 1964). These are

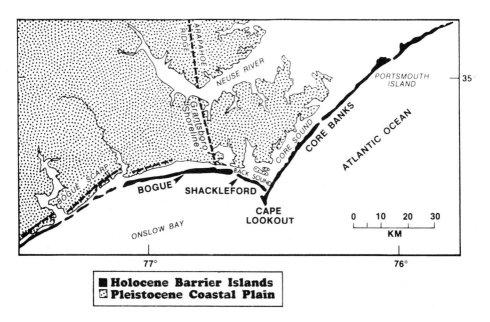

Fig.3. Map showing location and names of the Holocene barrier islands that form the "limbs" of the Cape Lookout cuspate foreland. Pleistocene shoreline trends are shown by the dashed and solid line patterns on the Coastal Plain. These Pleistocene shoreline features form a roughly cuspate orientation approximating the present day shoreline trend, suggesting that antecedant topography is a major control on Holocene shoreline configuration. Holocene barriers are black; Pleistocene deposits are stippled.

some of the lowest tidal ranges on the United States east coast (Nummedal et al., 1977). A significant aspect of tidal processes in the Cape Lookout cuspate foreland is that the 0.47 m mean tidal range for the northeast-trending barrier limb (Core and Portsmouth Banks) is roughly half the 0.89 m range for the southwest-trending limb (Boque and Shackleford Banks, Fig.3). The difference in tidal range is at least partly responsible for many of the variations in morphology and sedimentation patterns observed for the two barrier limbs. While the tidal range is relatively low, tidal processes should not be discounted as unimportant in the Cape Lookout area. Tidal currents have been measured in excess of 115 cm s^{-1} (Sarle, 1977), and are responsible for transport and deposition of fine- and coarse-grained sediment especially in the tidal inlets, tidal deltas and lagoons. The size of the tidal prism is an important factor in controlling the magnitude and duration of tidal currents, and in determining flood or ebb dominance of sediment transport in tidal inlets and backbarrier environments.

Although tides and tidal prism are important, the most significant coastal processes affecting Holocene sedimentation in the area are waves and storms. Mean annual wave height is 1.7 m with wave heights exceeding 2.0 m approximately 30% of the year (Fig.4; Nummedal et al., 1977). These values are among the highest for the U.S. east coast and are primarily responsible for the wave-dominated barrier shoreline morphology (Fig.5).

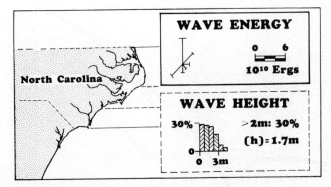

Fig.4. Deep-water wave-energy flux diagram and wave height—frequency histogram for portions of the North Carolina continental shelf and shoreline. Energy flux is in units of 10^{10} ergs m^{-1} s^{-1}. Only the onshore wave-energy flux components are shown. Wave height histogram indicates that wave heights in excess of 2 m occur 30% of the time. Average wave height (h) is 1.7 m for the study area (modified from Nummedal et al., 1977).

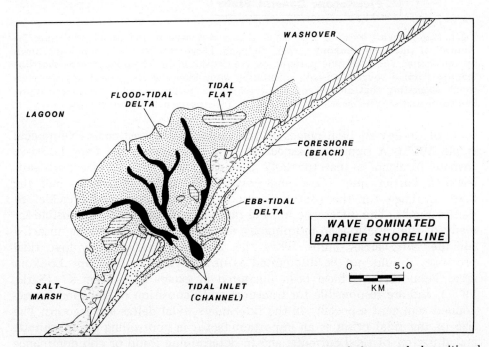

Fig.5. Diagram depicting barrier-island and tidal-inlet morphology and depositional environments in a wave-dominated (high average wave height, low mean tidal range) setting. This model is primarily based on observations of the study area.

An evaluation of the nearshore wave climate for North Carolina indicates that significantly more wave energy is focused on the Core and Portsmouth Banks barrier limb (Figs.3 and 4). These islands are directly affected by the northeast, east and southeast onshore components of wave-energy flux which account for approximately 75% (5.8×10^{10} ergs m^{-1} s^{-1}) of the total onshore wave energy for North Carolina (Fig.4; Nummedal et al., 1977). Bogue and Shackleford Banks barrier-limb is relatively sheltered with respect to wave energy because of its shoreline orientation. These islands are influenced only by the southerly onshore component which accounts for 25% (1.9×10^{10} ergs m^{-1} s^{-1}) of total onshore wave energy flux for the study area.

Extratropical storms and hurricanes have an important impact on sedimentation patterns in the Cape Lookout area. Storm-related processes redistribute coarse-grained sediment on the ocean margin of tidal inlets, tidal deltas and lagoons (Sarle, 1977). Large volumes of sand are transported across low-lying barriers during storms and deposited in the form of washover fans (Godfrey and Godfrey, 1976). The North Carolina coast has a history of 150 recorded hurricanes since 1585. An average of 1.64 hurricanes per year affect the Cape Lookout area (Crutcher and Quayle, 1974) with recorded storm surge up to 2.3 m above mean sea level (U.S. Corps of Engineers, 1976).

Barrier morphology

The barrier islands show a morphology typical of a wave-dominated, microtidal environment (Fig. 5; Hayes, 1979). The barriers are relatively long, linear and narrow and backed by wide shallow, open, lagoons. Tidal inlets are rare and ephemeral but migrate laterally at very high rates (Moslow and Heron, 1978). Tidal inlets on the higher-energy northeast barrier limb (Core and Portsmouth Banks) are associated with large flood-tidal deltas and small ebb deltas (Fig.6). The most prominent aspects of barrier morphology of Core and Portsmouth Banks are the extensive washover fans, fringing salt marsh and wide open lagoons.

In morphological contrast to the northeast limb the lower-energy southwest limb of Shackleford and Bogue Banks, while still falling within the wave-dominated shoreline morphology model, possess some unique features. Parallel and recurved beach ridges are common to both Shackleford and Bogue Banks and washover fans are much less common (Fig.7). The tidal inlets of the southwest lower-energy limb are associated with more prominent ebb-tidal deltas. Also, adjacent back-barrier lagoons are partially filled with salt marsh and tidal flat sediments.

CONTROLS ON HOLOCENE SEDIMENTATION

While the hydrographic regime (waves, tides and storms) is an important control on sedimentation patterns in the study area, other natural controls

Fig.6. Low-altitude oblique air photo of Core Banks and New Drum Inlet. Core Sound is to the right (west), Atlantic Ocean is to the left. Note the series of coalescing sand lobes that form the large flood-tidal delta on the sound side of the inlet throat. On the ocean side of the inlet, shoaling waves are breaking over the ebb-tidal delta. Note that the barriers are narrow, relatively flat, and featureless.

are equally important. As with any depositional system, sediment supply is important. The Cape Lookout area is a sand-rich environment with most of the sediment supplied from the Coastal Plain and Piedmont rivers or from the shoreface and inner shelf (Pierce, 1969; Mixon and Pilkey, 1976).

Holocene sedimentation in the area records several depositional processes. Each produced a specific shoreline response (Fig.8), the combination of which yielded the present observed sedimentary sequences. Perhaps the most important process has been the Holocene rise of sea level (Fig.2), a process that has resulted in the shoreline transgression most evident in the Core—Portsmouth barrier limb. Tidal-inlet migration has reworked large portions of Core, Portsmouth and Shackleford Banks. Shoreface sedimentation has resulted in a seaward progradation of Bogue Banks over the past 4000 yrs. Storms are important in washover sedimentation, inlet formation, and lagoonal infilling. Flood-dominated inlet sedimentation is responsible for partial infilling of shallow water lagoons. Flood-tidal deltas weld onto updrift portions of adjacent barriers resulting in abnormally wide, thick accumulations of flood-delta sediments (Fig.8).

Fig.7. Low-altitude air photo of Bogue Banks, Bogue Inlet and Bear Island (foreground). View is to the east. Note the series of shore-parallel beach ridges and greater width of the barrier at the western end of Bogue Island. This morphology contrasts with that typically found along the higher energy barrier limb as shown in Fig.6.

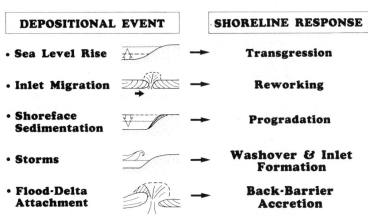

Fig.8. Major depositional events and resulting shoreline responses effecting Holocene sedimentation in the study area.

SEDIMENTARY SEQUENCES AND FACIES RELATIONSHIPS

The two foreland limbs have different facies that can be related to the processes of transgression, regression and inlet filling. Although many facies are common to both limbs, the general vertical and horizontal facies-associations of each limb are quite different.

Higher-energy transgressive limb

Core Banks and Portsmouth Island form the transgressive limb (Fig.3). Both barriers have a common transgressive stratigraphic sequence although the southwest and northeast ends reflect other processes. Four prominent facies are associated with the transgressive limb, each of which represents a number of sedimentary environments (Table I).

TABLE I

Facies characteristics of the high-energy transgressive limb and the lower-energy regressive limb of the Cape Lookout foreland

Depositional environment	Lithology	Shells and organics	Sedimentary structures	Large-scale features
Overwash and foreshore	Clean, moderately sorted, fine to medium sand	Whole and abraded shells in layers; variable assemblage (low diversity)	Horizontal and planar laminations	Caps inlet and barrier sequences
Shoreface	Well-sorted, fine to medium sand and silt	Abundance of sand-sized shell material *Gemma gemma*, *Arcopecten* sp., *Olivella* sp.	Cross-bedded (upper half) and burrowed (lower half) sequence	Coarsening-upward sequence; increase in mud content towards base
Backbarrier (lagoon, tidal flat, salt marsh)	Well-sorted, fine to medium silty sand and sandy clay	Organic rich: *Spartina* sp. and other plant material; *Ensis* sp., *Crassostrea* sp., *Crepidula* sp. (mollusks)	Burrowed; thin parallel clay laminations	Capped by salt marsh; increasing mud and organic content upwards
Flood-tidal delta	Moderately sorted, medium to coarse silty sand	Coarse shell frags. common; Echinoderm frags. common	Gently dipping cross-laminae; burrowed	Interbedded with backbarrier facies; cyclic fining-upward sequences
Inlet margin	Clean, well-sorted, fine to medium sand	Mollusks rare; low diversity	Planar and horizontal laminations	Caps fining-upward inlet sequence
Inlet channel	Moderately sorted, medium to coarse sand and shell; pebbly sand common	Mixed mollusk assemblage of shelf and backbarrier sps., shells common and abraded	Cross-bedded (trough and planar?)	Thickest unit of inlet sequence; fines upward
Inlet floor	Poorly sorted, coarse to pebbly sand and shell	Large, worn and abraded shell frags. common	Rip-up clasts; graded bedding	Basal scour lag

Overwash and foreshore. The beach and berm, depositional environments of the foreshore have been combined with overwash facies in cross-section Figs.9 and 11, and in the composite vertical section, Fig.10. The facies characteristics are described in Table I. Sedimentary structures and other diagnostic criteria associated with each facies in vertical sequences are shown in Table II.

Although not shown on the cross-sections, dune sands may locally overlie the overwash unit. Isolated wind-shadow or barchan dunes are scattered along the northeast limb of the foreland. Hosier (1973) attributes this to winds that parallel the island so that sand moves up and down rather than across the barriers. The highest dunes are 7 m and occur as isolated mounds just south of Ocracoke Inlet.

Backbarrier. The two dominant backbarrier depositional environments are the lagoon and flood-tidal delta, but others can be identified including tidal flat, salt marsh and distal overwash (Table I). The backbarrier facies underlies the overwash-foreshore deposits and commonly comprises the lower half of the transgressive barrier sequence.

Shoreface. At the Cape Lookout apex of Core Banks, the overwash-foreshore sequence is underlain by a coarsening-up sequence of well-sorted, fine-grained shoreface sediments (Fig.9 and Table I). The shoreface facies is relatively rare within the higher-energy transgressive barrier limb.

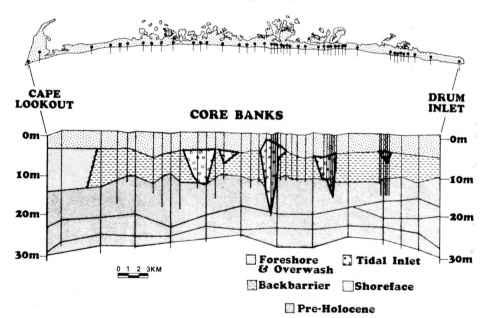

Fig.9. Shore-parallel cross section of Core Banks from Cape Lookout (south) to New Drum Inlet (north). Note shoreface sediments beneath Cape Lookout and five isolated inlet-fill sand bodies. Greatest volume of sediment is a transgressive sequence of foreshore and overwash overlying backbarrier facies, a sequence typical for most of the higher-energy barrier limb (modified from Moslow and Heron, 1979).

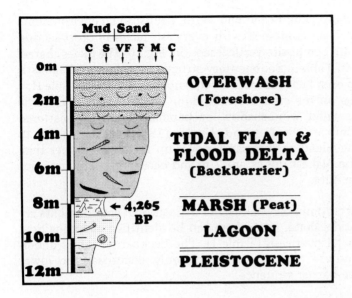

Fig.10. Composite vertical sequence of Holocene and uppermost Pleistocene sediments for Core Banks and Portsmouth Island. An explanation key to sedimentary features shown in all sequences is given in Table II.

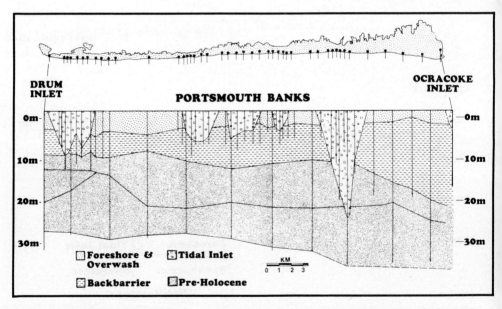

Fig.11. Shore-parallel cross-section of Portsmouth Island from New Drum Inlet (south) to Ocracoke Inlet (north). Note the extremely thick Holocene inlet-fill sequence at northern end of cross-section. The thick backbarrier sequence adjacent to Ocracoke Inlet is primarily a sequence of flood-tidal delta and lagoonal sediments (modified from Herbert, 1978).

TABLE II

Key to lithologies and physical and biogenic sedimentary structures shown in vertical sedimentary sequences (Figs.10, 14, 15, and 16)

LITHOLOGIES	PHYSICAL SEDIMENTARY STRUCTURES
Sand	Horizontal Laminations
Pebbly Sand and Gravel	Planar Laminations
Plant Material	Trough Cross-bedding
Shell Material	Mud Lenses
Rip-up Clasts	Parallel Laminations

BIOGENIC SEDIMENTARY STRUCTURES
Rooting
Sand-filled Burrows
Mud-filled Burrows

Tidal inlet. Ten historical relict inlets and one modern but artificial inlet have deposited expansive inlet-fill sand bodies within the transgressive limb (Figs.9 and 11). Fifteen to 20% of Core Banks and nearly 40% of Portsmouth Island is underlain by inlet-fill.

The stratigraphic sequence of the five relic inlets on Core Banks has been documented by Moslow and Heron (1978) and the facies details of the three sub-environments are shown in Table I. Though not very common, the inlet-fill deposit can consist of a series of stacked fining-up sequences, as observed beneath Portsmouth Island (Herbert, 1978).

The transgressive barrier sequence. Transgressive barriers do not form an easily recognizable thick vertical sequence of sediments. On the northeast limb, transgressive barriers are underlain by about 10 m of lagoon, marsh, tidal flat, flood delta and overwash-foreshore sands with some silts and muds (Fig.10). Lagoon, marsh, tidal flat and flood delta facies are not really characteristic of the barrier island per se although they may be associated with a barrier shoreline. The transgressive barrier proper consists of 2—3 m of overwash-foreshore sands as linear bodies. Associated facies occur as non-linear or arcuate-shaped sand bodies (that is, flood-tidal deltas), overlying widespread lagoonal silts and muds. Thus, even though the transgressive barrier has a typical vertical sequence, recognition of ancient barriers would be difficult based on observation of this sequence alone. The presence of fining-upward inlet sequences associated with the other barrier-related facies would be the clue to identifying ancient wave-dominated transgressive barriers.

Lower-energy depositional limb

Bogue Banks and Shackleford Banks (Fig.3) are prograding and inlet-modified barriers of the lower-energy southwest limb. Both barriers are morphologically different from the higher-energy northeast limb. Whereas Core Banks and the southwest part of Portsmouth Island form a long, narrow, low barrier system (Fig.6), Bogue and Shackleford Banks are generally wide with prominent beach ridges and often extensive dune fields (Fig.7). The topography reflects the complex progradation and inlet migration evolution of these barriers. Three prominent facies are associated with the lower-energy limb of the Cape Lookout foreland.

Shoreface. The shoreface facies is combined with the foreshore facies in stratigraphic cross-section of Bogue Banks (Fig.12). Shoreface and foreshore sediments are a coarsening-upward sequence of fine- to coarse-grained sand with an increasing percentage of silt and clay towards the base. These facies thicken in a seaward direction beneath the barrier and erosionally overlie backbarrier deposits. The sedimentary characteristics of shoreface and foreshore facies are shown in Table I.

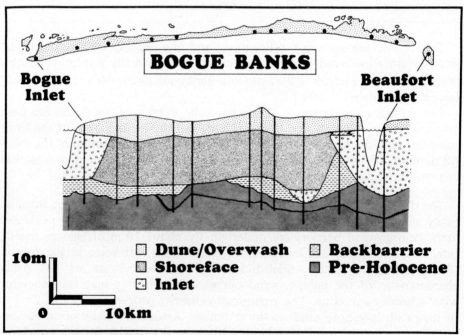

Fig.12. Shore-parallel cross section of Bogue Banks from Bogue Inlet (west) to the tip of Shackleford Banks adjacent to Beaufort Inlet (east). In sharp contrast to the transgressive barrier limb, the majority of the Holocene subsurface of Bogue Banks is a thick sequence of shoreface deposits erosionally overlying a thin sequence of basal backbarrier deposits (modified from Steele, 1980).

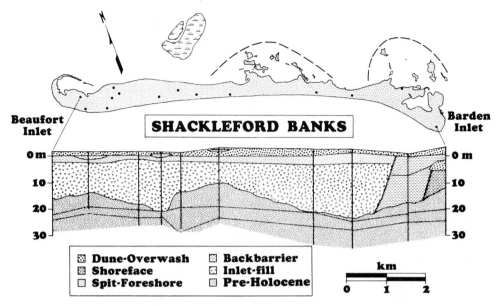

Fig.13. Shore-parallel cross-section of Shackleford Banks from Beaufort Inlet (west) to Barden Inlet (east). Most of the Holocene subsurface is a thick, laterally extensive sequence of tidal inlet-fill. The dashed lines forming an arcuate pattern on the sound side of the island enclose relict flood-tidal deltas (from Susman and Heron, 1979).

Tidal inlet. As much as 25 m of inlet fill underlies most of Shackleford Banks (Fig.13 and Table I). This inlet fill does not represent deposition from a single inlet that migrated westerly toward the present Beaufort Inlet as implied by Susman and Heron (1979). Instead, there are several stacked inlet sequences deposited by a minimum of two migrating inlets. Inlet sequences are crossbedded, poorly sorted, fine- to very coarse-grained sand and shell that are capped by dune and foreshore sands (Fig.14).

Flood-tidal delta. Vibracores from the relict flood-tidal delta behind Shackleford Banks in Back Sound show a three-part fining-upward sequence very similar to the inlet fill sequence except that the uppermost unit is usually an organic (that is salt marsh) mud (Table I). Flood-tidal delta sands thin in a landward direction and are interbedded with lagoon and tidal flat muds.

Vertical sequences. The general vertical sequence of prograding Bogue Banks is shown in Fig.15. The basal tidal flat/lagoon facies was deposited during an early Holocene transgressive phase. Regression started about 4000—5000 B.P. The general coarsening-up nature of this sequence is typical of prograding barriers (Bernard et al., 1970; Moslow and Colquhoun, 1981).

The vertical sequence of inlet-fill on Shackleford Banks is shown in Fig.14. The fining up trend is characteristic of inlet-fill. Flood-tidal delta sequences also fine up, and are capped by muddy, marsh deposits. A distinct difference in sand body geometry is observed between the somewhat isolated bodies

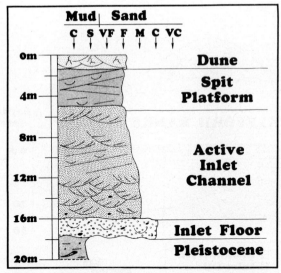

Fig.14. Composite sedimentary sequence of Holocene inlet-fill that is representative of an inlet-modified barrier island, such as Shackleford Banks (modified from Susman and Heron, 1979).

of inlet-fill in the higher-energy limb (Figs.9 and 11) and the extensive lateral body of inlet-fill in Shackleford (Fig.13).

DISCUSSION

Vertical sequence variability

Perhaps the most striking aspect of the subsurface stratigraphy of the Cape Lookout cuspate foreland is the extreme variability in facies relationships and vertical sedimentary sequences (Fig.16). The Holocene sedimentary deposits of the higher-energy transgressive barrier limb (Core and Portsmouth Banks) are a mixture of: (1) coarsening-upward, transgressive barrier-island sequences; (2) isolated, fining-upward inlet-fill sequences; and (3) interbedded, fining- and coarsening-upward sequences of flood-tidal delta, inlet-fill and lagoonal sediments (northern Portsmouth Island; Fig.16). In contrast, the lower-energy depositional limb consists of: (1) laterally extensive, fining-upward sequences of stacked inlet-fill and flood-delta sediments (entirety of Shackleford Banks); and (2) a coarsening-upward foreshore-shoreface sequence, also laterally extensive (majority of Bogue Banks). The Cape Lookout apex and adjacent subaqueous shoals are typified also by coarsening-upward sequences of primarily upper and lower shoreface sediments (Fig.16). Therefore, while the surficial geomorphology and observed depositional environments contrast somewhat for most of the Cape Lookout coastal area, this contrast does not reflect the variability of sedimentary sequences and facies relationships that occur in the Holocene stratigraphy.

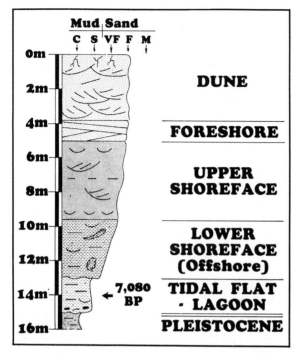

Fig.15. Composite vertical sequence of Holocene sediments from Bogue Banks. The coarsening-upward shoreface and foreshore sequence was deposited by seaward progradation of the barrier island. Sedimentary structures shown here are inferred from beach trenches, dune scarps and core data of shoreface sequences beneath Cape Lookout. Basal Holocene tidal flat and lagoon sediments were deposited during an earlier transgressive phase of island evolution (modified from Steele, 1980).

While not serving as an all-encompassing depositional model, Fig.16 does outline the variability and types of vertical sedimentary sequences that one can expect to find in a wave-dominated barrier island shoreline. This aspect of subsurface complexity is important to the sedimentologist examining other wave-dominated barriers or their ancient equivalents.

Facies preservation potential

The facies correlations and cross-sections developed from closely spaced drill holes in the study area have been used to calculate the relative percentages of Holocene sedimentary facies presently found in the subsurface (Fig.17). Previous subsurface studies in wave- and tide-dominated shorelines elsewhere have identified the more preservable barrier-island related sedimentary facies (Hoyt and Henry, 1967; Kumar and Sanders, 1974; Hayes, 1980; Belknap and Kraft, 1981). In general, sediments deposited in a subaqueous environment, primarily those deposited below wave base or protected from wave erosion, should have a much better potential for preservation in the coastal environment. This generalization is applicable to the Cape Lookout barrier islands, where tidal-inlet, backbarrier and shoreface facies account for approximately 80% of the Holocene sediments presently

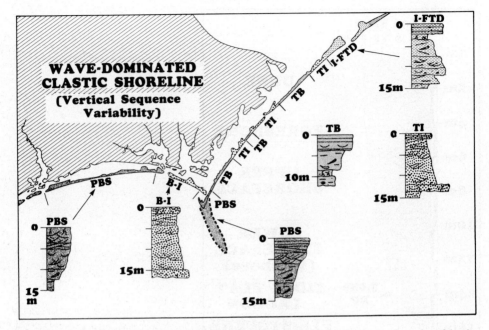

Fig.16. Diagram showing the variability in vertical sedimentary sequences found within the Holocene barrier islands of the Cape Lookout cuspate foreland. Vertical sequences are keyed by shaded patterns to those portions of the shoreline where they are most prevalent. Transgressive barrier (*TB*), tidal inlet (*TI*) and flood-tidal delta (*I-FTD*) sequences comprise the higher-energy barrier limb. Prograding barrier-shoreface (*PBS*) and barrier-inlet (*B-I*) sequences are found within the barrier islands of the lower-energy limb.

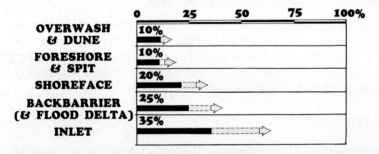

Fig.17. Diagram showing the relative amounts of the sedimentary facies presently found in the Holocene barrier and backbarrier subsurface. The solid black bars indicate the percentages existing at this moment in time as determined from cross-section analysis. The stippled bars in the diagram depict the possible volumetric increase of a sedimentary facies with time.

found in the subsurface (Fig.17). Tidal inlet deposits account for the greatest percentage (35%) of Holocene sediments. This should not be surprising because inlet channels in a wave-dominated shoreline migrate laterally over great distances reworking the previously deposited barrier sediments, sometimes scouring through the entire Holocene section, depositing thick inlet-fill sequences. It is likely therefore, that as the inlets continue to migrate along the shoreline, an even higher percentage of the Holocene sediment package will be represented by the tidal inlet facies.

The backbarrier facies (flood-tidal delta, lagoon and tidal flat deposits) account for 25% of the Holocene subsurface. Rapid sedimentation and protection from wave erosion in backbarrier environments allows for enhanced accumulation, rapid burial and therefore greater preservation potential. High rates of sedimentation along the seaward margin of Bogue Banks and the Cape Lookout apex have preserved thick shoreface sequences. The shoreface facies accounts for approximately 20% of the Holocene subsurface (Fig.17). Those sediments deposited in intertidal or subaerial environments (foreshore, spit platform, overwash and dune) have a low preservation potential, and combined, account for approximately 20% of the Holocene subsurface.

CONCLUSION

The wave-dominated Cape Lookout cuspate foreland consists of two barrier-island limbs and a massive cape-shoal complex that display a variety of vertical sedimentary sequences and facies relationships within the Holocene subsurface. The diverse nature of the Holocene sedimentary record has been a function of five major depositional processes and shoreline responses: (1) sea-level rise and barrier-island transgression; (2) tidal inlet migration and barrier reworking; (3) shoreface sedimentation and barrier progradation; (4) storm washover and lagoonal infilling; and (5) flood-tidal delta attachment and accretion at updrift portions of barrier islands. Sediment (that is, sand) supply, hydrographic regime, structural setting and antecedent topography are important controls of Holocene sedimentation and facies preservation.

Inlet-fill and transgressive barrier island sequences form the bulk of the higher-energy, storm-dominated limb of the cuspate foreland (Core and Portsmouth Banks). While inlet-fill is a prevalent facies, a significant portion of the lower-energy depositional barrier island limb (Shackleford and Bogue Banks) is a thick regressive sequence of prograding shoreface sediments. Likewise, the massive Cape Lookout shoal complex is mostly a coarsening-upward shoreface sequence. The variety and types of sedimentary sequences found within the Holocene subsurface of the Cape Lookout cuspate foreland should be similar to those found in other modern wave-dominated shorelines or their ancient equivalents.

The tidal inlet, backbarrier (including flood-delta) and shoreface facies account for approximately 80% of the Holocene subsurface deposits. These facies have the highest preservation potential among those observed in the wave-dominated barrier shoreline.

ACKNOWLEDGEMENTS

The majority of the research incorporated into this manuscript was funded by grants from the National Park Service to the Duke University

Department of Geology. Other partial funding was provided by the Cities Service Oil Company and the American Association of Petroleum Geologists. Logistical support was provided by the Duke University Marine Laboratory and the University of North Carolina Marine Science Consortium. Albert C. Hine is especially thanked for his support of the research conducted on Bogue Banks. Identification of mollusk assemblages and their related environments of deposition were identified by Blake Blackwelder and William C. Miller, III. Figures were prepared by the Cities Service Research Graphics Department and the Geology Department of Louisiana State University. Bruce Wilkinson and Albert Hine are thanked for their helpful review of the manuscript.

This manuscript is a synthesis of several M.S. thesis research projects conducted from 1974 to 1980 at the Duke University Department of Geology. Contributions of the authors are as follows: S.D. Heron, Jr. coordinated all research and helped prepare the manuscript; T.F. Moslow examined the subsurface stratigraphy of southern Core Banks and Cape Lookout, and helped prepare the manuscript; W.M. Berelson performed a vibracore study of Back Sound sediments; northern Core Banks and Portsmouth Island was studied by J.R. Herbert; K.R. Susman conducted a study of Shackleford Banks; Bogue Banks was examined by G.A. Steele III.

REFERENCES

Belknap, D.F. and Kraft, J.C., 1981. Preservation potential of transgressive coastal lithosomes on the U.S. Atlantic shelf. Mar. Geol., 42: 429—442.

Berelson, W.M., 1979. Barrier island evolution and its effect on lagoonal sedimentation: Shackleford Banks, Back Sound, and Harkers Island: Cape Lookout National Seashore. Thesis, Duke University, Durham, N.C., 132 pp. (unpublished).

Berg, R.G. and Davies, D.K., 1968. Origin of Lower Cretaceous Muddy Sandstone at Bell Creek Field, Montana. Bull. Am. Assoc. Pet. Geol., 52: 1888—1898.

Bernard, H.A., LeBlanc, R.J. and Major, C.F., 1962. Recent and Pleistocene geology of southeast Texas. In: E.H. Rainwater and R.P. Zingula (Editors), Geology of the Gulf Coast and Central Texas and Guidebook Excursions. Houston Geol. Soc., Field Trip Guidebook, pp.175—224.

Bernard, H.A., Major, C.E., Parrott, B.S. and LeBlanc, R.J., 1970. Recent sediments of southeast Texas. In: A Field Guide to the Brazos Alluvial and Deltaic Plains and the Galveston Barrier Island Complex. Univ. of Texas, Bur. Econ. Geol. Guidebook 11, pp.1—16.

Blackwelder, B.W. and Cronin, T.M., 1981. Atlantic Coastal Plain geomorphology illustrated by computer-generated block diagrams. U.S. Geol. Surv., Misc. Field Stud. Map MF-1242.

Blackwelder, B.W., MacIntyre, I.G. and Pilkey, O.H., 1982. Geology of continental shelf, Onslow Bay, North Carolina, as revealed by submarine outcrops. Bull. Am. Assoc. Pet. Geol., 66: 44—56.

Cleary, W.J. and Hosier, P.E., 1979. Geomorphology, washover history, and inlet zonation: Cape Lookout, N.C., to Bird Island, N.C. In: S.P. Leatherman (Editor), Barrier Islands. Academic Press, New York, N.Y., pp.237—271.

Crutcher, H.D. and Quayle, R.G., 1974. Mariners worldwide climate guide to tropical storms at sea. Naval Weather Service Command, U.S. Dept. Commerce, Washington, D.C.

Davies, J.L., 1964. A morphogenic approach to world shorelines. Z. Geomorphol., 8: 27—42.

Dolan, R. and Ferm, J.C., 1968. Crescentic landforms along the Atlantic coast of the United States. Science, 159: 627—629.

Fisher, J.J., 1962. Geomorphic expression of former inlets along the Outer Banks of N.C. Thesis, University of North Carolina, Chapel Hill, N.C., 120 pp. (unpublished).

Fisher, J.J., 1967. Development pattern of relict beach ridges, Outer Banks barrier chain, N.C. Diss., University of North Carolina, Chapel Hill, N.C., 325 pp. (unpublished).

Fisk, G.N., 1959. Padre Island and Laguna Madre flats, coastal south Texas. Proc. Second Coastal Studies Institute Conf., Louisiana State University, Baton Rouge, La., pp.103—152.

Godfrey, P.J. and Godfrey, M.M., 1976. Barrier island ecology of Cape Lookout National Seashore and vicinity, North Carolina. National Park Service Scientific Monograph Series, no. 9, 160 pp.

Goff, T.W., 1977. Sedimentation in a modern tidal inlet: Bogue Inlet, North Carolina. Thesis, Univ. of North Carolina, Chapel Hill, N.C., (unpublished).

Hayes, M.O., 1979. Barrier island morphology as a function of tidal and wave regime. In: S.P. Leatherman (Editor), Barrier Islands. Academic Press, New York, N.Y., pp.1—27.

Hayes, M.O., 1980. General morphology and sediment patterns in tidal inlets. Sediment Geol., 26: 139—156.

Hayes, M.O. and Kana, T.W., 1976. Terrigenous clastic depositional environments. Tech. Rep. 11-CRD, Dept. of Geology, Univ. of South Carolina, Columbia, S.C.

Herbert, J.R., 1978. Post-Miocene stratigraphy and evolution of northern Core Banks, North Carolina. Thesis, Duke University, Durham, N.C., 166 pp. (unpublished).

Hosier, P.E., 1973. The effects of oceanic overwash on the vegetation of Core and Shackleford Banks, North Carolina. Diss., Duke University, Durham, N.C., 230 pp. (unpublished).

Howard, J.D. and Reineck, H.E., 1972. Georgia coastal region, Sapelo Island, U.S.A.: sedimentology and biology. V. III. Conclusions. Senckenbergiana Marit., 4: 81—123.

Hoyt, J.H. and Henry, V.J., 1967. Influence of inlet migration on barrier island sedimentation. Geol. Soc. Am. Bull., 78: 77—86.

Hoyt, J.H. and Henry, V.J., 1971. Origin of capes and shoals along the southeastern coast of the United States. Geol. Soc. Am. Bull., 82: 59—66.

Knowles, C.E., Langfelder, J. and McDonald, R., 1973. A preliminary study of storm induced beach erosion for North Carolina. Rep. 73-75, Center for Marine Coastal Studies, North Carolina State University, Raleigh, N.C.

Kraft, J.C., 1971. Sedimentary facies patterns and geologic history of a Holocene marine transgression. Geol. Soc. Am. Bull., 82: 2131—2158.

Kraft, J.C., Allen, E.A. and Maurmeyer, E.M., 1978. The geologic and paleogeomorphological evolution of a spit system and its associated coastal environments: Cape Henlopen Spit, Delaware. J. Sediment. Petrol., 48(1): 211—226.

Kumar, N. and Sanders, J.E., 1974. Inlet sequences: A vertical succession of sedimentary structures and textures created by the lateral migration of tidal inlets. Sedimentology, 21: 291—323.

Leatherman, S.P., 1979. Barrier Islands from the Gulf of St. Lawrence to the Gulf of Mexico. Academic Press, New York, N.Y., 325 pp.

McCubbin, D.G., 1982. Barrier island and strand-plain facies. In: P.A. Scholle and D. Spearing (Editors), Sandstone Depositional Environments. Am. Assoc. Pet. Geol., Tulsa, Okla., pp.247—279.

Mixon, R.B. and Pilkey, O.H., 1976. Reconnaissance geology of the submerged and emerged Coastal Plain province, Cape Lookout area, North Carolina. U.S. Geol. Surv., Prof. Pap., 859, 41 pp.

Moslow, T.F., 1977. Quaternary evolution of Core Banks, North Carolina from Cape Lookout to New Drum Inlet. Thesis, Duke University, Durham, N.C., 132 pp. (unpublished).

Moslow, T.F. and Colquhoun, D.J., 1981. Influence of sea level change on barrier island evolution. Oceanis, 7: 439—454.

Moslow, T.F. and Heron Jr., S.D. 1978. Relict inlets: preservation and occurrence in the Holocene stratigraphy of southern Core Banks, North Carolina. J. Sediment. Petrol., 48: 1275—1286.

Moslow, T.F. and Heron Jr., S.D. 1979. Quaternary evolution of Core Banks, North Carolina: Cape Lookout to New Drum Inlet. In: S.P. Leatherman (Editor), Barrier Islands. Academic Press, New York, N.Y., pp.211—236.

Moslow, T.F. and Heron Jr., S.D. 1981. Holocene depositional history of a microtidal cuspate foreland cape: Cape Lookout, North Carolina. Mar. Geol., 41: 251—270.

Moslow, T.F. and Tye, R.S., 1985. Recognition and characterization of Holocene tidal inlet sequences. In: G.F. Oertel and S.P. Leatherman (Editors), Barrier Islands. Mar. Geol., 63 (in press).

Nummedal, D., Oertel, G.F., Hubbard, D.K. and Hine, A.C., 1977. Tidal inlet variability — Cape Hatteras to Cape Canaveral. Proc. Conf. on Coastal Sediments 77, A.S.C.E., Charleston, S.C., pp.543—562.

Pierce, J.W., 1969. Sediment budget along a barrier island chain. Sediment. Geol., 3: 5—16.

Pierce, J.W. and Colquhoun, D.J., 1970. Holocene evolution of a portion of the North Carolina coast. Geol. Soc. Am. Bull., 81: 3697—3714.

Sarle, L.L., 1977. Processes and resulting morphology of sand deposits within Beaufort Inlet, Carteret County, North Carolina. Thesis, Duke University, Durham, N.C., 149 pp. (unpublished).

Schwartz, R.K., 1975. Nature and genesis of some storm washover deposits. Tech. Memo. No. 61. U.S. Army Corps of Engineers, 60 pp.

Shepard, F.P. and Wanless, H.R., 1971. Our Changing Coastlines. McGraw-Hill, New York, N.Y., 579 pp.

Steele, G.A., 1980. Stratigraphy and depositional history of Bogue Banks, North Carolina. Thesis, Duke University, Durham, N.C., 201 pp. (unpublished).

Susman, K.R., 1975. Post-Miocene subsurface stratigraphy of Shackleford Banks, Carteret County, North Carolina. Thesis, Duke University, Durham, N.C., 85 pp. (unpublished).

Susman, K.R. and Heron Jr., S.D., 1979. Evolution of a barrier island, Shackleford Banks, Carteret County, North Carolina. Geol. Soc. Am. Bull., 90: 205—215.

United States Corps of Engineers, 1976. Morehead City harbor, North Carolina, final environmental statement. U.S. Army Engineer District, Wilmington, N.C.

U.S. Department of Commerce, 1982, Tide Tables, East Coast of North and South America, Admin., Natl. Ocean Survey, Rockville, Md., 222 pp.

RECONSTRUCTION OF PALEO-WAVE CONDITIONS DURING THE LATE PLEISTOCENE FROM MARINE TERRACE DEPOSITS, MONTEREY BAY, CALIFORNIA

WILLIAM R. DUPRÉ

Department of Geosciences, University of Houston, Houston, TX 77004 (U.S.A.)

(Received February 24, 1983; revised and accepted July 29, 1983)

ABSTRACT

Dupré, W.R., 1984. Reconstruction of paleo-wave conditions during the Late Pleistocene from marine terrace deposits, Monterey Bay, California. In: B. Greenwood and R.A. Davis, Jr. (Editors), Hydrodynamics and Sedimentation in Wave-Dominated Coastal Environments. Mar. Geol., 60: 435—454.

The Santa Cruz coastal terrace fringes much of the northern Monterey Bay region, California. It consists mainly of a regressive sequence of high-energy, barred nearshore marine sediments deposited during the last (Sangamonian) highstand of sea level. This sequence can be sub-divided into several depth-dependent facies on the basis of paleo-current data and vertical sequence of sedimentary structures. These include a lower shoreface facies deposited in 10—16 m water depth, an upper shoreface facies (including both a storm-dominated assemblage and a surf zone assemblage) deposited in 0—10 m water depth, and a foreshore facies deposited in the swash zone, up to 3.5 m above high tide.

The *magnitudes* of the storm events responsible for depositing these sediments were estimated by calculating paleo-wave heights using a variety of criteria (e.g., critical threshold equations, breaker depths, berm heights). In addition, the climate and paleogeography during the deposition of these sediments were essentially the same as today, allowing the use of present-day wave statistics to estimate the *frequency* of these storm events. The largest storms formed offshore-flowing currents (e.g., rip, wind-forced, and possibly storm-surge ebb currents) that resulted in the deposition of approximately 30% of the sediments seaward of the surf zone; however, the magnitude and frequency of these events are unknown. The remaining 70% of the sediment beyond the surf zone was deposited in response to smaller storm waves which were, on the average, at least 1.6 m high; such waves presently occur no more than 15% of the time. Sediments deposited during "fairweather" conditions (i.e., the remaining 85% of the time) have a low preservation potential, and are generally not preserved in this facies. In contrast, surf zone sediments were deposited by a variety of processes associated with waves whose maximum offshore heights were probably ≤2.2 m; such waves presently occur up to 92% of the time. Sediments within the swash zone were deposited by waves up to 3 m high, the largest of which presently occur approximately 2% of the time.

Most of the sediments were deposited by storms of intermediate magnitude and frequency; different facies, however, appear to preferentially record events of different recurrence intervals. In particular, surf zone sediments were deposited under relatively small storm and post-storm conditions, whereas sediments deposited farther offshore record increasingly larger, less frequent storm events. Relatively rare events (e.g., the 100 or 1000 yr events) do not appear to have significantly affected sedimentation in these nearshore environments.

0025-3227/84/$03.00 © 1984 Elsevier Science Publishers B.V.

INTRODUCTION

The recognition of characteristic vertical sequences of sedimentary structures associated with prograding, wave-dominated shorelines has been an active area of research in the past few years (e.g., Clifton, 1969; Campbell, 1971; Ryer, 1977; Vos and Hobday, 1977; Roep et al., 1979; Hunter et al., 1979; Howard and Reineck, 1981; Clifton, 1981). These and other studies of modern and ancient marine sediments have led to an increasing awareness of the significance of storm deposits preserved in the stratigraphic record. In fact, it appears evident that storm deposits are preferentially preserved in many depositional settings (e.g., Wolman and Miller, 1960; Kumar and Sanders, 1974; Kreisa, 1981). The magnitude and frequency of the storm events responsible for these storm deposits are, however, less certain.

The purpose of this paper is two-fold. The first is to describe the vertical sequence of sedimentary structures in Late Pleistocene marine terrace deposits in the Monterey Bay region, California (Fig.1), emphasizing the characteristics and relative abundance of storm deposits. The second is to calculate the magnitude and frequency of the storm conditions under which these sediments were deposited.

The magnitudes of paleo-storm events are difficult to estimate, with investigators typically relying on either ripple characteristics (e.g., Tanner, 1971; Komar, 1974; Allen, 1979; Miller and Komar, 1980), or critical

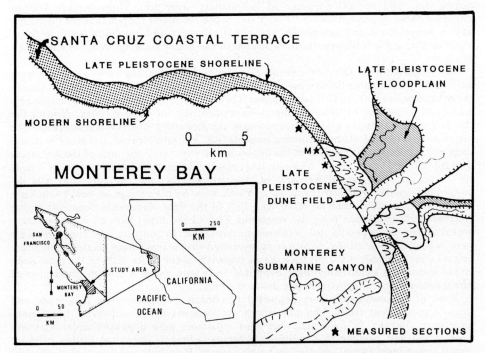

Fig.1. Location of study area. M indicates Manresa measured section (see Fig.2). Hachured lines indicate fluvial and marine scarps.

threshold criteria (e.g., Komar and Miller, 1974; Clifton, 1976; Dingler, 1979). The use of ripple morphology is a useful tool, particularly in low to moderate wave-energy environments. They are of less use in high-energy environments where large-magnitude storm events rarely produce ripples except at the very top of the deposits, only to be usually removed by the next storm event (Clifton, 1976; Howard and Reineck, 1981). Thus the use of velocity—grain size diagrams (e.g., Komar and Miller, 1974; Clifton, 1976, fig.4) to determine critical threshold boundaries often provides the best criteria for estimating the magnitude of ancient storm conditions. Other criteria such as breaker depth and berm height may also be used, and are discussed elsewhere in this paper.

Estimates of the frequency of paleo-storm events are even more tenuous, relying either on the assumption that the ancient wave regime was similar to that of the present (e.g., Bourgeois, 1980), or that the recurrence interval is equal to the number of storm deposits divided by the time interval of the unit as a whole. The latter method produces typical storm frequencies on the order of 100's of years for Holocene storm sands (e.g., Hayes, 1967; Morton, 1981; Nelson, 1982), and on the order of 1000—40,000 yrs for ancient storm sandstones (e.g., Brenchley et al., 1979; Hamblin and Walker, 1979; Kreisa, 1982).

Each method of estimating the recurrence interval of waves has potential problems. The use of modern wave statistics to estimate ancient wave conditions becomes increasingly unreliable with increasingly older deposits where wind patterns, storm tracks, and shelf and nearshore bathymetry may differ significantly from the present. The second method requires estimating the time interval represented by a part of the stratigraphic record, an estimate that is often unreliable. In addition, it should be noted that the recurrence interval of a storm event capable of forming a storm deposit is not necessarily the same as the number of storm deposits per unit time, as such an estimate cannot take into account storm deposits which were subsequently reworked or removed by erosion. For that reason such estimates tend to overestimate the actual recurrence interval. Lastly, the results of this method are very much an "average" storm frequency, without the ability to discern different recurrence intervals for different storm deposits. This may be particularly significant in nearshore sediments where the preservation potential of different frequency events may vary in different parts of the nearshore zone (Hunter et al., 1979).

Calculations of the magnitude and frequency of storm events are probably most accurate only where modern wave statistics can be extrapolated to the ancient wave conditions with some degree of confidence. This appears to be a valid assumption for the Late Pleistocene terrace deposits in the Monterey Bay region of California (Fig.1). The youngest of these sediments were deposited during the last (Sangamonian) interglacial period, during which time the climate was approximately the same as the present, albeit slightly warmer (Flint, 1971, p. 439). In addition, the configuration of the shoreline was essentially the same as it is today, thus wave refraction patterns are assumed to be the same as well. It seems reasonable, therefore, to assume that the nearshore wave spectrum during the deposition of these marine sediments was essentially the same as the present. Wave statistics for the Monterey Bay region were summarized by Arnal (1973) using Synoptic Meteorological Observations (SSMO) data for the period 1939—1971, and were used here as representative of the present-day wave climate. It should be noted that these data were obtained for a relatively small area (1° × 1°) near Monterey Bay, and differ from the SSMO data summarized by Harris (1972) for a 10° × 10° Marsden square for the area. The wave climate of the central California region is much higher than that in the relatively more sheltered Monterey Bay area, thus the use of the more generalized wave statistics would have resulted in significant errors with the generation of much larger recurrence intervals than those calculated in this study.

SANTA CRUZ COASTAL TERRACE DEPOSITS

Much of the central California coastal zone is rimmed by a series of uplifted marine terraces. The youngest of these is the Santa Cruz coastal terrace formed during the rise and fall of sea level associated with the last (Sangamonian) interglacial period (Dupré, 1975a; Dupré et al., 1980). These deposits are relatively well exposed in sea cliffs along the northern part of Monterey Bay, providing an opportunity to compare the preserved progradational sequence with that predicted on the basis of studies of modern nearshore marine facies (e.g., Clifton, 1969; Davidson-Arnott and Greenwood, 1976; Hunter et al., 1979; Howard and Reineck, 1981). Several sections were measured, however only one is described here as all showed basically the same vertical sequence of sedimentary facies illustrated in Fig.2. Associated non-marine facies are not described in this paper, but are discussed elsewhere (Dupré, 1975a; Dupré et al., 1980).

Lower shoreface deposits

The base of the terrace deposits is marked by a roughly planar erosional surface (Fig.3A) formed by marine planation mainly during a late stage of rising sea level (see Bradley and Griggs, 1976, for a more detailed discussion of the origin of the wave-cut platform). Some irregular scour depressions with up to a meter of relative relief are locally present, as are erosional ripples (cf. Reineck and Singh, 1975, p. 41) with wave length (L) = 30 cm and wave height (H) = 5 cm. Transgressive marine deposits, where present, are thin (typically less than 30 cm thick), and consist of landward-dipping (Fig.3B) ripple cross-stratified gravels (L = 1.2 m, H = 10 cm) and relict concretions of eolian deposits that appear to represent palimpsest sediments of an ancestral Monterey Bay.

The transgressive lag deposit grades upward into the lower shoreface deposits (Fig.3C). These deposits are approximately 1.5 m thick, and consists of highly bioturbated, parallel-laminated, fine to very fine grained sand

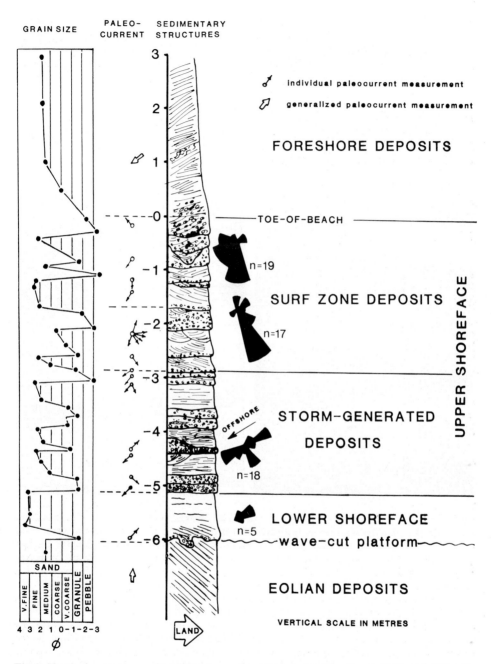

Fig.2. Vertical sequence of sedimentary structures and grain-size variations within the Santa Cruz coastal terrace, as exposed at the Manresa measured section. Individual and generalized paleocurrent data are from the Manresa section, however the rose diagrams include paleocurrent data from nearby measured sections as well (see Fig.1 for locations). Grain size determined by sieve analysis.

Fig.3. Outcrop photographs of the Santa Cruz coastal terrace sediments, northern Monterey Bay, California (arrow indicates direction of land when sediments were being deposited). A. Outcrop photograph of the Manresa measured section (see Fig.2). Dashed line delineates the wave-cut platform, separating eolian (E) deposits from overlying marine (M) deposits. B. Wave-cut platform separating eolian deposits (E) from overlying marine deposits (M). Note the onshore-dipping, cross-stratified gravels at the base of the lower shoreface deposits. C. Intensely bioturbated lower shoreface deposits of sand and scattered pebbles, overlain by graveliferous storm deposits. D. Landward-dipping, cross-stratified sand and gravel near the base of the storm-dominated sediments. Note the truncated vertical Asterosoma (?) burrows. E. Parallel-laminated sand deposited in the upper flow regime, grading upwards into onshore-dipping, cross-laminated sand, overlain by storm deposited gravels. Note the Asterosoma (?) burrows. F. Macraronichus segregatis burrows within the lower shoreface deposits.

with scattered pebbles. High-angle cross-stratification is not present, however some of the sets of parallel-laminated sand have landward dips of 5—15°. Scattered pebbles (long axis up to 2 cm) within the fine sand (Fig.3C), combined with the complete lack of ripple-drift stratification, suggests deposition under relatively high velocity, upper flow regime conditions (see also Hunter and Clifton, 1982, p.136). This unit is also characterized by extensive bioturbation (locally up to 100%), although body fossils are very rare. Recognizable trace fossils include polecypod burrows (up to 30 cm deep), thin, vertical burrows (Fig.3D, E) similar to *Asterosoma* (cf., Howard, 1972), and extensive, lined, worm-like feeding traces of *Macraronichus segregatis* (Fig.3F) described by Clifton and Thompson (1978).

Upper shoreface deposits

This unit is approximately 5 m thick, and consists of complexly interbedded sand and pebble gravel which can be further subdivided into a storm-dominated assemblage and a surf zone assemblage (Fig.2).

The storm-dominated deposits are approximately 2.5 m thick, and consist of parallel-laminated and cross-stratified, fine- to medium-grained sand and interbedded pebble gravel. The base of the gravel beds is marked by a sharp erosional surface (e.g., Figs.3D and 4A) which is planar where exposed perpendicular to the paleo-shoreline, but which is locally scoured with up to a meter of relative relief where exposed parallel to the paleo-shoreline. The gravel is typically 0.2—2 cm in diameter (long axis), moderately well sorted, and occurs in beds which commonly range from 10 to 30 cm thick. The gravel beds may be structureless, cross-stratified, or graded; most have a sandy matrix which appears to have formed by infiltration from overlying sandy units. Paleo-current data indicate most of the gravel beds were deposited by offshore-moving currents (some of which were channelized) during periods of maximum storm activity. The tops of some of the gravels have onshore-dipping ripple cross-laminations indicating reworking by onshore waves during periods of smaller storms or post-storm recovery.

The gravels are overlain by fine- to medium-grained sand which occurs in beds 0.1—1.2 m thick. Most of the sand is parallel-laminated, although medium-scale sets of cross-stratified coarse-grained sand approximately 30 cm thick (Fig.4A) formed by landward-migrating megaripples are also present. The sets show an overall fining-upward sequence due to avalanching down the slipface forsets. These forsets dip up to 30° in the landward direction, have an angular base, and show a distinct brinkpoint (Fig.4A), similar to that produced by migrating megaripples described by Clifton et al. (1969). These units occur as single sets of cross-stratified sand (i.e. no cosets), suggesting a large spacing between adjacent megaripples and/or relatively slow rates of deposition.

Most of the sand shows evidence of onshore directions of sediment transport, however some offshore movement of sand occurred as well. In some cases the gravel grades upwards into the parallel-laminated sands, suggesting

Fig.4. Outcrop photographs of the Santa Cruz coastal terrace sediments, northern Monterey Bay, California (arrow indicates direction of land when sediments were being deposited). A. Cross-laminated sand deposited by an onshore-migrating, lunate megaripple (LMR) with well developed brink point. Note the overlying graveliferous storm deposits. B. Parallel-laminated to hummocky cross-stratified sand overlain by a thick, graveliferous storm deposit. C. Complexly cross-stratified sand and gravel within the surf zone deposits (viewed parallel to the paleo-shoreline). D. Cosets of cross-stratified pebbly sand deposited by longshore-migrating megaripples (viewed parallel to the paleo-shoreline). E. Rip channel complex (RCC) within the inner part of the surf zone assemblage, overlain by foreshore deposits. F. Poorly stratified pebble gravels deposited at the "toe-of-beach".

deposition during the waning current of a single storm event (cf., Kumar and Sanders, 1974). In other cases, the contact is abrupt, and appears to record separate storm events of different magnitudes. Hummocky cross-stratification (Fig.4B) is present in only one area, where it grades both laterally and vertically into parallel laminations. The absence of "fairweather" deposits is indicated by the lack of any bioturbation or ripple cross-stratification.

The storm-dominated sediments are overlain by a 3.5 m thick surf zone assemblage of sediments which can be further subdivided into an outer and inner zone on the basis of dominantly longshore and offshore sediment transport, respectively. The outer surf zone assemblage is approximately 1.5 m thick, and mainly consists of cross-stratified coarse sand and pebble gravel (Fig.4C). In some places the gravel occurs as isolated linear ripple form sets ($L = 0.6-1$ m, $H = 5-7$ cm); elsewhere they are present at the base of 40—60 cm thick cosets of cross-stratified sand (Fig.4D) deposited by the migration of two- and three-dimensional megaripples ($H = 30$ cm, $L = 3$ m). The cross-stratified sediments all show a dominantly longshore direction of sediment transport, indicating formation in a longshore trough landward of the breaker zone (cf., Davidson-Arnott and Greenwood, 1974, 1976; Hunter et al., 1979). Both two- and three-dimensional megaripples of similar dimensions ($H = 25$ cm, $L = 2.5$ m) can be seen in longshore troughs along the northern parts of Monterey Bay today. In contrast to cross-stratified sands deposited farther offshore, the strata deposited within the surf zone typically form cosets of cross-stratified sand, and appear to have been deposited by rapidly migrating megaripples with relatively small wavelengths (approximately 3 m). Where one megaripple climbed over another, the base of the upper set of cross laminations is typically concave to tangential, rather than angular. In addition, the forsets have more shallow dips (typically 20°), and show no evidence of a distinct brinkpoint.

These sediments are overlain by up to 2 m of inner surf zone deposits which are characterized by coarse pebble gravels and interbedded, fine to coarse sand. The gravel beds range in thickness from a few centimeters to 0.75 m; they are generally thicker and more abundant at the top of the section. The gravels may be structureless, horizontally stratified, or form sets of cross-stratification up to 75 cm thick (Fig.4E). Where present, paleocurrent data is predominantly offshore, suggesting the gravels were deposited by offshore-flowing currents within the inner rough zone of Clifton et al. (1969). The coarsest gravels (long axis up to 10 cm) are found at the top of this section, which is interpreted to represent "toe-of-beach" deposits (Fig.4F) from which estimates of paleo-water depth can be made. Some thin interbedded zones of planar laminated sands (up to 30 cm thick) are also present within this zone, locally dipping up to 25° landward. These probably record the partial preservation of small bars migrating through the surf zone.

Foreshore deposits

This unit consists of up to 5 m of inclined parallel-laminated, well-sorted, medium- to coarse-grained sand and pebble gravel. The coarsest sediment occurs at the base of this unit, where layers of imbricated pebbles are locally common. Bioturbation is also restricted to the lowermost 0.5—0.75 m of this unit, where it may approach 100% in sandy sediments. The seaward-dipping laminated sands show abundant low-angle truncations (Fig.5A),

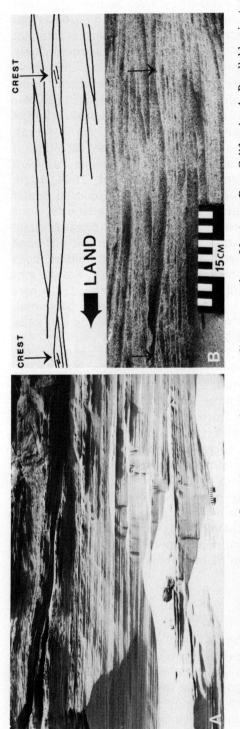

Fig.5. Swash zone deposits within the Santa Cruz coastal terrace sediments, northern Monterey Bay, California. A. Parallel-laminated swash zone (foreshore) deposits. Note the gentle seaward dips. B. Discontinuous laminae deposited by migrating antidunes within the swash zone.

typical of swash laminations (Thompson, 1937); inverse grading within these laminations is also relatively common (Clifton, 1969). Most of the laminations are characterized by extreme lateral continuity, however, some form discontinuous lenses (typically 1—2 cm thick and 15—30 cm long) in which some internal laminations dip at a low angle landward (Fig.5B). This is characteristic of laminations deposited by antidunes on the beach (Hayes and Kana, 1976). Thin laminations of heavy minerals are locally present, typically merging landward to form placer deposits up to 0.6 m thick. These deposits are typical of those formed within the swash zone of a high-energy, microtidal shoreline.

Vertical sequence of sedimentary structures

The Santa Cruz coastal terrace deposits are the result of sedimentation during a glacio-eustatic rise and fall of sea level. The transgressive marine deposits are almost completely lacking, however, because of their selective removal during the erosional transgression which accompanied the rise in sea level (Fisher, 1961; Dupré, 1975a; Ryer, 1977). Essentially all of the preserved sediments are those deposited during the interglacial highstand and subsequent lowering of sea level. The resultant regressive sequence of sedimentary structures (Fig. 6) is essentially that predicted for a moderate- to high-energy, barred nearshore marine environment (Davidson-Arnott and Greenwood, 1974, 1976; Hunter et al., 1979).

There are, however, some exceptions worthy of note. The gradual coarsening-upwards trend of mean grain size is almost completely obscured by the repeated introduction of coarse-grained sands and gravels during storm events (Fig.2). The graveliferous sediments were introduced into the shoreface by winter floods along the Pleistocene Pajaro River which was approximately 5 km to the southeast of the Manresa measured section (Fig.1). The gravels were transported laterally by longshore currents within the surf zone and then beyond the surf zone during storm events. In addition, the general lack of ripple and lunate megaripple cross-stratification which characterizes the modern zone of wave buildup (outer rough of

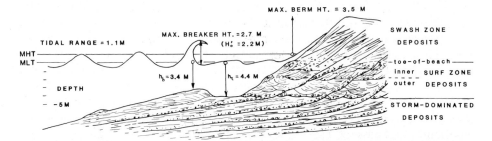

Fig.6. Facies associations and environmental reconstruction of the Santa Cruz coastal terrace deposits (h_b = depth of bar; h_t = depth of trough). Water depths were measured from mean low tide (MLT); elevation of the berm was measured from mean high tide (MHT).

Clifton et al., 1971), is evidence of the low preservation potential of these fair-weather features. Storm deposits, characterized by sharp erosional bases (with or without a basal gravel), overlain by parallel-laminated sand, have been preferentially preserved stratigraphically below the surf zone deposits, as previously noted by several authors (e.g., Clifton, 1976; Howard and Reineck, 1981). The lack of any significant hummocky cross stratification in these wave-dominated storm deposits is also noteworthy, however it probably simply reflects the relatively shallow depth (mainly <10 m) in which they were deposited. The highly bioturbated, lower shoreface deposits represent a zone where the rate of biogenic reworking is greater than the rate of storm deposition. It appears to correspond to the "transitional zone" which presently occurs in water depths of 9—18 m off the coast of southern California (Howard and Reineck, 1981).

RECONSTRUCTION OF PALEO-WAVE CONDITIONS

The main variables controlling nearshore marine sedimentation along wave-dominated coastlines are: (1) rate of sediment supply; (2) grain size; (3) water depth; and (4) wave height and period (e.g. Komar, 1974; Clifton, 1976). Grain size and associated sedimentary structures are observable on the outcrop; the other variables must either be assumed or calculated, as discussed below.

Calculation of paleo-water depths

The reconstruction of paleo-depth conditions is relatively easy if we can assume no change in the relative position of sea level during deposition of the sediments, and no compaction following deposition. Under such conditions, the water depth is approximated by the vertical distance a deposit occurs below sediment deposited at sea level (cf., Klein, 1974). Thus storm-generated sediments located 3 m below the top of the "toe-of-beach" deposits would have been deposited 3 m below low tide. Such a thickness/sea-level curve is shown on Fig. 7 (curve A).

The sediments of the Santa Cruz terrace are relatively young sands and gravels with no visible signs of significant compaction. In addition, the amount of tectonic uplift was negligible over the relatively short interval of time during which the sediments were deposited. The assumption that sea level did not change during the deposition of the sediment seems less valid, however. Where exposed, the marine sediments are approximately 1.5 km seaward of the maximum transgressive shoreline (Dupré, 1975b). If we assume no change in sea level occurred during the deposition of these sediments, then the maximum water depth 1.5 km offshore would equal the maximum thickness of the sediments deposited below sea level (i.e., 6 m). Studies by Bradley and Griggs (1976) show that the gradient of the wave-cut platform of the Santa Cruz coastal terrace is essentially the same as that of the modern day shelf (0.01—0.015), where the maximum water depth

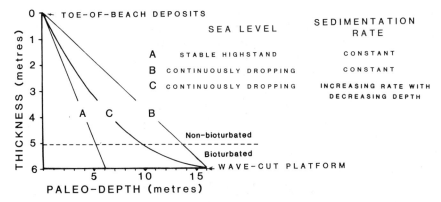

Fig.7. Selected curves relating stratigraphic thickness below toe-of-beach deposits with paleo-water depths. Curve C was used for the calculation of water depths in this paper. Dashed line indicates contact between highly bioturbated (lower shoreface) and non-bioturbated (upper shoreface) deposits.

1.5 km offshore today is approximately 16 m. This suggests that much of the deposition must have occurred during the late interglacial lowering of sea level, with a drop of up to 10 m having occurred during the deposition of the sediments (Fig.8). This is consistent with the fact that there has been almost no seaward progradation of the shoreline during the last 5000 yrs of the present highstand of sea level, presumably because of the relatively high wave regime of the Pacific Coast and low sediment input (Dupré, 1975a). Why progradation should have preferentially occurred during the early stages of sea level lowering is unclear; however, it could be the result of the increased rate of sediment influx due to fluvial entrenchment which accompanied the sea level lowering (Dupré, 1975a), increased wave attenuation due to the decreased gradient of the newly emergent shelf (Bradley and Griggs, 1976), or some element of both.

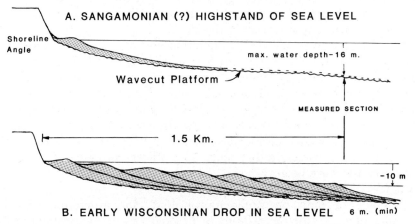

Fig.8. Configuration of the shelf and water depths during deposition of the Santa Cruz coastal terrace. Overlying eolian deposits are not shown for the purpose of clarity.

If the sediments at the base of the section were deposited in approximately 16 m water depth, and if the rates of sedimentation and sea-level drop were constant, then a linear thickness/water-depth curve can be made (Fig.7, curve B) to estimate maximum water depths. It seems more likely, however, that the rate of sedimentation was not constant, but rather was greatest near the shoreline, as implied by the non-linear curve (C) in Fig.7. This type of curve is also preferred on the basis of the predicted depth of the highly bioturbated, lower shoreface deposits. The transition from non-bioturbated to highly bioturbated sediment in southern California presently occurs at approximately 9 m water depth (Howard and Reineck, 1981). This transition is predicted to have occurred in water depths of 9—10 m using the non-linear curve (curve C), as opposed to 14 m water depths using the maximum water depth curve (curve B). For these reasons, I have chosen the non-linear curve (curve C) to determine paleo-water depths.

Calculation of paleo-wave heights

Several methods exist for the determination of ancient wave conditions from the stratigraphic record. Perhaps the most accurate of these are the equations which relate wave height (H) to maximum orbital velocity (U_{max}). Two such equations are provided below:

$$H = T \sinh (2\pi h/L) U_{max}/\pi \qquad (1)$$

$$H = T \sinh (2\pi h/L) (U_{max} - \Delta U_{max}/2)/\pi \qquad \text{(Clifton, 1976)} \qquad (2)$$

The first equation is based on Airy wave theory, whereas the second is a modified Stokes (second-order) equation which assumes negligible mass transport. Preliminary calculations using probable wave periods, wave heights, and water depths indicated that the waves responsible for the storm layers in the terrace deposits were relatively shallow-water waves, thus strictly speaking the Airy wave theory is inapplicable (see U.S. Army Coastal Engineering Research Center, 1975, fig. 2-7). In fact, the use of eq. 1 produces relatively small differences (see Table I), typically less than 5%, nonetheless, the Stokes equation was used in this study.

In order to calculate wave heights using either equation, it is necessary to know or to approximate the water depth (h), wave period (T), and maximum orbital velocity (U_{max}). Paleo-water depths were estimated using curve C (Fig.7), as discussed previously. The wave period was assumed to be 10 s, inasmuch as approximately 85% of the waves in Monterey Bay today have a period of 8—12 s (based on SSMO data summarized by Arnal, 1973). Only 8.5% of the waves have periods less than 8 s. It should be noted that the use of 8 or 12 s waves (rather than 10 s waves) produces relatively small differences for wave heights in shallow water (e.g., 3% deviation in 5 m water depths).

In the absence of data on ripple morphology, the minimum critical threshold value of U_{max} for the movement of sand can be determined as a function of wave period, grain size, and associated bedform (Komar, 1974;

TABLE I

Estimate of minimum storm wave heights (assuming $T = 10$ s), using eqs. 1 (Airy) and 2 (Stokes), for 10 selected storm deposits (A—J). U_{max} = maximum orbital velocity; ΔU_{max} = maximum orbital velocity asymmetry; h = water depth; H = wave height; H'_o = offshore wave height (ignoring effects of refraction and frictional attenuation)

	U_{max} (m s^{-1})	ΔU_{max} (m s^{-1})	h (m)	H_{airy} (m)	H_{stokes} (m)	H'_o (m)
A	0.7	0.05	11	1.7	1.7	1.7
B	1.25	0.05	8	2.5	2.4	2.4
C	1.0	0.2	7	1.7	1.6	1.6
D	1.0	0.2	6.8	1.7	1.6	1.6
E	1.1	0.05	6.5	1.9	1.9	1.6
F	1.0	0.05	5	1.5	1.5	1.3
G	1.0	0.05	4	1.3	1.2	1.0
H	2.0	0.4	3.5	2.5	2.2	1.9
I	2.0	0.4	3.4	2.5	2.2	1.9
J	0.9	0.05	3.2	1.1	1.0	0.8

Average H'_o = 1.6 m.

Clifton, 1976; Dingler, 1979). For the purpose of this study, mean grain size was determined by sieving; bedforms (e.g., lunate megaripples, upper flow regime plane beds) were inferred on the basis of preserved sedimentary structures. Given these data, critical values of U_{max} (and ΔU_{max} in the case of the Stokes equation) were determined using graphs provided by Clifton (1976).

The wave height as calculated by this procedure is a minimum wave height at the site where the sediment was deposited. In fact, this minimum height is typically greater than the minimum offshore wave height (H_o) because of the shoaling effect. This can cause some confusion unless nearshore wave heights are converted to their offshore wave equivalents for the purpose of comparisons. H'_o represents the offshore wave height ignoring the effects of wave refraction and frictional attenuation. It can be calculated from the ratio H/H'_o provided in tables in the Corps of Engineers Shoreline Protection Manual (U.S. Army Coastal Engineering Research Center, 1975, vol. 3). The assumption of no significant changes in wave height due to refraction appears to be valid, as most of the paleo-waves were from the southwest, and southwesterly waves today have a refraction coefficient (K_r) of approximately 1 (i.e., no refraction; Johnson, 1953). The effect of decreasing wave heights due to frictional attenuation was considered negligible, in part because of the relatively steep offshore gradients (0.01—0.015); however, this could represent a source of error (see Goldsmith, 1976).

Ten storm deposits were studied in order to calculate the height (and frequency of occurrence) of waves responsible for their deposition (Table I). These storm deposits are representative of the approximately 70% of the sediment beneath the surf zone assemblage that was deposited in response to shoreward propagating wave trains. The remaining 30% of the sediment deposited beyond the surf zone consists of pebble gravels and sands deposited by offshore-flowing storm currents (e.g., rip currents, wind-forced currents, and perhaps storm-surge ebb currents).

The minimum offshore wave height (H'_o) as calculated for the ten storm deposits averaged 1.6 m (range 0.8—2.4 m); such waves presently occur in Monterey Bay no more than 15% of the time (Arnal, 1973). Presumably these waves represent relatively frequent, small storms and post-storm recovery periods. The smaller waves which occurred during the remaining 85% of the time (i.e. "fairweather conditions") were probably capable of reworking the tops of the deposits, however their effects were not preserved because of subsequent reworking by storm waves. The coarser-grained storm sediments deposited by offshore-flowing storm currents presumably reflect larger storm events, however there is presently no method by which the height of the waves associated with these storm-generated currents can be derived.

The methods described above allow an estimate of the minimum size and recurrence interval of waves responsible for the deposition of most of the sediments deposited beyond the surf zone. It is also possible to estimate the wave conditions under which more shallow water sediments were deposited based on: (1) thickness of the surf zone assemblage; and (2) thickness of the swash zone assemblage.

The maximum thickness of the surf zone deposits is a function of the maximum depth of scour within the surf zone (either in the longshore trough or rip channels). In the case of the Santa Cruz coastal terrace deposits (Fig.2), the maximum thickness is 2.8 m, which corresponds to a maximum depth of scour in the longshore trough of 4.4 m below low tide (using curve C, Fig.7). If the associated longshore bar (Fig.5) during storms was approximately 1 m higher than the adjacent trough (cf., Bascom, 1964; Davidson-Arnott and Greenwood, 1976; Hunter et al., 1979; Sallenger et al., 1983), then the maximum depth of the bar crest during storms was approximately 3.4 m. The calculated depths to the trough (4.4 m) and to the bar (3.4 m) are similar to the observed depths to the storm-generated trough (3.2 m below mean sea level) and bar (4.4 m) measured by Sallenger et al. (1983) in the southern Monterey Bay region. The ratio of trough depth (h_t) to bar depth (h_b) is virtually identical (Pleistocene = 1.29; Modern = 1.31), and is within the range of many modern and experimental bar systems (see summary by Greenwood and Davidson-Arnott, 1975, p. 144).

It is possible to estimate the height of the breaking waves by using the depth of the bar crest to approximate the depth at which the waves are breaking. The relationship between the height of breaking waves (H_b) and breaker depth (h_b) for solitary waves is commonly approximated by the following:

$$\gamma_b = H_b/h_b = 0.78 \quad \text{(Weigel, 1964)} \tag{3}$$

Assuming h_b = 3.4 m below low tide, we can use this value to estimate that the sediment in the surf zone was deposited in response to breaking waves no larger than 2.7 m high. This is equivalent to a maximum offshore wave height (H'_o) of 2.2 m, which presently occurs approximately 8% of the time. The actual value of γ_b depends on wave shape, offshore wave steepness, and beach slope, however, with experimental values ranging from 0.7 to 1.2 (Galvin, 1972; Komar, 1976). Such values serve to limit the likely range of maximum wave conditions (i.e. H'_o = 2—3.4 m), with the 2.2 m waves being used as a representative value.

The thickness of the swash zone deposits also provides an indication of the height of waves responsible for their deposition. The maximum thickness is a function of the paleo-tidal range and the maximum height of the berm above high tide. The mean tidal range during the last interglacial is assumed to have been the same as the present (1.1 m), thus the maximum berm height can be estimated as being equal to the total thickness of the swash zone deposits (4.5 m) minus the paleo-tidal range (Fig.5). This suggests a maximum berm height of approximately 3.5 m, which can be related to associated offshore wave heights by the following equation:

Max. berm ht. = 1.3 $H_o K_r$ (Bascom, 1964)   (4)

If the characteristic refraction coefficient (K_r) = 1 (as discussed earlier), then the swash zone sediments were deposited in response to waves whose offshore heights were no greater than approximately 3 m; the largest of these waves presently occur approximately 2% of the time.

CONCLUSIONS

The Santa Cruz coastal terrace deposits consist of a regressive sequence of marine sediments deposited in a high-energy, barred nearshore marine environment during the end of the last interglacial highstand of sea level. This sequence can be subdivided into several depth-dependent facies on the basis of paleocurrent data and vertical sequence of sedimentary structures. The lower shoreface deposits consist of highly bioturbated, parallel-laminated sands deposited in water depths of approximately 10—16 m. These are overlain by a sequence of interbedded sand and gravel deposited in response to storm waves in water depths of approximately 5—10 m. These are, in turn, overlain by a surf zone assemblage of sediments consisting of complexly cross-stratified sands and gravels deposited in water depths of 0—5 m. The top of the marine section consists of sediments deposited in the swash zone, where storm berms were built up to 3.5 m above high tide.

An analysis of the magnitude and frequency of storm conditions responsible for the deposition of these deposits indicates that they preferentially record storm events of intermediate frequency and magnitude (cf., Wolman and Miller, 1960). In addition, it appears that different facies within the high-energy, nearshore environment record events of different recurrence intervals. Sediments within the surf zone were deposited under relatively small storm and post-storm conditions, whereas sediments farther offshore were deposited by relatively larger, less frequent storm events (cf., Hunter et al., 1979).

ACKNOWLEDGEMENTS

This study was initiated as part of my dissertation at Stanford University, under the supervision of E.I. Rich; J.C. Ingle, Jr. and W.R. Dickinson also served as committee members. Additional work was done as part of a regional study of the Quaternary sediments in the Monterey Bay region, under the supervision of E.E. Brabb (U.S. Geological Survey). Many other people aided me throughout the course of this study, however I would like to especially thank H.E. Clifton (U.S. Geological Survey), who showed me that many sediments record the processes by which they were formed, and Brian Greenwood (Univ. of Toronto), for his many useful editorial comments. I would also like to acknowledge partial funding from the University of Houston Geology Foundation.

REFERENCES

Allen, J.R.L., 1979. A model for the interpretation of wave ripple-marks using their wavelength, textural composition, and shape. J. Geol. Soc. London, 136: 673—682.

Arnal, R.E., 1973. Marine geology of Monterey Bay near Moss Landing. In: Sand Transport Studies in Monterey Bay, California, Moss Landing Tech. Rep. 73-5, pp.III-1—III-21.

Bascom, W., 1964. Waves and Beaches. Garden City, Doubleday, 267 pp.

Bourgeois, J., 1980. A transgressive shelf sequence exhibiting hummocky stratification: The Cape Sebastian sandstone (Upper Cretaceous), southwestern Oregon. J. Sediment. Petrol., 50: 681—702.

Bradley, W.C. and Griggs, G.B., 1976. Form, genesis and deformation of Central California wave-cut platforms. Geol. Soc. Am. Bull., 87: 433—449.

Brenchley, P.J., Newall, G. and Stanistreet, I.G., 1979. A storm surge origin for sandstone beds in an epicontinental platform sequence, Ordovician, Norway. Sediment. Geol., 22: 185—217.

Campbell, C.V., 1971. Depositional model, Upper Cretaceous Gallup beach shoreline, Ship Rock area, northwestern New Mexico. J. Sediment. Petrol, 41: 395—409.

Clifton, H.E., 1969. Beach lamination: nature and origin. Mar. Geol., 7: 553—559.

Clifton, H.E., 1976. Wave-formed sedimentary structures — a conceptual model: In: R.A. Davis, Jr. and R.L. Ethington (Editors), Beach and Nearshore Sedimentation. Soc. Econ. Paleontol. Mineral., Spec. Publ., 24: 126—148.

Clifton, H.E., 1981. Progradational sequences in Miocene shoreline deposits, southeastern Caliente range, California. J. Sediment. Petrol., 51: 165—184.

Clifton, H.E. and Thompson, J.K., 1978. *Macraronichnus segregatis*: a feeding structure of shallow marine polychaetes. J. Sediment. Petrol., 48: 1293—1302.

Clifton, H.E., Hunter, R.E. and Phillips, R.L., 1971. Depositional structures and processes in the non-barred high-energy nearshore. J. Sediment. Petrol., 41: 651—670.

Davidson-Arnott, R.G.D. and Greenwood, B., 1974. Bedforms and structures associated with bar topography in the shallow-water wave environment, Kouchibouguac Bay, New Brunswick, Canada. J. Sediment. Petrol., 44: 698—704.

Davidson-Arnott, R.G.D. and Greenwood, B., 1976. Facies relationships on a barred coast, Kouchibouguac Bay, New Brunswick, Canada. In: R.A. Davis, Jr. and R.L. Ethington (Editors), Beach and Nearshore Sedimentation. Soc. Econ. Paleontol. Mineral., Spec. Publ., 24: 149—168.

Dingler, J.R., 1979. The threshold of grain motion under oscillatory flow in a laboratory wave channel. J. Sediment. Petrol., 49: 287—294.

Dupré, W.R., 1975a. Quaternary history of the Watsonville lowlands, north-central Monterey Bay region, California. Unpubl. Ph.D. Diss., Stanford University, Palo Alto, Calif.

Dupré, W.R., 1975b. Maps showing geology and liquefaction potential of the Quaternary deposits in Santa Cruz County, California. U.S. Geol. Surv., Misc. Field Studies Map, MF-643, 2 sheets, scale 1:62,500.

Dupré, W.R., Clifton, H.E. and Hunter, R.E., 1980. Modern sedimentary facies of the open Pacific coast and Pleistocene analogs from Monterey Bay, California. In: M. E. Fields et al. (Editors), Quaternary Depositional Environments of the Pacific Coast. Pacific Coast Paleogeography Symposium 4, Pacific Section, Soc. Econ. Paleontol. Mineral., Los Angeles, Calif., pp.105—120.

Fisher, A.G., 1961. Stratigraphic record of transgressing seas in light of sedimentation on the Atlantic coast of New Jersey. Bull. Am. Assoc. Pet. Geol., 45: 1656—1666.

Flint, R.F., 1971. Glacial and Quaternary Geology. Wiley, New York, N.Y., 892 pp.

Galvin Jr., C.J., 1972. Waves breaking in shallow water. In: R.E. Meyer (Editor), Waves on Beaches and Resulting Sediment Transport. Academic Press, New York, N.Y., pp.413—456.

Goldsmith, V., 1976. Continental shelf wave climate models: Critical links between shelf hydraulics and shoreline processes. In: R.A. Davis, Jr. and R.L. Ethington (Editors), Beach and Nearshore Sedimentation. Soc. Econ. Paleontol. Mineral., Spec. Publ., 24: 24—47.

Greenwood, B. and Davidson-Arnott, R.G.D., 1975. Marine bars and nearshore sedimentary processes, Kouchibouguac Bay, New Brunswick, Canada. In: J. Hails and A. Carr (Editors), Nearshore Sediment Dynamics and Sedimentation. Wiley, London, pp.123—150.

Hamblin, A.P. and Walker, R.G., 1979. Storm-dominated shallow marine deposits: the Fernie—Kootenay (Jurassic) transition, southern Rocky Mountains. Can. J. Earth Sci., 16: 1673—1690.

Harris, D.L., 1972. Wave statistics for coastal regions. In: D.J.P. Swift, D.B. Duane and O.H. Pilkey (Editors), Shelf Sediment Transport: Process and Pattern. Dowden, Hutchinson and Ross, Stroudsburg, Pa., pp.99—125.

Hayes, M.O., 1967. Hurricanes as geologic agents: case studies of hurricane Carla, 1961, and Cindy, 1963. Rep. Inv. 61, Univ. Tex., Austin, Bureau Econ. Geol., 54 pp.

Hayes, M.O. and Kana, T.W., 1976. Terrigenous Clastic Depositional Environments. Tech. Rep. 11-CRD, Coastal Research Div., Univ. S. Carolina, Columbia, S.C., 301 pp.

Howard, J.D., 1972. Trace fossils as criteria for recognizing shorelines in the stratigraphic record. In: J.K. Rigby and W.K. Hamblin (Editors), Recognition of Ancient Sedimentary Environments. Soc. Econ. Paleontol. Mineral., Spec. Publ., 16: 215—225.

Howard, J.D. and Reineck, H.-E., 1981. Depositional facies of high-energy beach-to-offshore sequence: Comparison with low energy sequence. Bull. Am. Assoc. Pet. Geol., 65: 807—830.

Hunter, R.E. and Clifton, H.E., 1982. Cyclic deposits and hummocky cross-stratification of probable storm origin in Upper Cretaceous rocks of the Cape Sebastian area, southwestern Oregon. J. Sediment. Petrol., 52: 127—143.

Hunter, R.E., Clifton, H.E. and Phillips, R.L., 1979. Depositional processes, sedimentary structures, and predicted vertical sequences in barred nearshore systems, southern Oregon coast. J. Sediment. Petrol., 49: 711—726.

Ingle Jr., J.C., 1966. The Movement of Beach Sand. (Developments in Sedimentology, 5) Elsevier, Amsterdam, 221 pp.

Johnson, J.W., 1953. Engineering aspects of diffraction and refraction. Trans. Am. Soc. Civ. Eng., 118: 617—648.

Klein, G. deV., 1974. Estimating water depths from analysis of barrier island and deltaic sedimentary sequences. Geology, 2: 408—412.

Komar, P.D., 1974. Oscillatory ripple marks and the evaluation of ancient wave conditions and environments. J. Sediment. Petrol., 44: 169—180.

Komar, P.D., 1976. Beach Processes and Sedimentation. Prentice-Hall, Englewood Cliffs, N.J., 429 pp.

Komar, P.D. and Miller, M.C., 1974. Sediment threshold under oscillatory waves. Am. Soc. Civ. Eng., Proc. 14th Conf. on Coastal Engineering, pp.756—775.

Kreisa, R.D., 1981. Storm-generated sedimentary structures in subtidal marine facies with examples from the middle and upper Ordovician of southwestern Virginia. J. Sediment. Petrol., 51: 823—848.

Kumar, N. and Sanders, J.E., 1974. Characteristics of shoreface storm deposits: modern and ancient examples. J. Sediment. Petrol., 46: 145—162.

Miller, M.C. and Komar, P.D., 1980. A field investigation of the relationship between oscillatory ripple spacing and the near-bottom water orbital motions. J. Sediment. Petrol., 50: 183—191.

Morton, R.A., 1981. Formation of storm deposits by wind-forced currents in the Gulf of Mexico and the North Sea. In: S.-D. Nio, R.T.E. Shuttenhelm and Tj.C.E. van Weering (Editors), Holocene Marine Sedimentation in the North Sea Basin. Int. Assoc. Sedimentol., Spec. Publ., 5: 385—396.

Nelson, C.H., 1982. Modern shallow-water graded sand layers from storm surges, Bering shelf: a mimic of Bouma sequences and turbidite systems. J. Sediment. Petrol., 52: 537—545.

Reineck, H.E. and Singh, I.B., 1975. Depositional Sedimentary Environments. Springer, New York, N.Y., 439 pp.

Roep, Th.B., Beets, D.J., Dronkert, H. and Pagnier, H., 1979. A prograding coastal sequence of wave-built structures of Messinian age, Sorbas, Alameria, Spain. Sediment. Geol., 22: 135—163.

Ryer, T.A., 1977. Patterns of Cretaceous shallow-marine sedimentation, Coalville and Rockport areas, Utah. Geol. Soc. Am. Bull., 88: 177—188.

Sallenger Jr., A.H., Howard, P.C., Fletcher III, C.H. and Howd, P.A., 1983. A system for measuring bottom profile, waves and currents in the high-energy nearshore environment. Mar. Geol., 51: 63—76.

Tanner, W.F., 1971. Numerical estimates of ancient waves, water depths and fetch. Sedimentology, 16: 71—88.

Thompson, W.O., 1937. Original structures of beaches, bars, and dunes. Geol. Soc. Am. Bull., 48: 723—752.

U.S. Army Coastal Engineering Research Center, 1975. Shore Protection Manual. U.S. Govt. Printing Office, Washington, D.C., 3 vols.

Vos, R.G. and Hobday, D.K., 1977. Storm beach deposits in the late Palaeozoic Ecca Group of South Africa. Sediment. Geol., 19: 217—232.

Weigel, R.L., 1964. Oceanographical Engineering. Prentice Hall, Englewood Cliffs, N.J., 532 pp.

Wolman, M.G. and Miller, J.P., 1960. Magnitude and frequency of forces in geomorphic processes. J. Geol., 68: 54—74.

RECONSTRUCTION OF ANCIENT SEA CONDITIONS WITH AN EXAMPLE FROM THE SWISS MOLASSE

PHILIP A. ALLEN

Department of Geology, University College, P.O. Box 78, Cardiff CF1 1XL (United Kingdom)

(Received March 15, 1983; revised and accepted August 29, 1983)

ABSTRACT

Allen, P.A., 1984. Reconstruction of ancient sea conditions with an example from the Swiss Molasse. In: B. Greenwood and R.A. Davis, Jr. (Editors), Hydrodynamics and Sedimentation in Wave-Dominated Coastal Environments. Mar. Geol., 60: 455—473.

Ancient sea conditions can be estimated from the grain size, spacing and steepness of preserved ripple-marks. The element of greatest uncertainty in such reconstructions is the relationship between near-bed orbital diameter of water particles and the ripple spacing. This relationship is simple for vortex ripples of high steepness but is problematical for the low-steepness forms known as post-vortex, rolling-grain or anorbital ripples.

The existence field for wave ripples is between the threshold velocity for sediment movement and the onset of sheet flow, most low-steepness forms occurring close to the bed planation threshold. A range of maximum period of formative waves can be obtained using combinations of orbital diameter and orbital velocity, assuming linear wave theory to be a reasonable approximation.

Probable wave heights, wave lengths and water depths can be investigated using the transformation of wave parameters in shallowing waters and the constraints on wave dimensions provided by the wave-breaking condition. Given reasonable estimates of wave height, crude estimates of wave power allow a comparison of ancient wave-influenced sequences with modern counterparts.

Wave ripple-marks preserved in the Upper Marine Molasse of western Switzerland have been investigated. Results, which are in agreement with regional geology, suggest deposition in a seaway of approximately 100 km width, where moderate period waves ($T = 3$—6 s) were generated. The depositional facies belts were adjusted to the prevailing waves, tides and fluvial outflows.

INTRODUCTION

Since Harms (1969), Tanner (1971) and Komar (1974), with varying degrees of rigour, proposed the use of preserved wave ripple-marks in reconstructing ancient wave conditions, surprisingly few ancient sequences have been analysed in this way. Further encouragements both from theoretical and empirical (Clifton, 1976; Allen, 1979; Dingler, 1979; Miller and Komar, 1980a) and field studies (Newton, 1968; Cook and Gorsline, 1972; Stone and Summers, 1972; Dingler and Inman, 1977; Miller and Komar, 1980b) have not yet resulted in a flourish of case-studies of ancient marine

0025-3227/84/$03.00 © 1984 Elsevier Science Publishers B.V.

or lacustrine sedimentary basins. Allen (1981a) analysed a Devonian lacustrine basin-fill and Homewood and Allen (1981) studied the marine sediments of the "Upper Marine Molasse" of Switzerland, but to the author's knowledge very few oceanic coastline deposits have been comprehensively treated.

There are, of course, formidable assumptions and approximations which must be made in quantifying ancient wave conditions (Allen, 1981b), but with knowledge of the pitfalls, important insights into palaeoenvironments and processes can be acquired. Without knowledge of the assumptions and approximations, spurious results will be obtained. The purpose of this paper is to describe the methods used by Homewood and Allen (1981) in their analysis of the Upper Marine Molasse of western Switzerland and to synthesize some of the more important elements of particular uncertainty in the reconstruction of ancient sea conditions. This contribution therefore acts as something of a cookbook which may encourage workers to analyse or re-analyse quantitatively their ancient wave-influenced sequences.

METHODS

Are they wave ripple-marks?

Boersma (1970) distinguished a number of features thought to be characteristic of wave-generated ripple cross-lamination, including irregular or catenary-arcuate lower bounding surfaces, bundled upbuilding of cross-sets within ripple cross-laminated lenses, chevron structures, cross-stratal offshoots and form-discordancy. Tanner (1967) and Reineck et al. (1971) likewise summarized wave ripple-marks in terms of ripple steepness, symmetry and crestal arrangement in plan. Wave ripple-marks characteristically have straight crestlines which commonly bifurcate, low ripple indices and symmetrical to near-symmetrical profiles.

There have been relatively few studies of the natural variability in morphology and structure of wave ripple-marks. A very wide spectrum of ripple morphologies has been described from experimental studies (Bagnold, 1946; Manohar, 1955; Dingler, 1974; Sleath, 1976), ranging from the very flat varieties which have variously been termed "rolling grain ripples" (Bagnold, 1946; Sleath, 1976; Allen, 1979) and "post-vortex ripples" (Dingler, 1974; Dingler and Inman, 1976) to the very steep varieties termed 'vortex ripples' (Bagnold, 1946; Sleath, 1976) or 'orbital ripples' (Clifton, 1976). The identification of these ripple types in natural environments has not always been unequivocal.

The relationship between orbital diameter of water particles and ripple spacing

Miller and Komar (1980a) summarized laboratory experiments seeking to determine the relationship between near-bed orbital diameter and ripple spacing. Oscillating beds (Bagnold, 1946; Manohar, 1955; Kalkanis, 1964; Sleath, 1975, 1976), oscillating water tunnels or U-tubes (Carstens et al., 1969; Chan et al., 1972; Mogridge, 1973; Brebner and Reidel, 1973; Lofquist, 1977, 1978) and wave flumes (Yalin and Russell, 1962; Horikawa and Watanabe, 1967; Mogridge and Kamphuis, 1973; Dingler, 1974; Dingler and Inman, 1977) have been used to study oscillatory flows. From these data, for small orbital diameters, the relationship between near-bed orbital diameter (d_0) and ripple spacing (λ) is very simple (Fig.1):

$$\lambda = 0.65 \, d_0 \tag{1}$$

(Miller and Komar, 1980, p.178), the flume and U-tube data providing the best fit to this curve. Equation 1 has a weak Reynolds number dependence demonstrated independently by Sleath (1976) and Japanese workers (e.g. Hom-Ma et al., 1965), but can be neglected here for simplicity. As near-bed orbital diameter increases eq.1 no longer holds, becoming invalid as a function of grain size:

$$\lambda_{\max} = 0.0028 \, D^{1.68} \tag{2}$$

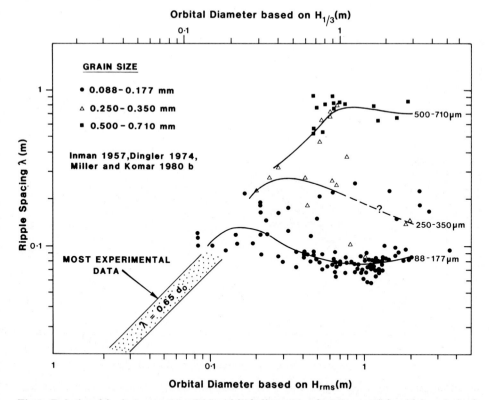

Fig.1. Relationship between near-bed orbital diameter of water particles (d_0) and ripple spacing (λ) with data sources indicated. Note the poor correlation between d_0 and λ for d_0 and $\lambda \gtrsim 0.2$ m. $H_{1/3}$ is significant wave height, H_{rms} is root mean square of wave height. Modified from Miller and Komar (1980b).

(Miller and Komar, 1980, p.180) where D is the median grain-diameter in microns and λ_{max} is in centimetres.

Allen (1979) showed that ripple steepness is an important factor influencing wave reconstructions. He plotted the data of Inman (1957), Kennedy and Falcon (1965), Manohar (1955), Inman and Bowen (1963), Hom-Ma et al. (1965), Carstens et al. (1969), Lofquist (1978) and Sleath and Ellis (1978) on a dimensionless ripple spacing (λ/D) versus dimensionless orbital diameter (d_0/D) diagram with ripple steepness as a further variable (Fig.2). A range of dimensionless orbital diameters could be obtained for a given ripple steepness. Equation 1 then merely represented one limit on Allen's fig.2 (1979, p.676) for vortex ripples, the other limit being determined for post-vortex or rolling grain ripples (Sleath, 1975, 1976) as:

$$0.036 < \lambda/d_0 < 0.059 \tag{3}$$

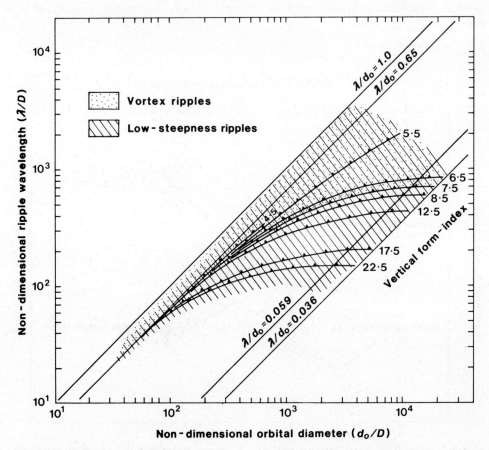

Fig.2. Occurrence of wave ripple-marks as a function of dimensionless ripple spacing (λ/D) and dimensionless near-bed orbital diameter of water particles (d_0/D). The lower limit of ripple occurrence is given by $\lambda/d_0 = 0.65$ or 1.0 and the upper limits for given ripple steepnesses (vertical form indices) are given by the series of curves. Note that the curves represent the upper limit of d_0/D and are not isopleths. After Allen (1979, p.676).

Allen (1981) suggested that the use of a wide range of ripple steepness inevitably led to an unacceptably wide range of estimated orbital diameters, making wave reconstructions hazardous.

The relationship between ripple spacing and grain and flow parameters can be expressed as:

$$\lambda = F(d_0, \nu, D, \rho_s, \rho, g, \omega) \qquad (4)$$

Since $U_0 = \omega d_0/2$, Sleath (1976) was able to express the relationship between orbital diameter and ripple spacing for "rolling grain ripples" as a function of four dimensionless variables:

$$\frac{d_0}{2\lambda} = F[\beta D, R, \rho_s/\rho, (\rho_s - \rho)gD/\rho\omega\nu] \qquad (5)$$

where $R = U_0/(\omega\nu)^{1/2}$ is a form of wave Reynolds number and $\beta = (\omega/2\nu)^{1/2}$. It can be seen that three of these dimensionless groups contain ω and one contains U_0. Only prior knowledge of U_0 and an iterative solution of Sleath's equations (1976, p. 78) would allow a solution to be made.

Clearly, the best results will come from those ripple-marks for which an unambiguous relationship between orbital diameter and ripple spacing exists. Such vortex ripple-marks are defined by the incidence of flow separation over a steep crest. Bagnold (1946), J.F.A. Sleath (pers. commun., 1979) and Allen (1979) suggested limiting ripple steepnesses (expressed as a vertical form index, VFI; Bucher 1919) of 8.0, 8.3 and 7.5, respectively, and Dingler and Inman (1977) stated an average value of 6.7 for vortex ripples. It is worth emphasising that ripple steepness is a function of several fluid, sediment and flow-related variables but that ripple steepness alone is the predominant control on the existence of vortex ripples. Extreme caution must be exercised in analysing low-steepness ripple-marks for wave reconstructions and it is recommended that attention is focused on vortex ripple-marks possessing vertical form indices of less than 7.5 and certainly less than 10.

The critical velocities for wave ripple formation

The critical threshold for entrainment under waves is given by a modified Shields parameter, and is determined by grain and fluid density (ρ_s, ρ), fluid viscosity (μ), grain diameter (D) and near-bed orbital diameter of water particles (d_0). For grain sizes of less than 0.5 mm, Komar and Miller's (1973) relation is:

$$\rho U_t^2/(\rho_s - \rho)gD = 0.21 (d_0/D)^{1/2} \qquad (6)$$

which is based exclusively on Bagnold's (1946) data and corresponds to laminar boundary layers. For grain sizes greater than 0.5 mm, Komar and Miller (1973) suggested the expression:

$$\rho U_t^2/(\rho_s - \rho)gD = 0.46 \pi (d_0/D)^{1/4} \qquad (7)$$

derived from the data of Rance and Warren (1969) which applies to turbulent boundary layers.

Ripple steepness is intimately related to processes in the wave boundary layer. It is the near-bed curvature-related drift velocities (Sleath, 1975, 1976) which are directly responsible for wave ripple-mark formation (Bagnold, 1946; Kaneko and Honji, 1979), whereas it is the purely oscillatory (simple harmonic) component which causes most grain movement. Allen (1979, fig.1, p. 675) plotted the purely oscillatory component, U_{max}, against grain diameter for a range of steepness values (Fig.3). Together with the analysis of Komar and Miller (1976), Allen's compilation shows that wave ripple-marks occur at orbital velocities well above those at the threshold condition, up to the point at which ripples are destroyed and sheet flow commences (Fig.3).

Komar and Miller (1976) gave the critical relative stress for ripple disappearance as a function of grain size alone, as:

$$\theta_c = 0.413 \, D^{-0.396} \tag{8}$$

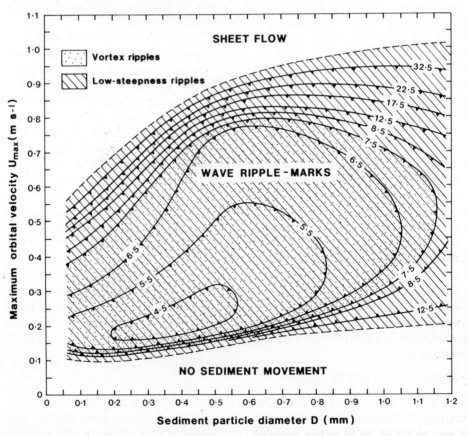

Fig.3. Occurrence of wave ripple-marks as a function of maximum near-bed orbital velocity of water particles (U_{max}) and grain size of sediment (D_{50}). The curves for ripple steepness (vertical form index) enclose the existence-field of ripple-marks with a given steepness. After Allen (1979, p.675).

or in terms of grain Reynolds number as:

$$\theta_c = 4.40\,(U_{max}D/\nu)^{-0.333} \tag{9}$$

both based on Manohar's (1955) data (Fig.4). The simpler method is to go directly to Allen's (1979) fig.1 which is based on a far wider range of data.

Calculation of wave period

Using linear wave theory, which is adequate for these purposes (Le Méhauté et al., 1969), it is possible to relate near-bottom orbital diameters and velocities to wave period (T) with the simple expression:

$$U_{max} = \pi d_0/T \tag{10}$$

Equation 10 indicates that although the entire range of conditions from the threshold velocity to the destruction velocity should be studied, it is the threshold condition (minimum U_0) which gives the maximum wave period. Maximum wave periods are of greater interest in reconstructing ancient sea conditions. If one experiments with the limits of the range of d_0 obtained above, a range of T for both the threshold and destruction condition is obtained.

The rationale behind calculating T is that the period of gravity waves is determined by wind strength, duration and fetch. Placing reasonable limits on wind speed and duration (for example, for typical fair-weather and typical storm conditions), it is possible to use wave forecasting methods to derive fetch limits for calculated wave periods (Sverdrup and Munk, 1946; refined by Bretschneider, 1970, or Darbyshire and Draper, 1963). In fetch-limited seas the analysis of Neumann (1953) is particularly useful (Fig.5).

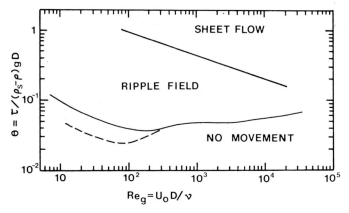

Fig. 4. Occurrence of wave ripple-marks as a function of a Bagnold (1963)-Shields (1936) relative stress (θ) and grain Reynolds number (Re_g). The ripple existence-field was based on the laboratory data of Manohar (1955), Horikawa and Watanabe (1967) and Carstens et al. (1969) and the field data of Inman (1957). The bed planation threshold was constructed from the data of Manohar (1955) and supported by the theoretical criterion of Bagnold (1956, 1966). After Komar and Miller (1975, p.701).

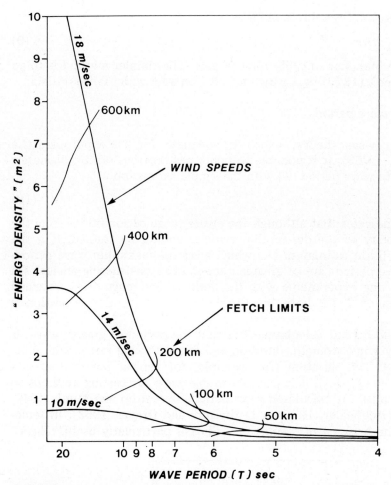

Fig.5. The energy density spectra for sea conditions where fetch is limiting, after Neumann (1953).

Calculated fetches for ancient waves are an important asset in palaeogeographical studies. Homewood and Allen (1981) used fetch length calculated from Miocene wave ripple-marks to confirm traditional views, based on the distribution of marine facies, on the size of the peri-Alpine sea in Switzerland. Allen (1981) used fetch data to postulate the extent of a Devonian lacustrine basin where outcrops were incomplete.

Simulation of wavelength, wave height and water depth

The orbital diameter of water particles near the bed is the result of a wave of period T, height H and wavelength L acting in water depth h. It follows that it is impossible to obtain a unique solution to H, L and h; it is only possible to obtain combinations of parameters. This is why estimation of H, L and h is well suited to computer simulation (see Komar, 1974, for instance).

Because deep-water waves are unaffected by water depth, it is possible to construct deep-water wavelength based purely on wave period, as follows:

$$L_\infty = gT^2/2\pi \tag{11}$$

which is a simplification of the general case:

$$L = (gT^2/2\pi)\tanh\left(\frac{2\pi h}{L}\right) \tag{12}$$

As the wave travels into shallower water its form changes so that orbital diameter of water particles near the bottom is given by:

$$d_0 = H/\sinh(2\pi h/L) \tag{13}$$

It is necessary therefore to study the transformations which take place in a wave of known period T as it moves into shallower water. This can be done in various ways, but a relatively painless method is to calculate h/L_∞ for the primary field of interest for water depth and use the graphs for Airy wave transformations provided by Wiegel (1964) (Fig.6). Alternatively Eckart (1952) gave the approximation:

$$L = L_\infty \text{ abs}[\tanh(2\pi h/L_\infty)] \tag{14}$$

The variation of H is of particular interest since it affects near-bed orbital diameter. Assuming that in order to form the observed ripple-marks the waves possessed finite near-bottom orbital diameters (i.e., the formative waves were not deep-water waves), eq.13 can be used to obtain wave height.

Wave breaking

There is another limit which is critical to the validity of the reconstructed wave parameters, that of wave breaking. Miche (1944) gives the limiting steepness for waves in water of finite depth as:

$$(H/L)_{\text{lim}} = 0.142\tanh(2\pi h/L) \tag{15}$$

and in progressively shallower water (McCowan, 1894) waves break at:

$$(H/h)_{\text{lim}} = 0.78 \tag{16}$$

Equation 16 is sensitive to beach slope (Ippen and Kulin, 1955). For small slopes ($\tan \beta < 0.07$):

$$0.72 < (H/h)_{\text{lim}} < 1.18 \tag{17}$$

and for a reasonable beach slope of 0.003:

$$(H/h)_{\text{lim}} = 0.88 \tag{18}$$

which is in close agreement with eq.16. Only rarely do geologists have data on ancient beach slope, so it is normal to simply implement eqs. 15 and 16.

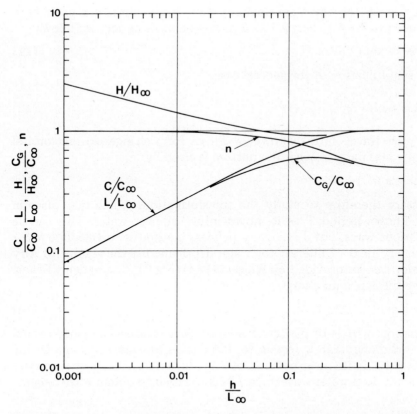

Fig.6. The shoaling transformations for Airy waves as functions of the ratio of water depth to deep-water wavelength, h/L_∞. C, C_g and C_∞ are wave celerity (phase velocity), group velocity and deep-water celerity. H and H_∞ are wave height and deep-water wave height. L and L_∞ are surface wavelength and deep-water wavelength. n is a shoaling coefficient relating wave celerity to group velocity. After Wiegel (1964).

Energy flux (wave power)

The motion of waves produces a transfer of energy over the sea surface which is of great interest to physical oceanographers and coastal engineers. A wave train possesses a total energy made up of two components. The potential energy component is caused by water particles being displaced from the still water level, and the kinetic energy component accounts for the orbital motions of water particles. The total energy is given by:

$$E = \rho g H^2/8 \tag{19}$$

(Teleki, 1972, p.38; Madsen, 1976, p.72) where E is a surface energy density per unit width of wave crest. The rate at which this energy is propagated in the direction of wave advance is the energy flux (or wave power), and is directly related to the velocity of the wave train (group velocity C_g) rather than the phase velocity of individual waveforms. The energy flux per unit length of wave crest is:

$$P = \rho g H^2 C_g/8 \tag{20}$$

Group velocity can be obtained from Wiegel's (1964) Airy wave transformations as the dimensionless ratio C_g/C_∞ (Fig. 6) by use of the expression for wave phase velocity (celerity):

$$C = L/T = (gT/2\pi)\tanh(2\pi h/L) \tag{21}$$

where $C = C_\infty = gT/2\pi$ in deep water.

Energy flux for reconstructed wave conditions should decrease in shallowing waters, the ratio of deep-water energy flux to shallow-water energy flux providing an index of power attenuation. Homewood and Allen (1981) used estimates of energy flux to provide modern analogues to the Miocene Sea of Switzerland by comparison with the energy flux (wave power) data of Coleman and Wright (1975) and Wright and Coleman (1973).

THE UPPER MARINE MOLASSE OF SWITZERLAND

The wedge of Tertiary clastic sediments north of the Swiss Alps is traditionally subdivided into four units (Matter et al., 1980; and Fig.7A). The lowermost, of Oligocene age, is termed the Lower Marine Molasse and represents offshore mudstones with storm sandstones and culminates in a wave-dominated shoreline sequence. The overlying unit, the Lower Freshwater Molasse comprises predominantly fluviatile clastic sediments with some playa and lacustrine sediments. The third unit is the Upper Marine Molasse of Miocene (Burdigalian) age, consisting of wave- and tide-dominated shallow marine sandstones and the conglomerates of fringing fan-deltas. The uppermost unit, the Upper Freshwater Molasse is composed of alluvial fan and fluviatile clastics and lacustrine deposits.

The Upper Marine Molasse was deposited in a peri-Alpine depression north of the Alps (Fig.7B) which extended eastward to the Austro-Vienna basin and southwestward into France. Although wave-formed structures occur throughout the Upper Marine Molasse, the present study concerns the area in the vicinity of Fribourg where road cuttings and river gorges provide spectacular sections through the marine sand bodies.

In the Fribourg area Homewood (1978, 1981) described four facies belts in the Upper Marine Molasse (Fig.7C). The *proximal facies belt*, restricted to the south (Hoffman, 1960), is composed of thick fan-delta deposits which represent the major feeder systems of sediments from the Alpine hills to the marine seaway. A *coastal facies belt* contains abundant tidal sandwaves (Allen and Homewood, 1984) interbedded with intertidal sandflat deposits and distributary and tidal channel sandstones. The *nearshore facies belt* is constructed of thick, elongate subtidal shoals with shoal crevasse deltas and intershoal swales. The *offshore facies belt* contains sandy and pebbly coquinas deposited as giant-sized flow-transverse tidal bedforms.

Wave ripple-marks are very common in the coastal, nearshore and offshore facies belts. Homewood and Allen (1981, pp.2540—2543) summarized

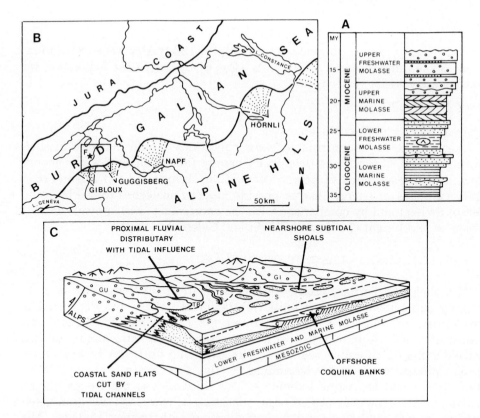

Fig.7. A. The four lithostratigraphic subdivisions of the Swiss Molasse after Matter et al. (1980). B. Palaeogeographic map of the Upper Marine Molasse of Switzerland with location of major fan-deltas, after Hofmann (1960), Rigassi, in Matter et al. (1980) and Lemcke (1981). F marks city of Fribourg. C. Diagrammatic reconstruction of distribution of facies belts in the Fribourg area during Burdigalian (Miocene) times (not to scale). GI = Gibloux fan; GU = Guggisberg fan; TB = transverse bars in tidally influenced distributaries; TS = tidal sandwaves in coastal belt; S = elongate subtidal shoals in nearshore belt; C = coquina banks in offshore belt. After Homewood and Allen (1981).

the major findings of a study of ancient wave and tide conditions from these facies belts.

Reconstructed sea conditions

The wave ripple-marks measured in the field possess the symmetries and steepnesses shown in Fig.8. Care was taken to omit ripple-marks which were of questionable origin, in particular those resulting from probable combined flows of waves superimposed on tidal currents. Such ripple-marks were somewhat more asymmetric and were commonly associated with the flanks of tidal sandwaves. A large number of ripple-marks have steepnesses (large VFI) that make estimation of ancient wave conditions hazardous because of the wide range of possible orbital diameters. Fifty-three out of 150 ripple-

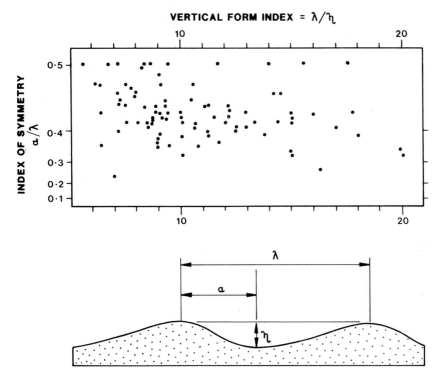

Fig.8. Plot of an index of symmetry against ripple steepness (vertical form index) for wave ripple-marks from the Upper Marine Molasse near Fribourg. After Homewood and Allen (1981).

marks are clearly of the vortex type, where eq. 1 can be used with confidence, and only three of these 53 ripple-marks had a spacing greater than λ_{max} in eq.2 (Fig.9). Bearing in mind the original scatter of data from eq.1 and difficulties of accurate field measurements, it is justifiable to simply use the threshold condition for vortex ripples in estimating wave period from orbital diameter and orbital velocity (Fig.10).

For each locality a table was constructed giving lengths and heights of formative waves over a range of water depths. Unreasonable combinations of H, L and h were then eliminated according to wave breaking criteria. In this way, an impression of the maximum water depths at which the wave ripple-marks formed was obtained. An example for one locality is given in Table I.

Water depths under formative waves were in most cases less than 25 m, but in extreme cases very high waves near the breaking limit may theoretically have been responsible for the wave ripple-marks in deeper waters, perhaps up to 60 m. Such large water depths are unlikely from the facies associations. Furthermore, the association of wave ripple-marks with tidal sandwaves showing shallow-stage run-off patterns and miniripples (as at Illens, map co-ord. 574.50/176.50, Swiss topographic maps No. 252) and

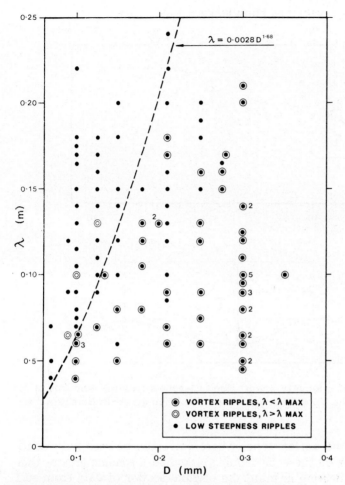

Fig.9. Plot of median grain size against ripple spacing showing the ripple spacing at which the linear relationship between orbital diameter and ripple spacing (eq.1) breaks down (dashed line). Double black circles, vortex ripples, VFI < 7.5; double open circles, vortex ripples with $\lambda > \lambda_{max}$; small dots, low-steepness ripples, VFI > 7.5. Number of superimposed data points also indicated for vortex ripples.

swash bars or flood ramps (as at Fribourg, 578.90/184.70 Map 242) suggests that water depths were shallow, and most wave ripple-marks may have been produced under modest fair-weather waves.

The variability of reconstructed wave conditions between localities is not great, but it may be explained by the relative exposure or sheltering of sub-environments from wave attack and to the viscissitudes of depth during the tidal cycle. In the case of the offshore facies comprising coquina banks, the reconstructed wave periods are generally about 3 s and water depths must have been less than 10 m for formative waves. Since these ripple-marks originate from the facies most distal from the shoreline in the south, it is inconceivable that more proximal facies were deposited in water depths

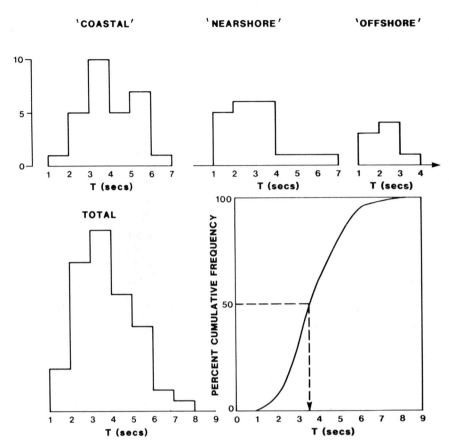

Fig.10. Histograms of calculated maximum wave periods of formative waves for the coastal, nearshore and offshore facies belts and histogram and cumulative frequency curve for total. Assumes the threshold condition and eq.1, utilizing vortex ripple-mark data only.

TABLE I

Example of work-sheet for each locality showing wave transformations in shallowing water depths. Asterisk at $h = 20$ m indicates unstable wave conditions. Terms defined in text and caption to Fig.6; h, L and H are in metres

Water depth (h)	(h/L_∞)	(L/L_∞)	(H/H_∞)	Shoaling coefficient (n)	(C_g/C_∞)	Surface wavelength (L)	Surface wave height (H)
0.5	0.019	0.33	1.30	0.97	0.33	8.58	0.06
1.0	0.038	0.48	1.10	0.93	0.42	12.48	0.08
5.0	0.192	0.85	0.91	0.68	0.59	22.10	0.31
10.0	0.385	0.99	0.99	0.53	0.53	25.74	0.91
20.0	0.769	1.00	1.00	0.50	0.50	26.00	9.8*

Locality: Côtes vers le Lac, Estavayer — Yverdon Road. Mean of maximum wave periods: 4.08 s. Deep-water wave length: 26.0 m.

substantially greater than this. The apparent landward increase in period of formative waves (Fig.10) may be due to a steadily increasing fetch for the northerly or westerly winds, as the Jura coast became more distant and the Alpine shore was approached.

The calculated values of wave power (or energy flux) fall within a wide range of approximately 10^8-10^6 erg s^{-1}. It must be stressed that calculated wave power values are highly sensitive to estimated water depth since h determines the height of formative waves. Nevertheless, bearing in mind the probable errors, it is possible to compare the wave power at the coast of the Burdigalian Sea in the Fribourg area to that of the Danube, Ebro, Niger and Nile deltas (Table II; Coleman and Wright, 1975). In contrast, the wave-dominated deltas such as the São Francisco and Senegal are moulded by considerably more powerful waves than those estimated for the Burdigalian Sea. Delta morphology is also a function of river input (discharge effectiveness index) and tidal range. Homewood and Allen's (1981) plots using wave power, discharge effectiveness index and tidal range as the three controlling parameters suggested that the closest affinities of the Burdigalian coastal systems in the Fribourg area lie with the present-day deltas of the Niger and possibly Burdekin and Klang.

Comparing the wave power calculated for the offshore coquina bank facies to that of the coastal swash bar or flood ramp facies, representing the passage from the open Burdigalian seaway to the Alpine coast, there appears to have been a wave attenuation of between 50:1 and 100:1.

CONCLUSIONS

Further case-studies of wave-influenced sedimentary sequences are required to broaden the presently inadequate data base. Substantial progress

TABLE II

Year-average wave powers for seven of the world's major deltas and estimated values for the Burdigalian Sea of western Switzerland (Fribourg area). Wright and Coleman's (1973) data have been converted to metric c.g.s. units. The Fribourg molasse values are derived from formative waves in 10 m water depth and in depths of less than 2 m for shoreline wave powers

Delta	Year-average wave power at 10 m contour erg s^{-1} × 10^7	Year-average wave power at shoreline erg s^{-1} × 10^7
Mississippi	190	0.041
Danube	49	0.037
Ebro	172	0.155
Niger	107	2.01
Nile	128	10.17
São Francisco	598	30.37
Senegal	285	114.72
Fribourg Molasse	10—50	0.1—1.0

in the quantification of ancient wave conditions can only be made possible by advances in our knowledge of the relationship between near-bed orbital diameter of water particles and ripple-mark spacing. In this respect, linkage of characteristic ripple geometry or internal structure with flow parameters is necessary.

The reconstructed sea conditions of the Miocene (Burdigalian) Sea in the Fribourg area of western Switzerland are of moderate period waves (3—6 s). Fetch lengths for such waves were of the order of 100 km, placing some constraint on the minimum size of the seaway. The estimated wave power of the Burdigalian Sea indicates an affinity with sea conditions off deltas such as the Ebro and Danube, located in semi-restricted seas, but tidal range was substantially larger than in these two examples.

ACKNOWLEDGEMENTS

I am grateful to John Allen, Peter Homewood and Darwin Spearing for their useful comments on the manuscript. Data were collected with the financial support of the Swiss National Science Foundation, Project 2.242-0.79.

REFERENCES

Allen, G.P., Laurier, D. and Thouvenin, J., 1979. Etude Sédimentologique du Delta du Mahakam. Notes et Mémoires No.15, Compagnie Française des Pétroles, Paris.
Allen, J.R.L., 1979. A model for the interpretation of wave ripple-marks using their wavelength, textural composition and shape. J. Geol. Soc. London, 136: 673—682.
Allen, P.A., 1981a. Wave-generated structures in the Devonian lacustrine sediments of SE Shetland, and ancient wave conditions. Sedimentology, 28: 369—379.
Allen, P.A., 1981b. Some guidelines in reconstructing ancient sea conditions from wave ripple marks. Mar. Geol., 43: M59—M67.
Allen, P.A. and Homewood, P., 1983. Mechanics and evolution of a Miocene tidal sand-wave. Sedimentology, 31: 63—82.
Bagnold, R.A., 1946. Motion of waves in shallow water. Interactions between waves and sand bottoms. Proc. R. Soc. London, Ser. A, 187: 1—15.
Boersma, J.R., 1970. Distinguishing features of wave-ripple cross-stratification and morphology. Ph.D. Thesis, University of Utrecht, Utrecht, 65 pp. (unpubl.).
Brebner, A. and Reidel, P.H., 1973. A new oscillating water tunnel. J. Hydraul. Res., 11: 107—121.
Bretschneider, C.L., 1966. Wave generation by wind, deep and shallow water. In: A.T. Ippen (Editor), Estuary and Coastline Hydrodynamics. McGraw Hill, New York, N.Y., pp.133—196.
Bucher, W.H., 1919. On ripples and related sedimentary surface forms and their palaeogeographic interpretations. Am. J. Sci., 47: 149—210, 241—269.
Carstens, M.R., Nielson, F.M. and Altinbilek, H.D., 1969. Bedforms generated in the laboratory under an oscillatory flow: analytical and experimental study. U.S. Army Coastal Eng. Res. Centre, Tech. Memo. 28.
Chan, K.W., Baird, M.H.I. and Round, G.F., 1972. Behaviour of beds of dense particles in a horizontally oscillating liquid. Proc. R. Soc. London, Ser. A, 330: 537—559.
Clifton, H.E., 1976. Wave-formed sedimentary structures — a conceptual model. In: R.A. Davies and R.L. Ethington (Editors), Beach and Nearshore Sedimentation. Soc. Econ. Paleontol. Mineral., Spec. Publ., 24: 126—148.

Coleman, J.M. and Wright, L.D., 1975. Modern river deltas: variability of processes and sand bodies. In: M.L. Broussard (Editor), Deltas, Models for Exploration. Houston Geol. Soc., pp.99—149.
Cook, D.O. and Gorsline, D.S., 1972. Field observations of sand transport by shoaling waves. Mar. Geol., 13: 31—56.
Darbyshire, M. and Draper, L., 1963. Forecasting wind-generated sea waves. Engineering (London), 195: 482—484.
Dingler, J.R., 1974. Wave-formed ripples in nearshore sands. Ph.D. Thesis, Univ. of California, San Diego, Calif., 136 pp.
Dingler, J.R., 1979. The threshold of grain motion under oscillatory flow in a laboratory wave channel. J. Sediment. Petrol., 49: 287—294.
Dingler, J.R. and Inman, D.L., 1977. Wave-formed ripples in nearshore sands. Proc. 15th Conf. Coastal Engineering, pp.2109—2126.
Eckart, C., 1952. The propagation of gravity waves from deep to shallow water. U.S. Nat. Bur. Stand. Gravity Waves, Circ., 521: 165—173.
Harms, J.C., 1969. Hydraulic significance of some sand ripples. Geol. Soc. Am. Bull., 80: 363—396.
Hoffman, F., 1960. Materialherkunft, Transport und Sedimentation im Schweizerischen Molassebecken. Jahrb. St. Gallischen Naturwissensch. Ges., 76: 5—28.
Hom-Ma, N., Horikawa, K. and Hajima, R., 1965. A study on suspended sediment due to wave action. Coastal Eng. Jpn., 8: 85—103.
Homewood, P., 1978. Exemples de séquences de faciès dans la molasse fribourgeoise et leur interpretation. Soc. Fribourg Sci. Nat. Bull., 67: 73—82.
Homewood, P., 1981. Faciès et environnements de depôt de la Molasse de Fribourg. Eclogae. Geol. Helv., 74: 29—36.
Homewood, P. and Allen, P.A., 1981. Wave-, tide- and current-controlled sandbodies of Miocene Molasse, western Switzerland. Bull. Am. Assoc. Pet. Geol., 65: 2534—2545.
Horikawa, K. and Watanabe, A., 1967. A study of sand movement due to wave action. Coastal Eng. Jpn., 10: 39—57.
Inman, D.L., 1957. Wave generated ripples in nearshore sands. U.S. Technol. Memo. Beach Erosion Board, 100, 42 pp.
Inman, D.L. and Bowen, A.J., 1963. Flume experiments on sand transport by waves and currents. Proc. 8th Conf. Coastal Engineering, 1: 137—150.
Ippen, A.T. and Kulin, G., 1955. The shoaling and breaking of the solitary wave. Proc. 5th Conf. Coastal Engineering, pp.27—49.
Kalkanis, G., 1964. Transportation of bed material due to wave action. U.S. Army Corps Engrs, Coastal Eng. Res. Centre, Tech. Memo., 2, 38 pp.
Kaneko, A. and Honji, H., 1979. Initiation of ripple marks under oscillating water. Sedimentology, 26: 101—113.
Kennedy, J.F. and Falcon, M., 1965. Wave generated sediment ripples. M.I.T. Dept. Civ. Eng. Hydro. Lab. Rept. 86, 55 pp.
Komar, P.D., 1974. Oscillatory ripple marks and the evaluation of ancient wave conditions and environments. J. Sediment. Petrol., 44: 169—180.
Komar, P.D., 1976. Beach Processes and Sedimentation. Prentice-Hall, Englewood Cliffs, N.J., 429 pp.
Komar, P.D. and Miller, M.C., 1973. The threshold of sediment movement under oscillatory water waves. J. Sediment. Petrol., 43: 1101—1110.
Komar, P.D. and Miller, M.C., 1976. The initiation of oscillatory ripple marks and the development of plane-bed at high shear stresses under waves. J. Sediment. Petrol., 45: 697—703.
Le Méhauté, B., Divoky, D. and Lin, A., 1969. Shallow water waves: a comparison of theories and experiments. Am. Soc. Civ. Eng., Proc. 11th Conf. Coastal Engineering, pp.86—96.
Lofquist, K.E.B., 1977. A positive displacement oscillatory water tunnel. Coastal Eng. Res. Centre, Misc. Rep., 77-1, 26 pp.

Lofquist, K.E.B., 1978. Sand ripple growth in an oscillatory-flow water tunnel. Tech. Pap. U.S. Coastal Eng. Res. Centre, 78-5, 101 pp.
Madsen, O.S., 1974. Wave climate of the continental margin (its mathematical description). In: D.J. Stanley and D.J.P. Swift (Editors), The New Concepts of Continental Margin Sedimentation — Sediment Transport and its Application to Environmental Management. Short Course Lecture Notes, Am. Geol. Inst., pp.42—108.
Manohar, M., 1955. Mechanics of bottom sediment movement due to wave action. U.S. Army Corps Engrs, Beach Erosion Bd., Tech. Memo., 75, 121 pp.
Matter, A. et al., 1980. Flysch and Molasse of central and western Switzerland. In: R. Trumpy (Editor), Geology of Switzerland, a Guide Book. Exc. 126A, 26th Int. Geol. Congress, Paris, Schweiz. Geol. Komm., pp. 261—293.
McCowan, J., 1894. On the highest wave of permanent type. Philos. Mag., 5: 351—357.
Miche, R., 1944. Undulatory movements of the sea in constant and decreasing depth. Ann. Ponts Chauss, May—June, July—August, 25—78, 131—164, 270—292, 369—406.
Miller, M.C. and Komar, P.D., 1980a. Oscillation sand ripples generated by laboratory apparatus. J. Sediment. Petrol., 50: 173—182.
Miller, M.C. and Komar, P.D., 1980b. A field investigation of the relationship between oscillation ripple spacing and the near-bottom orbital motions. J. Sediment. Petrol., 50: 183—191.
Mogridge, G.R., 1973. Bedforms generated by wave action. DME/NAE Q. Bull., 1973(2), 41 pp.
Mogridge, G.R. and Kamphuis, J.W., 1973. Experiments on bedform generation by wave action. Am. Soc. Civ. Engr., Proc. 13th Conf. Coastal Engineering, pp.1123—1142.
Neumann, G., 1953. On ocean wave spectra and a new way of forecasting wind-generated sea. U.S. Army Corps Eng., Beach Erosion Board, Tech. Mem., 43, 42 pp.
Newton, R.S., 1968. Internal structure of wave-formed ripple marks in the nearshore zone. Sedimentology, 11: 275—292.
Rance, P.J. and Warren, N.F., 1969. The threshold of movement of coarse material in oscillatory flow. Am. Soc. Civ. Eng., Proc. 11th Conf. Coastal Engineering, pp.487—491.
Reineck, H.E., Singh, I.B. and Wunderlich, F., 1971. Einteilung der Rippeln und anderer mariner Sandkorper. Senckenbergiana Marit., 3: 93—101.
Sleath, J.F.A., 1975. A contribution to the study of vortex ripples. J. Hydraul. Res., 13: 315—328.
Sleath, J.F.A., 1976. On rolling grain ripples. J. Hydraul. Res., 14: 69—80.
Sleath, J.F.A. and Ellis, A.C., 1978. Ripple geometry in oscillatory flow. Univ. Cambridge Dept. Engr. Rept. A/Hydraulics/TR2, 15 pp.
Stone, R.O. and Summers, H.J., 1972. Study of subaqueous and subaerial sand ripples. Univ. S. Calif. Dept. Geol. Sci., Final Rept., 72-1.
Sverdrup, H.U. and Munk, W.J., 1946. Empirical and theoretical relations between wind, sea and swell. EOS, Trans. Am. Geophys. Union, 28: 823—927.
Tanner, W.F., 1967. Ripple mark indices and their uses. Sedimentology, 9: 89—104.
Tanner, W.F., 1971. Numerical estimates of ancient waves, water depth and fetch. Sedimentology, 16: 71—88.
Teleki, P.G., 1972. Wave boundary layers and their relation to sediment transport. In: D.J.P. Swift, D.B. Duane and O.H. Pilkey (Editors), Shelf Sediment Transport: Process and Pattern. Dowden, Hutchinson and Ross, Stroudsbourg, Pa., pp.21—60.
Wiegel, R.L., 1964. Oceanographical Engineering. Prentice Hall, Englewood Cliffs, N.J., 532 pp.
Wright, L.D. and Coleman, J.M., 1973. Variations in morphology of major river deltas as functions of ocean wave and river discharge regimes. Bull. Am. Assoc. Pet. Geol., 57: 370—398.
Yalin, S. and Russell, R.D.H., 1962. Similarity in sediment transport due to waves. Proc. 8th Conf. Coastal Engineering, pp.151—171.